D1281321

Random Integral Equations

This is Volume 96 in
MATHEMATICS IN SCIENCE AND ENGINEERING
A series of monographs and textbooks
Edited by RICHARD BELLMAN, *University of Southern California*

The complete listing of books in this series is available from the Publisher upon request.

Random Integral Equations

A. T. BHARUCHA-REID

CENTER FOR RESEARCH IN PROBABILITY
DEPARTMENT OF MATHEMATICS
WAYNE STATE UNIVERSITY
DETROIT, MICHIGAN

ACADEMIC PRESS New York and London 1972

ACADEMIC PRESS, INC.
111 Fifth Avenue, New York, New York 10003

United Kingdom Edition published by
ACADEMIC PRESS, INC. (LONDON) LTD.
24/28 Oval Road, London NW1

LIBRARY OF CONGRESS CATALOG CARD NUMBER: 72-77332

AMS (MOS) 1970 Subject Classification: 60H20

PRINTED IN THE UNITED STATES OF AMERICA

Dedicated to

RODABÉ, KURUSH, AND RUSTAM

Contents

Chapter 3—**Random Equations: Basic Concepts and Methods of Solution**

Chapter 4—**Random Linear Integral Equations**

Chapter 5—**Eigenvalue Problems for Random Fredholm Equations**

Chapter 6—**Random Nonlinear Integral Equations**

Chapter 7—**Itô Random Integral Equations**

Preface

At the present time the theory of random equations is a very active area of mathematical research; and applications of the theory are of fundamental importance in the formulation and analysis of various classes of operator equations which arise in the physical, biological, social, engineering, and technological sciences. Of the several classes of random equations which have been studied, random integral equations (and random differential equations formulated as integral equations) have been studied rather extensively. This book is intended as an introductory survey of research on random integral equations and their applications.

Research on random integral equations has, in the main, proceeded along two lines. There are the fundamental studies on random integral equations associated with Markov processes, these studies being initiated by Itô in 1951; and there are the studies on classical linear and nonlinear integral equations with random right-hand sides, random kernels, or defined on random domains, these studies being initiated by Špaček in 1955. In this book we attempt to present a complete account of the basic results that have been obtained in both of the above rather broad areas of research.

The material in this book is presented in seven chapters. In Chapters 1 and 2 we present a survey of those basic concepts and theorems of probability theory in Banach spaces required for the formulation and study of random equations in Banach spaces. Chapter 3 is an introduction to the theory of random equations. In this chapter the material presented in Chapters 1 and 2 are utilized to discuss the basic concepts and to formulate various methods of solving random equations. Chapters 1–3 are intended as an introduction to probabilistic functional analysis, and they can be read independently of the remaining chapters. In Chapter 4 linear Fredholm and Volterra integral equations with random right-hand sides and/or random kernels are considered. Chapter 5 is devoted to the formulation and analysis of eigenvalue problems for some random Fredholm equations. In Chapter 6 we consider random nonlinear integral equations, in particular Hammerstein and Volterra equations. Finally, in Chapter 7 we study Itô random integral equations.

This book is intended primarily for probabilists, applied mathematicians, and mathematical scientists interested in probabilistic functional analysis and the theory of random equations and its applications. Readers are assumed to have some knowledge of probabilistic measure theory, the basic classes of stochastic processes, and the elements of functional analysis in Banach spaces. Since there are many excellent texts and reference works available which cover the background material needed, no effort has been made to make this volume self-contained.

Acknowledgments

This book could not have been written without direct and indirect assistance from a number of individuals and institutions. Firstly, I would like to express my gratitude to Joseph Kampé de Fériet and the late Antonín Špaček for introducing me to the field of random equations, and for their guidance and encouragement. Secondly, I would like to thank Richard Bellman for inviting me to prepare a book on random equations for his series Mathematics in Science and Engineering.

During the preparation of this book I received generous support from several institutions: Mathematics Research Center, The University of Wisconsin, Madison (U.S. Army Research Office Grant No. DA-31-124-ADO(D)-462), National Science Foundation (Grant No. GP-13741), and the Office of Research Administration, Wayne State University.

Many colleagues, students and friends have provided me with constructive suggestions and criticism. In particular, I would like to acknowledge the help given me by D. Kannan, V. Mandrekar, A. Mukherjea, B. L. S. Prakasa Rao, and C. P. Tsokos.

Finally, I would like to express my thanks to Hing Lau Ng for typing the manuscript and providing invaluable editorial assistance, and to the members of the book production department of Academic Press for their wonderful cooperation in producing this book.

Introduction

Mathematical equations play a central role in all branches of the applied mathematical sciences. In the formulation of an equation, or a system of equations, used to represent a given physical system or phenomenon, coefficients and/or parameters are often introduced which have some definite physical interpretation. For example, in the theory of diffusion (or heat conduction) we have the diffusion coefficient, in the theory of wave propagation the propagation coefficient; and in the theory of elasticity the modulus of elasticity. In concrete examples of all of the cases mentioned, the magnitudes of the coefficients or parameters are experimentally determined. When solving the associated mathematical equations, and in subsequent calculations, it is usually the *mean value* of a set of experimental determinations that is used as the value of the coefficient or parameter. In some instances this may provide an adequate or reasonable description, but in many instances the variance may be sufficiently large to warrant consideration. Hence, when we talk about physical constants or parameters, etc., we are not, in many instances, talking about constants at all, but random variables whose values are determined by some probability distribution or law. The same thing can be said about coefficients and nonhomogeneous terms (or forcing functions) of equations, which may be random variables or functions.

As is well known, in classical physical theories the role of initial conditions is idealized. That is to say, the initial conditions are assumed to be known. In reality, however, initial conditions are known only within certain ranges of values. Hence a more realistic formulation of initial-value problems would involve treating the initial data as random. Similarly, boundary-value problems should be treated under the assumption that the boundary data are random.

In a more general way, not only the initial or boundary conditions, but the operators describing the behavior of a system may not be known exactly. For example, in the case of difference and differential operators the coefficients might not be known exactly, and are therefore assumed to be of a known form, or approximations are employed. In the case of integral equations the kernel often is not known exactly. In many studies, workers use in their formulation what might be called mean coefficients or kernels; hence the solutions of the differential or integral equations often give only the behavior of average, mean, or expected values of the physical quantities used to describe the state of a system.

In view of the above, a more realistic formulation of the equations arising in applied mathematics would involve the study of random equations. A *random operator equation* of the form $L(\omega)x = y(t, \omega)$ is a *family of operator equations* depending on a parameter ω which ranges over a probability measure space $(\Omega, \mathfrak{A}, \mu)$. The probability measure μ determines the probability of an event (i.e., a subset of Ω), and therefore the probability of the corresponding equation of the family. The objective of the theory of random equations is the determination of the probability distribution (or law) of the *random solution* $x(t, \omega)$, or functionals of the solution; and/or the determination of various statistical properties of $x(t, \omega)$ such as its expectation, variance, and higher moments.

A random equation of the form considered above will arise, for example, in the representation of systems with random coefficients and forcing functions. However, it is necessary to consider within the framework of the theory of random equations the study of random solutions of deterministic equations which are generated by random initial and/or boundary conditions. In many instances problems of this type can be so transformed as to lead to a random integral equation.

It is of interest to remark that the distinction between a deterministic and probabilistic approach to the study of mathematical equations lies mainly in the nature of the questions they try to answer, and in the interpretation of the results. The advantages of a probabilistic approach are that (1) it permits from the initial formulation a flexibility and, therefore, a greater generality than that offered by a deterministic approach, and (2) it permits the incorporation of stochastic features in the equations, the inclusion of which may play an essential role in making the connection between mathematical equations and the real phenomena they attempt to describe.

The theory of random equations is still in the formative stage; however, there is a very extensive literature dealing with four basic classes of random equations, namely: (1) random algebraic equations, (2) random difference equations, (3) random differential equations, and (4) random integral equations. Other random equations can be considered (and indeed have been considered), for example, random differential–difference equations, random integrodifferential equations, and rather general random functional equations;

however, for the purposes of this exposition we regard the above four classes as the basic classes of random equations. We present below a list of survey articles and books which deal with the basic classes of random equations and their applications.

1. *Random algebraic equations:* Bharucha-Reid [4].
2. *Random difference equations:* Aoki [1], Åström [3], Bharucha-Reid [5], and Jazwinski [9].
3. *Random differential equations:* Arnold [2], Åström [3], Bucy and Joseph [5a], Gihman and Skorohod [7], Jazwinski [9], Khasminskiĭ [10], Kushner [11, 11a], Saaty [13], Srinivasan and Vasudevan [15], Stratonovich [16], Wong [18], and Wonham [19].
4. *Random integral equations;* Dynkin [6], Gihman and Skorohod [7], Itô and McKean [8], McKean [12], Skorohod [14], and Tsokos and Padgett [17].

It is clear from the above list of references that random differential equations have been the subject of rather extensive research. This, however, should not be surprising; for deterministic (or classical) differential equations are utilized as models for a wide variety of phenomena considered in the applied mathematical sciences and, consequently, play a dominant role in applied mathematics. Integral equations also constitute an important class of operator equations; and in applied mathematics they occur in their own right as mathematical models, and also as alternative formulations of models that lead to differential equations. As in the deterministic case, random integral equations arise as primary models of physical phenomena, and also as alternative formulations of random differential equations. The books listed above as dealing with random integral equations are, with the exception of [17], concerned with random differential equations of Itô type which arise in the theory of Markov processes. However, in order to give a precise formulation of this class of random differential equations it is necessary to formulate these random differential equations as random integral equations. In general, random integral equations provide a more appropriate formulation of random differential equations, for it is not necessary to introduce the notion of differentiability (in some sense) of random functions in order to give a mathematically correct formulation of the random differential equations.

The modern theories of linear and nonlinear equations belong to the domain of functional analysis. Within this framework, the formulation of an equation is considered as a correspondence between elements of a suitable linear space; and then, utilizing properties of the transformation (linear or nonlinear) that defines the correspondence and the properties of the space, the problem is to obtain conditions for the solvability of the equation.

The same general procedure has been followed in the development of the theory of random equations. The systematic study of random equations employing the methods of functional analysis was initiated by the Prague school of probabilists under the direction of the late Antonín Špaček. Their research was motivated by the importance of the role of random equations in the applied mathematical sciences, and as a concrete application of the results of research in probabilistic functional analysis. The formulation of a random equation from the functional analytic point of view considers the correspondence between elements of a given linear space as being defined by a deterministic or random transformation, and then, utilizing the properties of the transformations and the properties of the space on which the transformation is defined, the problem is to obtain conditions under which there exists a random function with values in the linear space which satisfies the equation with probability one.

As stated in the preface, the theory of random equations is still in its formative stage. However, even at the present time I do not think that its importance for applied mathematics can be denied; and I feel that within the not too distant future we will witness the development of a unified theory of random equations that will be based on, and utilize, the most modern concepts and tools of stochastic processes, functional analysis, and numerical analysis.

References

1. Aoki, M., "Optimization of Stochastic Systems." Academic Press, New York, 1967.
2. Arnold, L., "Stochastische Differentialgleichungen—Theorie und Anwendung." Oldenbourg, Munich, to be published.
3. Åström, K. J., "Introduction to Stochastic Control Theory." Academic Press, New York, 1970.
4. Bharucha-Reid, A. T., Random algebraic equations. *In* "Probabilistic Methods in Applied Mathematics" (A. T. Bharucha-Reid, ed.), Vol. 2, pp. 1–52. Academic Press, New York, 1970.
5. Bharucha-Reid, A. T., Random difference equations. *In* "Probabilistic Methods in Applied Mathematics" (A. T. Bharucha-Reid, ed.), Vol. 4. Academic Press, New York, to be published.

5a. Bucy, R. S., and Joseph, P. D., "Filtering for Stochastic Processes with Applications to Guidance." Wiley (Interscience), New York, 1968.

6. Dynkin, E. B., "Markov Processes," Vols. I and II, translated from the Russian. Academic Press, New York, 1965.
7. Gihman, I. I., and Skorohod, A. V., "Stochastic Differential Equations" (Russian). Izdat. Naukova Dumka, Kiev, 1968.
8. Itô, K., and McKean, H. P., "Diffusion Processes and Their Sample Paths." Springer-Verlag, Berlin and New York, 1965.
9. Jazwinski, A. H., "Stochastic Processes and Filtering Theory." Academic Press, New York, 1970.

10. Khasminskiĭ, R. Z., "Stability of Systems of Differential Equations under Random Perturbation of Their Parameters" (Russian). Izdat. "Nauka," Moscow, 1969.
11. Kushner, H. J., "Stochastic Stability and Control." Academic Press, New York, 1967.
11a. Kushner, H. J., "Introduction to Stochastic Control." Holt, New York, 1971.
12. McKean, H. P., Jr., "Stochastic Integrals." Academic Press, New York, 1969.
13. Saaty, T. L., "Modern Nonlinear Equations." McGraw-Hill, New York, 1967.
14. Skorohod, A. V., "Studies in the Theory of Random Processes," translated from the Russian. Addison-Wesley, Reading, Massachusetts, 1965.
15. Srinivasan, S. K., and Vasudevan, R., "Introduction to Random Differential Equations and Their Applications." Amer. Elsevier, New York, 1971.
16. Stratonovich, R. L., "Conditional Markov Processes and Their Application to the Theory of Optimal Control," translated from the Russian. Amer. Elsevier, New York, 1968.
17. Tsokos, C. P., and Padgett, W. J., "Random Integral Equations with Applications to Stochastic Systems." Springer-Verlag, Berlin and New York, 1971.
18. Wong, E., "Stochastic Processes in Information and Dynamical Systems." McGraw-Hill, New York, 1971.
19. Wonham, W. M., Random differential equations in control theory. *In* "Probabilistic Methods in Applied Mathematics" (A. T. Bharucha-Reid, ed.), Vol. 2, pp. 131–212. Academic Press, New York, 1970.

CHAPTER 1

Probability Theory in Banach Spaces: An Introductory Survey

1.1 Introduction

Let $(\Omega, \mathfrak{A}, \mu)$ be a probability measure space, and let $(\mathfrak{X}, \mathfrak{B})$ be a measurable space, where $\mathfrak{X}$ is a Banach space and $\mathfrak{B}$ is the σ-algebra of all Borel subsets of $\mathfrak{X}$. With reference to the notation and terminology used above, we recall the following definitions. The set Ω is a nonempty abstract set, whose elements ω are termed *elementary events*. $\mathfrak{A}$ is a σ-algebra of subsets of Ω; that is, $\mathfrak{A}$ is a nonempty class of subsets of Ω satisfying the following conditions: (i) $\Omega \in \mathfrak{A}$, (ii) if $A \in \mathfrak{A}$ and $B \in \mathfrak{A}$, then $A - B \in \mathfrak{A}$, and (iii) if $A_i \in \mathfrak{A}$, $i = 1, 2, \ldots$, then $\bigcup_{i=1}^{\infty} A_i \in \mathfrak{A}$. The elements of $\mathfrak{A}$ are called *events*. μ is a *probability measure* on $\mathfrak{A}$; that is, μ is a set function, with domain $\mathfrak{A}$, which is nonnegative, countably additive, and such that $\mu(A) \in [0,1]$ for all $A \in \mathfrak{A}$, with $\mu(\Omega) = 1$. We will assume throughout this book that μ is a *complete* probability measure; that is μ is such that the conditions $A \in \mathfrak{A}$, $\mu(A) = 0$, and $A_0 \subseteq A$ imply $A_0 \in \mathfrak{A}$.

In this chapter we present a survey of the basic concepts of probability theory in Banach spaces; hence we will be concerned with random variables defined on $(\Omega, \mathfrak{A}, \mu)$ with values in $(\mathfrak{X}, \mathfrak{B})$, where $\mathfrak{X}$ is a Banach space and $\mathfrak{B}$ is the σ-algebra of Borel subsets of $\mathfrak{X}$. We remark that when $\mathfrak{X} = R$ (the real line) or $\mathfrak{X} = R_n$ (Euclidean n-space), the theory to be presented in this chapter yields the results of classical probability theory.

Since the setting for the theory is a Banach space, Sect. 1.2 is devoted to some basic definitions and concepts from the theory of Banach spaces. In Sect. 1.3 we introduce the notion of a Banach space-valued random variable, and give a rather complete survey of the basic definitions, concepts, and theorems. Section 1.4 is concerned with Banach space-valued random functions; and in Sect. 1.5 we study probability measures on Banach spaces. The material in these two sections is of fundamental importance in the study of random equations in Banach spaces. Finally, in Sect. 1.6 we consider some limit theorems for Banach space-valued random variables.

Since this chapter is intended as an introductory survey of results from the theory of probability in Banach spaces which are useful in the development of a theory of random equations, we do not prove the theorems that are stated. This chapter is fully referenced; hence the interested reader can refer to the original papers or books for proofs.

1.2 Banach Spaces

A. Introduction

This section is devoted to some basic definitions and concepts from the theory of Banach spaces. We also give a list of some concrete Banach spaces which will be encountered in this book. The results presented in this section are all standard, and can be found in any of the many text and reference books on the theory of Banach spaces. We refer, in particular, to the books of Dunford and Schwartz [23], Goffman and Pedrick [37], Hille and Phillips [46], Kolmogorov and Fomin [69], Lorch [79], Simmons [111], Taylor [116], and Zaanen [130].

B. Banach spaces, some examples

***Definition* 1.1.** A nonempty set $\mathfrak{X}$ is a *real* (resp. *complex*) *Banach space* if (1) $\mathfrak{X}$ is a linear space over the real numbers R (resp. the complex numbers C); (2) $\mathfrak{X}$ has a *norm*, that is, there is a real-valued function $\|x\|$ defined on $\mathfrak{X}$ such that (a) $\|x\| \geqslant 0$ for all $x \in \mathfrak{X}$, with $\|x\| = 0$ if and only if $x = \theta$ (the null element of $\mathfrak{X}$), (b) $\|\alpha x\| = |\alpha| \, \|x\|$ for any scalar α, and all $x \in \mathfrak{X}$, and (c) $\|x+y\| \leqslant \|x\| + \|y\|$ for all $x, y \in \mathfrak{X}$; (3) $\mathfrak{X}$ is a complete metric space with metric $d(x,y) = \|x-y\|$.

A Banach space $\mathfrak{X}$ is said to be *separable* if it has a countable subset that is everywhere dense. Separable Banach spaces constitute an important class of Banach spaces, and they are of particular importance in the development of probability theory in Banach spaces.

We now list some concrete Banach spaces which will be encountered in this book.

1. The spaces R and C, the *real numbers* and the *complex numbers*, respectively, are the simplest of all Banach spaces. The norm of an element x in either R or C is, of course, defined by $\|x\| = |x|$, the absolute value of x.

2. The space R_n of all *n-tuples* $x=(x_1,x_2,\ldots,x_n)$ *of real numbers* is a real Banach space with respect to coordinatewise addition and scalar multiplication, and with the norm defined by $\|x\|=(\sum_{i=1}^n |x_i|^2)^{1/2}$. R_n is, of course, *Euclidean n-space.* The space C_n of all *n-tuples* $z=(z_1,z_2,\ldots,z_n)$ *of complex numbers* is a complex Banach space with respect to coordinatewise addition and scalar multiplication, and with the norm defined by $\|z\|=(\sum_{i=1}^n |z_i|^2)^{1/2}$. C_n is called *n-dimensional unitary space.* We remark that the *infinite-dimensional Euclidean space* R_∞, the set of all sequences $x=\{x_i\}$ of real numbers such that $\sum_{i=1}^\infty |x_i|^2$ converges, is a real Banach space with norm $\|x\|=(\sum_{i=1}^\infty |x_i|^2)^{1/2}$; and C_∞, the *infinite-dimensional unitary space*, is a complex Banach space.

3. For p real, $1 \leqslant p < \infty$, we denote by l_p^n or $l_p(n)$ the space of all n-tuples $x=(x_1,x_2,\ldots,x_n)$ of scalars. With norm $\|x\|=(\sum_{i=1}^n |x_i|^p)^{1/p}$ the l_p^n spaces are Banach spaces; and it is clear that the real and complex spaces l_2^n are the n-dimensional spaces R_n and C_n, respectively.

4. *The sequence spaces* l_p, $1 \leqslant p \leqslant \infty$. We denote by l_p the space of all sequences $x=\{x_i\}$ of scalars such that $\sum_{i=1}^\infty |x_i|^p < \infty$, and with addition and scalar multiplication defined coordinatewise. For $1 \leqslant p < \infty$, the norm of an element in l_p is given by $\|x\|_p=(\sum_{i=1}^\infty |x_i|^p)^{1/p}$; and with this norm the l_p spaces are Banach spaces. The space l_∞, which is the space of all bounded sequences of scalars, is a Banach space with norm $\|x\|_\infty = \sup|x_i|$. (We remark that the real and complex spaces l_2 are R_∞ and C_∞, respectively.)

5. *The space of continuous function* $C[a,b]$. The space of all bounded continuous scalar-valued functions $x(t)$, $t \in [a,b]$, $-\infty \leqslant a < b \leqslant \infty$, with the operations of addition and scalar multiplication as usually defined for scalar-valued functions, is a Banach space with norm defined by

$$\|x\| = \sup_{t\in[a,b]} |x(t)|.$$

This norm is usually referred to as the *sup* or *uniform norm.* The function space $C[a,b]$ is a Banach space of great importance in analysis; and it plays an important role in the theory of stochastic processes and its applications.

6. *The Lebesgue spaces* L_p, $1 \leqslant p \leqslant \infty$. The function space

$$L_p(S,\mathfrak{F},m)=L_p(S),$$

defined for any real number p, $1 \leqslant p < \infty$, and any positive measure space $(S,\mathfrak{F},m)$, consists of those scalar-valued measurable functions $x(s)$ on S for which $\int_S |x(s)|^p dm(s)$ is finite. The L_p $(1 \leqslant p < \infty)$ spaces are Banach spaces with norm defined by

$$\|x\|_p = \left(\int_S |x(s)|^p dm(s)\right)^{1/p}.$$

We remark that the elements of the L_p spaces are not actually functions, but equivalence classes of functions; that is, two functions belong to the same equivalence class if they differ only on a set of measure zero. The space $L_\infty(S)$ is defined for a positive measure space, and consists of all (equivalence classes of) essentially bounded, scalar-valued measurable functions $x(s)$. L_∞ is a Banach space with norm defined by $\|x\|_\infty = \operatorname{ess\,sup} |x(s)|$. The Lebesgue spaces L_p, especially L_2, are of importance in the theory of random functions; hence they have a number of applications to problems associated with random equations.

We remark that the space l_p is just the space $L_p(S, \mathfrak{F}, m)$ where S is the set of all integers, $\mathfrak{F}$ is the collection of all subsets of S, and $m(F)$ the number (finite or infinite) of elements of $F \in \mathfrak{F}$.

7. *The Orlicz spaces* L_Φ. Let $s = \varphi(r)$ and $r = \psi(s)$ be nondecreasing functions, inverse to each other, and satisfying the conditions $\varphi(0) = 0$, $\varphi(\infty) = \infty$ and $\psi(0) = 0$, $\psi(\infty) = \infty$. For given $\varphi(r)$ and $\psi(s)$, $r, s \geqslant 0$, the convex functions $\Phi(r)$ and $\Psi(s)$ defined by the Lebesgue integrals

$$\Phi(r) = \int_0^r \varphi(\tau)\, d\tau, \qquad \Psi(s) = \int_0^s \psi(\tau)\, d\tau$$

are complementary in the sense of Young; that is, if the functions $\Phi(r)$ and $\Psi(s)$ are complementary in the sense of Young, then $rs \leqslant \Phi(r) + \Psi(s)$ for arbitrary $r \geqslant 0$, $s \geqslant 0$, with equality if and only if at least one of the relations $s = \varphi(r)$, $r = \psi(s)$ is satisfied. Now, let D be a subset of R_n, and let m denote Lebesgue measure on D. The Orlicz space $L_\Phi = L_\Phi(D, m)$ is defined as the collection of complex-valued measurable functions $x(s)$ on D for which

$$\|x\|_\Phi = \sup \int_D |x(s)\, y(s)|\, dm(s),$$

where the supremum is taken over all y such that

$$\int_D \Psi(|y(s)|)\, dm(s) \leqslant 1.$$

Similarly, the Orlicz space L_Ψ is defined as the collection of all complex-valued measurable functions $x(s)$ on D for which

$$\|x\|_\Psi = \sup \int_D |x(s)\, y(s)|\, dm(s),$$

where the supremum is taken over all y such that

$$\int_D \Phi(|y(s)|)\, dm(s) \leqslant 1.$$

For a fixed function $\Phi(r)$, the Orlicz spaces L_Φ can be made into Banach spaces, with norm $\|x\|_\Phi$, if, as in the case of the L_p spaces, we consider the elements of L_Φ to be equivalence classes of functions. We remark that when the function $\Phi(r) = kr^p$, $k > 0$, $1 \leqslant p < \infty$, L_Φ contains the same functions as the L_p $(1 \leqslant p < \infty)$ spaces.

The Orlicz spaces are an important class of Banach function spaces; and in recent years they have played an important role in the study of linear and nonlinear integral equations (cf. Krasnosel'skiĭ and Rutickiĭ [70] and Zaanan [130]), and have found many applications in analysis and probability.

8. *Hilbert space H.* A linear space H is called an *inner-product space* (or *pre-Hilbert space*) if there is defined for all pairs of elements $x, y \in H$ a scalar-valued function on $H \times H$, denoted by (x, y), and called the *inner* or *scalar product* of x and y, such that

(i) $(x, y) = \overline{(y, x)}$,
(ii) $(\alpha x + \beta y, z) = \alpha(x, z) + \beta(y, z)$,
(iii) $(x, x) \geqslant 0$,

with $(x, x) = 0$ if and only if $x = \theta$. The norm of an element $x \in H$ is defined by $\|x\| = (x, x)^{1/2}$. If H is complete under the norm obtained from its inner product, then H is called a *Hilbert space*; and H is Banach space with norm as defined above.

Of the Banach spaces listed above, the finite-dimensional spaces R_n, C_n, and l_2^n are Hilbert spaces with $(x, y) = \sum_{i=1}^{n} x_i \bar{y}_i$. Also, l_2 with $(x, y) = \sum_{i=1}^{\infty} x_i \bar{y}_i$, and $L_2(S)$ with $(x, y) = \int_S x\bar{y}\, dm$ are Hilbert spaces. A Hilbert space is a very special type of Banach space in that it possesses enough additional structure to permit us to determine if two elements of the space are orthogonal (perpendicular). For example, let $L_2(\Omega)$ denote the Hilbert space of second-order real- (or complex-) valued random variables. Then, for $x, y \in L_2(\Omega)$, $(x(\omega), y(\omega)) = \mathscr{E}\{x(\omega)\overline{y(\omega)}\}$; and x and y are orthogonal if $\mathscr{E}\{x(\omega)\overline{y(\omega)}\} = 0$. L_2 spaces are of fundamental importance in the theory of square-integrable random functions and its applications.

As we remarked earlier, separable Banach spaces form an important class of Banach spaces which play an important role in the development of probability theory in Banach spaces and the theory of random equations. A fundamental result concerning separable Banach spaces is the following theorem of Banach and Mazur.

THEOREM 1.1. *If $\mathfrak{X}$ is a separable Banach space, then $\mathfrak{X}$ is isometrically isomorphic to a closed subspace of the Banach space $C[0,1]$.*

In view of the Banach–Mazur theorem, probability theory in separable Banach spaces is applicable to problems concerning continuous random functions.

Of the Banach spaces listed in this section which will be of particular interest to us in this book, the following are separable: (1) $l_p(n)$; (2) l_p $(1 \leqslant p < \infty)$; (3) the real space $C[a,b]$ is separable if $[a,b]$ is a finite closed interval of the real line; (4) $L_p(S)$ $(1 \leqslant p < \infty)$ is separable if the measure m is separable; (5) L_Φ is separable if the measure m is separable and there exists a constant $M > 0$ such that $\Phi(2r) \leqslant M\Phi(r)$, $r \geqslant 0$; (6) a Hilbert space H is separable if and only if H contains a maximal orthonormal system which is at most countable; and a separable Hilbert space is isometrically isomorphic to l_2.

C. Linear functionals. The adjoint space of a Banach space

We now consider linear functionals on a Banach space $\mathfrak{X}$, and the adjoint space of $\mathfrak{X}$.

***Definition* 1.2.** A *functional* f on a Banach space $\mathfrak{X}$ is a function on $\mathfrak{X}$ to the scalars. The value of f for an element $x \in \mathfrak{X}$ is denoted by $f(x)$. A functional f is said to be *linear* if (i) $f(x_1 + x_2) = f(x_1) + f(x_2)$, $x_1, x_2 \in \mathfrak{X}$, and (ii) $f(\alpha x) = \alpha f(x)$, $x \in \mathfrak{X}$, α a scalar; and f is said to be *bounded* if there exists a real constant $M \geqslant 0$ such that $|f(x)| \leqslant M\|x\|$ for all $x \in \mathfrak{X}$.

A basic result is the following: *A linear functional is bounded if and only if it is continuous.*

***Definition* 1.3.** Let $\mathfrak{X}^*$ denote the set of all bounded linear functionals on a Banach space $\mathfrak{X}$. $\mathfrak{X}^*$ is a Banach space, and is called the *adjoint* (*conjugate* or *dual*) space of $\mathfrak{X}$.

Since $\mathfrak{X}^*$ denotes the set of all bounded linear functionals on $\mathfrak{X}$, an element of $\mathfrak{X}^*$ is frequently denoted by x^*. Hence $x^*(x)$ denotes the value of x^* for an element $x \in \mathfrak{X}$.

The set of all bounded linear functionals on $\mathfrak{X}^*$ is called the *second adjoint space* of $\mathfrak{X}$; and it is denoted by $\mathfrak{X}^{**}$. A Banach space $\mathfrak{X}$ is said to be *reflexive* if $\mathfrak{X} = \mathfrak{X}^{**}$.

We also need the following notion. A set $\mathfrak{X}_0^* \subset \mathfrak{X}^*$ is said to be *total* on a set $M \subset \mathfrak{X}$ if $x \in M$ and $y \in M$ together with $x^*(x) = x^*(y)$ for every $x^* \in \mathfrak{X}_0^*$ imply $x = y$.

D. Topologies and convergence concepts

A number of topologies can be introduced in Banach spaces, and with each one there is associated a type of convergence.

***Definition* 1.4.** The topology induced in a Banach space $\mathfrak{X}$ by the metric $d(x,y) = \|x-y\|$ is called the *metric*, *norm*, or *strong* topology of $\mathfrak{X}$.

***Definition* 1.5.** A sequence $\{x_n\}$ of elements in a Banach space $\mathfrak{X}$ *converges strongly*, or *converges in the strong topology* to an element x if $\lim_{n\to\infty}\|x_n - x\| = 0$. x is called the *strong limit* of $\{x_n\}$.

In addition to the strong topology, another topology can be introduced in $\mathfrak{X}$ called the weak topology.

***Definition* 1.6.** Let x_0 be a fixed element of a Banach space $\mathfrak{X}$, and consider the neighborhood $S(x_0)$ defined as follows: $S(x_0) = S(x_0; x_1^*, \ldots, x_n^*; \epsilon) = \{x: |x(x_i^*) - x_i^*(x_0)| < \epsilon, i = 1, 2, \ldots, n\}$, where $x_1^*, \ldots, x_n^* \in \mathfrak{X}^*$ and $\epsilon > 0$. The topology defined by the neighborhoods $S(x_0)$, $x_0 \in \mathfrak{X}$, is called the *weak topology* of $\mathfrak{X}$.

***Definition* 1.7.** A sequence $\{x_n\}$ of elements in a Banach space $\mathfrak{X}$ *converges weakly*, or *converges in the weak topology*, to an element x if (i) the norms $\|x_n\|$ are uniformly bounded, that is, $\|x_n\| \leqslant M$, and (ii) $\lim_{n\to\infty} x^*(x_n) = x^*(x)$ for every $x^* \in \mathfrak{X}^*$.

A weak topology can also be introduced in the adjoint space $\mathfrak{X}^*$, called the weak $*$-topology of $\mathfrak{X}^*$; and we can consider weak convergence of functionals.

***Definition* 1.8.** Let x_0^* be a fixed element of $\mathfrak{X}^*$, and consider the neighborhood $S(x;)$ defined as follows:

$$S(x_0^*) = S(x_0^*; x_1, x_2, \ldots, x_n; \epsilon) = \{x^*: |x^*(x_i) - x_0^*(x_i)| < \epsilon, i = 1, 2, \ldots, n\},$$

where $x_1, x_2, \ldots, x_n \in \mathfrak{X}$ and $\epsilon > 0$. The topology defined by the neighborhoods $S(x_0^*)$, $x_0^*; \in \mathfrak{X}^*$, is called the *weak $*$-topology* of $\mathfrak{X}^*$.

***Definition* 1.9.** A sequence $\{x_n^*\}$ of bounded linear functionals *converges weakly*, or *converges in the weak $*$-topology*, to the bounded linear functional x^* if (i) the norms $\|x_n^*\|$ are uniformly bounded, and (ii) $\lim_{n\to\infty} x_n^*(x) = x^*(x)$ for every $x \in \mathfrak{X}$.

The following facts are useful: (1) *Weak and strong convergence are equivalent in R_n*. (2) *If a sequence $\{x_n\}$ of elements in a Banach space $\mathfrak{X}$ converges strongly to an element $x \in \mathfrak{X}$, then $\{x_n\}$ also converges weakly to x*. (3) *A sequence $\{x_n\}$ of elements in a Hilbert space H converges weakly to an element $x \in \mathfrak{X}$ if $\lim_{n\to\infty} (x_n, y) = (x, y)$ for every $y \in H$.*

1.3 Banach Space-Valued Random Variables

A. Introduction

Classical probability theory is concerned primarily with real-valued random variables and random functions. The need to study random variables and random functions with values in general topological spaces was pointed out in 1947 by Fréchet [30] (cf. also [31]); and in 1953 Mourier published her fundamental paper [84] which initiated the systematic study of random variables with values in a Banach space.† Another basic paper is that of Hanš [44].

In this section we will consider five topics: (1) the various definitions of Banach space-valued random variables, (2) convergence concepts, (3) the collection of random variables with values in a separable Banach space, (4) integration of Banach space-valued random variables, and (5) Banach algebra-valued random variables.

B. Definitions of Banach space-valued random variables

In classical probability theory random variables are introduced as real-valued measurable functions defined on a probability measure space. Analogously, Banach space-valued random variables are Banach space-valued measurable functions; however for Banach space-valued functions a number of concepts of measurability can be introduced which lead to different definitions of a Banach space-valued random variable.

Let $(\Omega, \mathfrak{A}, \mu)$ be a complete probability measure space, and let $(\mathfrak{X}, \mathfrak{B})$ be a measurable space where $\mathfrak{X}$ is a Banach space and $\mathfrak{B}$ is the σ-algebra of all Borel subsets of $\mathfrak{X}$.

***Definition* 1.10.** A mapping $x: \Omega \to \mathfrak{X}$ is said to be a *random variable with values in* $\mathfrak{X}$ if the inverse image under the mapping x of every Borel set B belongs to $\mathfrak{A}$; that is, $x^{-1}(B) \in \mathfrak{A}$ for all $B \in \mathfrak{B}$.

Definition 1.10 is equivalent to stating that a random variable with values in $\mathfrak{X}$ is a *Banach space-valued Borel measurable function*; and in the case where $\mathfrak{X}$ is the real line, the above definition coincides with the usual *descriptive* definition of a real-valued or ordinary random variable (cf. Loève [77, p. 150]).

We now introduce the notion of a strong random variable.

† Earlier contributions to the development of probability theory in Banach spaces are the paper of Glivenko [36] on the law of large numbers in function spaces, and the paper of Kolmogorov [67] on the Fourier transform, or characteristic functional, of Banach space-valued random variables.

***Definition* 1.11.** A mapping $x: \Omega \to \mathfrak{X}$ is said to be a *finitely valued random variable* if it is constant on each of a finite number of disjoint sets $A_i \in \mathfrak{A}$ and equal to θ on $\Omega - (\bigcup_{i=1}^{n} A_i)$, and a *simple random variable* if it is finitely valued and $\mu(\{\omega: \|x(\omega)\| > 0\}) < \infty$. x is said to be a *countably valued random variable* if it assumes at most a countable set of values in $\mathfrak{X}$, assuming each value different from θ on a set in $\mathfrak{A}$.

***Definition* 1.12.** A mapping x: $\Omega \to \mathfrak{X}$ is a *μ-almost separably valued random variable* if there exists a set $A_0 \in \mathfrak{A}$ such that $\mu(A_0) = 0$ and $x(\Omega - A_0)$ is separable.

Finally, we have:

***Definition* 1.13.** A mapping x: $\Omega \to \mathfrak{X}$ is said to be a *strong* (or *Bochner*) *random variable* if there exists a sequence $\{x_n(\omega)\}$ of countably valued random variables which converges to $x(\omega)$ almost surely; that is, there exists a set $A_0 \in \mathfrak{A}$, with $\mu(A_0) = 0$ such that

$$\lim_{n\to\infty} \|x_n(\omega) - x(\omega)\| = 0 \qquad \text{for every} \quad \omega \in \Omega - A_0.$$

Since $\mu(\Omega) = 1$, we can replace "countably valued" in Definition 1.13 by "simple."

Next we introduce the notion of a weak random variable.

***Definition* 1.14.** A mapping x: $\Omega \to \mathfrak{X}$ is said to be a *weak* (or *Pettis*) *random variable* if the functions $x^*(x(\omega))$ are real-valued random variables for each $x^* \in \mathfrak{X}^*$.

Mourier [84] has used the term *L-random element* for a weak random variable.

Definition 1.13 is a constructive definition of a Banach space-valued random variable, while Definitions 1.10 and 1.14 are descriptive definitions. We remark that general properties of Banach space-valued random variables are often easier to discover and theorems easier to prove when a descriptive definition is used.

The concepts of weak and strong random variables are connected by the following result (cf. Hille and Phillips [46, p. 72]).

THEOREM 1.2. *A mapping $x: \Omega \to \mathfrak{X}$ is a strong random variable if and only if it is a weak random variable and μ-almost separably-valued.*

Another definition of a Banach space-valued random variable is based on the concept of measurability introduced by Price [96].

***Definition* 1.15.** A mapping $x: \Omega \to \mathfrak{X}$ is said to be a *Price random variable* if and only if for every $x_0 \in \mathfrak{X}$ and positive real number r,

$$\{\omega: \|x(\omega) - x_0\| < r\} \in \mathfrak{A};$$

that is, the inverse images of all spherical neighborhoods are measurable.

We now restrict our attention to the important case where $\mathfrak{X}$ is a separable Banach space. In this case, an important corollary of Theorem 1.2 states that the concept of weak and strong random variable are equivalent (cf. Hille and Phillips [46, p. 73]). Furthermore, the σ-algebra generated by the class of all spherical neighborhoods of $\mathfrak{X}$ is equal to the σ-algebra $\mathfrak{B}$ of all Borel subsets of $\mathfrak{X}$. Finally, every strong random variable is measurable in the sense of Definition 1.10. From the above we can conclude that when $\mathfrak{X}$ is separable the four definitions of random variables with values in a Banach space are equivalent; hence we can talk about a *Banach space-valued random variable*, *$\mathfrak{X}$-valued random variable*, or *random element* in $\mathfrak{X}$ without reference to an associated concept of measurability.

Another consequence of assuming $\mathfrak{X}$ to be separable concerns the measurability of the norm $\|x(\omega)\|$. A subset D of $\mathfrak{X}^*$ is said to be a *determining set* for $\mathfrak{X}$ if

$$\|x\| = \sup_{x^* \in D} |x^*(x)| \qquad \text{for all} \quad x \in \mathfrak{X}.$$

We state the following result: *Let $\mathfrak{X}$ be an arbitrary Banach space. If (i) $x(\omega)$ is a weak random variable, and (ii) there exists a denumerable determining set D for $\mathfrak{X}$, then $\|x(\omega)\|$ is a nonnegative real-valued random variable.* We refer to Hille and Phillips [46, p. 72] for a proof of the above result. Now, it is known that if $\mathfrak{X}$ is separable, then $\mathfrak{X}$ possesses a denumerable determining set; hence when $\mathfrak{X}$ is a separable Banach space $\|x(\omega)\|$, $x \in \mathfrak{X}$, is a nonnegative real-valued random variable.

Before closing this subsection, we give the definition of a random variable with values in a Hilbert space; and then consider the notions of Gaussian random elements, equivalence, independence, and sequences of Banach space-valued random variables.

Since a Hilbert space is a Banach space, all of the definitions given earlier can be used to define Hilbert space-valued random variables; however the following definition (which is equivalent to that of a weak random variable) is frequently employed.

***Definition* 1.16.** Let H be a Hilbert space with inner product $(\cdot,\cdot)$. A mapping $x: \Omega \to H$ is said to be a *Hilbert space-valued random variable* if $(x(\omega), y)$ is a real- (or complex-) valued random variable for every $y \in H$.

When H is a separable Hilbert space, the above definition is, of course, equivalent to the other definitions of Banach space-valued random variables.

As is well known, normal or Gaussian random variables play a central role in classical probability theory. For random variables with values in a Banach space we have the following definition.

***Definition* 1.17.** A random variable $x(\omega)$ with values in a Banach space $\mathfrak{X}$ is said to be a *Gaussian* or *normal random variable* if $x^*(x(\omega))$ is a scalar-valued Gaussian random variable for every $x^* \in \mathfrak{X}^*$.

Let $x(\omega)$ and $y(\omega)$ be $\mathfrak{X}$-valued random variables defined on the same probability space.

***Definition* 1.18.** $x(\omega)$ and $y(\omega)$ are said to be *equivalent* if for every $B \in \mathfrak{B}$ the sets $\{\omega : x(\omega) \in B\}$ and $\{\omega : y(\omega) \in B\}$ coincide with probability one; that is, $\mu(\{\omega : x(\omega) \in B\} \triangle \{\omega : y(\omega) \in B\}) = 0$. If $\mathfrak{X}$ is separable, equivalence means $x(\omega)$ and $y(\omega)$ are equal with probability one; that is, $\mu(\{\omega : x(\omega) \neq y(\omega)\}) = 0$.

Sequences of Banach space-valued random variables are encountered in many problems in the theory of random equations. We give a few definitions which are the analogues of the classical definitions. Let $\{x_n(\omega)\}$ be a sequence of $\mathfrak{X}$-valued random variables.

***Definition* 1.19.** (i) $\{x_n(\omega)\}$ is a sequence of *independent* $\mathfrak{X}$-valued random variables if for every positive integer n and all sets $B_1, B_2, \ldots, B_n \in \mathfrak{B}$

$$\mu\left(\bigcap_{i=1}^{n} \{\omega : x_i(\omega) \in B_i\}\right) = \prod_{i=1}^{n} \mu(\{\omega : x_i(\omega) \in B_i\}).$$

(ii) $\{x_n(\omega)\}$ is a sequence of *identically distributed* $\mathfrak{X}$-valued random variables if for every $B \in \mathfrak{B}$ and for every pair of positive integers i,j

$$\mu(\{\omega : x_i(\omega) \in B\}) = \mu(\{\omega : x_j(\omega) \in B\}).$$

(iii) $\{x_n(\omega)\}$ is a *stationary* sequence of $\mathfrak{X}$-valued random variables if for every pair of positive integers n,j, and all sets $B_1, B_2, \ldots, B_n \in \mathfrak{B}$

$$\mu\left(\bigcap_{i=1}^{n} \{\omega : x_i(\omega) \in B_i\}\right) = \mu\left(\bigcap_{i=1}^{n} \{\omega : x_{i+j}(\omega) \in B_i\}\right).$$

C. Convergence concepts

In Sect. 1.2D we listed several modes of convergence for sequences of elements in a Banach space. We now consider the modes of convergence for

Banach space-valued random variables. Let $x(\omega)$ and $\{x_n(\omega)\}$ be $\mathfrak{X}$-valued random variables.

***Definition* 1.20.** The sequence $\{x_n(\omega)\}$ converges to $x(\omega)$ in Ω (i) *almost uniformly* if to every $\epsilon > 0$ there is a set $A_\epsilon \in \mathfrak{A}$ with $\mu(A_\epsilon) < \epsilon$ and to every $\delta > 0$ there is an integer $n(\epsilon, \delta)$ such that $\|x_n(\omega) - x(\omega)\| < \delta$ for $\omega \in \Omega - A_\epsilon$ and $n \geqslant n(\epsilon, \delta)$; (ii) *strongly almost surely* if there exists a set $A_0 \in \mathfrak{A}$ (with $\mu(A_0) = 0$) such that

$$\lim_{n\to\infty} \|x_n(\omega) - x(\omega)\| = 0 \qquad \text{for every} \quad \omega \in \Omega - A_0;$$

(iii) *weakly almost surely* if

$$\lim_{n\to\infty} x^*(x_n(\omega)) = x^*(\omega) \qquad \text{for every} \quad x^* \in \mathfrak{X}^* \quad \text{and every} \quad \omega \in \Omega - A_0;$$

(iv) *in measure* if for every $\epsilon > 0$ the outer measure μ^* of

$$\{\omega : \|x_n(\omega) - x(\omega)\| > \epsilon\} \to 0$$

as $n \to \infty$.

D. The collection of random variables with values in a separable Banach space

Let $\mathscr{V}(\Omega, \mathfrak{X})$ denote the collection of all random variables with values in a separable Banach space $\mathfrak{X}$. We now state several results which show that $\mathscr{V}$ is closed with respect to addition and scalar multiplication. Hence $\mathscr{V}$ is a linear space. We also show that $\mathscr{V}$ is closed under composition with Borel measurable functions, and that the weak limit of a sequence of elements of $\mathscr{V}$ is in $\mathscr{V}$.

THEOREM 1.3. *If $x, y \in \mathscr{V}$, then $x + y \in \mathscr{V}$.*

It is of interest to point out that Nedoma [90] has shown that if $\mathfrak{X}$ is not separable, then the sum of two $\mathfrak{X}$-valued random variables may not be an $\mathfrak{X}$-valued random variable. To be more precise, Nedoma proved the following result: *Let $\mathfrak{X}$ be an arbitrary Banach space, the cardinal number of which is greater than that of the continuum. Then the sum of two $\mathfrak{X}$-valued random variables need not be an $\mathfrak{X}$-valued random variable.*

THEOREM 1.4. *If $x \in \mathscr{V}$ and α is a real number, then $\alpha x \in \mathscr{V}$*

THEOREM 1.5. *Let x be a random variable with values in an arbitrary Banach space $(\mathfrak{X}_1, \mathfrak{B}_1)$, and let φ be a Borel measurable mapping of $(\mathfrak{X}_1, \mathfrak{B}_1)$ into an arbitrary Banach space $(\mathfrak{X}_2, \mathfrak{B}_2)$. Then $\varphi(x(\omega))$ is an $\mathfrak{X}_2$-valued random variable.*

If $\mathfrak{X}_1 = \mathfrak{X}_2 = \mathfrak{X}$ and $\mathfrak{X}$ is separable, the above result states that if $x \in \mathscr{V}$ and φ is a Borel measurable mapping of $\mathfrak{X}$ into itself, then $\varphi(x) \in \mathscr{V}$.

Finally, we state the following convergence theorem.

THEOREM 1.6. *Let $\{x_n(\omega)\}$ be a sequence of random variables in $\mathscr{V}$ which converges weakly almost surely to the element $x(\omega)$. Then $x(\omega) \in \mathscr{V}$.*

E. Integration of Banach space-valued random variables

1. ***Introduction.*** Let $x(\omega)$ be a real-valued random variable defined on $(\Omega, \mathfrak{A}, \mu)$. The *expectation* or *mean* of $x(\omega)$ is defined as the Lebesgue integral of $x(\omega)$ over Ω, if the integral exists; that is,

$$\mathscr{E}\{x(\omega)\} = (\mathrm{L})\int_{\Omega} x(\omega)\, d\mu.$$

We denote by $L_1(\Omega)$ the set of equivalence classes of real-valued random variables whose expectations exist. Hence $x(\omega) \in L_1(\Omega)$ if $\mathscr{E}\{x(\omega)\} < \infty$; and we remark that the mapping $x \to \mathscr{E}\{x\}$ is a linear functional on the real Banach space $L_1(\Omega)$. The expectations of the functions x^k $(k = 1, 2, \ldots)$ and $|x|^k$ $(k > 0)$ are called the kth *moments* and kth *absolute moments*, respectively, of the random variable $x(\omega)$. For every $p \in [1, \infty]$, we denote by $L_p(\Omega)$ the set of equivalence classes of real-valued random variables such that $\mathscr{E}\{|x(\omega)|^p\} < \infty$. $L_\infty(\Omega)$ is the set of all a.s. bounded real-valued random variables. Of particular interest is the second moment of $x(\omega)$ with respect to $\mathscr{E}\{x(\omega)\}$, which is called the *variance of* $x(\omega)$. We have

$$\begin{aligned}\mathrm{Var}\{x(\omega)\} &= \mathscr{E}\{(x(\omega) - \mathscr{E}\{x(\omega)\})^2\} \\ &= \mathscr{E}\{x^2(\omega)\} - (\mathscr{E}\{x(\omega)\})^2.\end{aligned}$$

Random variables with finite variances are called *second-order random variables*. Since $\mathscr{E}\{|x(\omega)|^2\} < \infty$, second-order random variables are elements of $L_2(\Omega)$.

Now, let $x(\omega)$ be a real-valued random variable defined on $(\Omega, \mathfrak{A}, \mu)$, and let $\mathfrak{A}_0$ be a σ-subalgebra of $\mathfrak{A}$. The *conditional expectation of* $x(\omega)$ *relative to* $\mathfrak{A}_0$, denoted by $\mathscr{E}\{x(\omega) | \mathfrak{A}_0\}$, is defined by the equality

$$\int_A \mathscr{E}\{x | \mathfrak{A}_0\}(\omega)\, d\mu = \int_A x(\omega)\, d\mu$$

for every $A \in \mathfrak{A}_0$, where $\mathscr{E}\{x | \mathfrak{A}_0\}$ is $\mathfrak{A}_0$-measurable.

We refer to the books of Doob [20], Loève [77], and Neveu [92] for detailed expositions of the expectation and conditional expectation of real-valued random variables and their properties.

In this subsection we consider the expectation and conditional expectation of Banach space-valued random variables. For real-valued random variables, the expectation was defined via the Lebesgue integral; hence in order to define in an analogous manner the expectation (and conditional expectation) of a Banach space-valued random variable we require generalizations of the Lebesgue integral for Banach space-valued measurable functions. The two generalizations we will consider are the Pettis and Bochner integrals, defined for weakly and strongly measurable functions, respectively.

2. ***Expectation of Banach space-valued random variables.*** Let $x(\omega)$ be an $\mathfrak{X}$-valued random variable defined on Ω.

***Definition* 1.21.** $x(\omega)$ is said to be *Pettis integrable* if and only if there is an element $m_A \in \mathfrak{X}$ corresponding to each $A \in \mathfrak{A}$ such that

$$x^*(m_A) = (\mathrm{L})\int_A x^*(x(\omega))\, d\mu$$

for all $x^* \in \mathfrak{X}^*$, where the integral is assumed to exist in the sense of Lebesgue. The *Pettis integral* $x(\omega)$ is defined as

$$m_A = (\mathrm{P})\int_A x(\omega)\, d\mu.$$

It is clear from the above definition that an $\mathfrak{X}$-valued random variable which is Pettis integrable is a weak random variable, but not necessarily a strong random variable. Hence for weak random variable we can introduce the following definition.

***Definition* 1.22.** Let $x(\omega)$ be a weak random variable. The *expectation* of $x(\omega)$, denoted by $\mathscr{E}_{\mathrm{w}}\{x(\omega)\}$, is defined as the Pettis integral of $x(\omega)$ over Ω; that is

$$\mathscr{E}_{\mathrm{w}}\{x(\omega)\} = (\mathrm{P})\int_\Omega x(\omega)\, d\mu.$$

The expectation of weak random variables is frequently referred to as the *weak expectation* or the *Mourier expectation*, since the above definition of the expectation was introduced by Mourier [84]. The Mourier expectation has been explicitly determined by Nasr [89] for random variables with values in a number of concrete Banach spaces, and by Bharucha-Reid [9] for Orlicz space-valued random variables.

The following properties of the weak expectation are immediate consequences of Definition 1.22.

1. $\mathscr{E}_{\mathrm{w}}\{x(\omega)\}$ is uniquely defined.

2. If $\mathscr{E}_w\{x_1(\omega)\}$ and $\mathscr{E}_w\{x_2(\omega)\}$ exist, then $\mathscr{E}_w\{x_1(\omega)+x_2(\omega)\}$ exists and $\mathscr{E}_w\{x_1(\omega)+x_2(\omega)\}=\mathscr{E}_w\{x_1(\omega)\}+\mathscr{E}_w\{x_2(\omega)\}$.
3. If $x(\omega)=x_0$ (a constant) a.s., then $\mathscr{E}_w\{x(\omega)\}$ exists and $\mathscr{E}_w\{x(\omega)\}=x_0$.
4. If $\mathscr{E}_w\{x(\omega)\}$ exists, then $\mathscr{E}_w\{\alpha x(\omega)\}$ exists and $\mathscr{E}_w\{\alpha x(\omega)\}=\alpha\mathscr{E}_w\{x(\omega)\}$.
5. If $\mathscr{E}\{\|x(\omega)\|\}<\infty$ (that is, $\|x(\omega)\|\in L_1(\Omega)$), then $\mathscr{E}_w\{x(\omega)\}$ exists and $\|\mathscr{E}_w\{x(\omega)\}\|\leqslant\mathscr{E}\{\|x(\omega)\|\}$.

Another property of the Pettis integral, which is applicable in the study of random equations, is contained in the following result (cf. Hille and Phillips [46, p. 78]).

THEOREM 1.7. *Let L be a bounded linear operator on $\mathfrak{X}$ to itself, and let $x(\omega)$ be a weak random variable with values in $\mathfrak{X}$. If $\mathscr{E}_w\{x(\omega)\}$ exists and is equal to m, then $\mathscr{E}_w\{L[x(\omega)]\}$ exists and*

$$(\mathrm{P})\int_\Omega L[x(\omega)]\,d\mu=\mathscr{E}_w\{L[x(\omega)]\}=L[m].$$

We next consider the definition of the expectation of Banach space-valued random variables via the Bochner integral.

***Definition* 1.23.** A simple random variable $x(\omega)$ is said to be *Bochner integrable* if and only if $\|x(\omega)\|\in L_1(\Omega)$. By definition

$$(\mathrm{B})\int_A x(\omega)\,d\mu=\sum_{i=1}^{\infty}x_i\,\mu(A_i\cap A),$$

where $x(\omega)=x_i$ on $A_i\in\mathfrak{A}$, $i=1,2,\ldots$.

We remark that the integral is well defined for all $A\in\mathfrak{A}$, hence for Ω itself. This follows from the fact that the above series is absolutely convergent since

$$\sum_{i=1}^{\infty}\|x_i\|\,\mu(A_i\cap A)=(\mathrm{L})\int_A\|x(\omega)\|\,d\mu.$$

Therefore,

$$\left\|(\mathrm{B})\int_A x(\omega)\,d\mu\right\|\leqslant(\mathrm{L})\int_A\|x(\omega)\|\,d\mu$$

for simple random variables. Also, since the series is absolutely convergent, we have

$$\begin{aligned}x^*\Big(\int_A x(\omega)\,d\mu\Big)&=\sum_{i=1}^{\infty}x^*(x_i)\,\mu(A_i\cap A)\\&=(\mathrm{L})\int_A x^*(x(\omega))\,d\mu\end{aligned}$$

for every $x^* \in \mathfrak{X}^*$. It follows from the above that the Bochner and Pettis integrals of simple random variables coincide.

***Definition* 1.24.** $x(\omega)$ is said to be *Bochner integrable* if and only if there exists a sequence of simple random variables $\{x_n(\omega)\}$ converging almost surely to $x(\omega)$ and such that

$$\lim_{n\to\infty} \int_\Omega \|x_n(\omega) - x(\omega)\| \, d\mu = 0.$$

By definition

$$(\mathrm{B})\int_A x(\omega)\, d\mu = \lim_{n\to\infty} (\mathrm{B})\int_A x_n(\omega)\, d\mu$$

for every $A \in \mathfrak{A}$ and $A = \Omega$.

It is clear from the above definition that every Bochner integrable random variable is a strong random variable. Furthermore, it is easy to show that every Bochner integrable random variable is also Pettis integrable, and that the integrals have the same value.

***Definition* 1.25.** Let $x(\omega)$ be a strong random variable. The *expectation* of $x(\omega)$, denoted by $\mathscr{E}_s\{x(\omega)\}$, is defined as the Bochner integral of $x(\omega)$ over Ω; that is,

$$\mathscr{E}_s\{x(\omega)\} = (\mathrm{B})\int_\Omega x(\omega)\, d\mu.$$

The above expectation is often referred to as the *strong expectation*; and, in view of the remark above, the existence of the strong expectation implies the existence of the weak expectation, and $\mathscr{E}_s\{x(\omega)\} = \mathscr{E}_w\{x(\omega)\}$.

The class of random variables which have strong expectation can be characterized using the following result (cf. Hille and Phillips [46, p. 80]).

THEOREM 1.8. *A necessary and sufficient condition that $\mathscr{E}_s\{x(\omega)\}$ exists is that $x(\omega)$ be a strong random variable and that $\mathscr{E}\{\|x(\omega)\|\} < \infty$.*

Let $B_1 = B_1(\Omega, \mathfrak{X})$ denote the class of all $\mathfrak{X}$-valued random variables which are Bochner integrable. It is clear that B_1 is a complex linear space, and that the Bochner integral is a linear operator on B_1. Furthermore, B_1 is a Banach space with norm

$$[x]_1 = (\mathrm{L})\int_\Omega \|x(\omega)\|\, d\mu = \mathscr{E}\{\|x(\omega)\|\}$$

if we consider the elements to be equivalence classes of random variables, where equivalence is in the sense of Definition 1.18. We remark that when

$\mathfrak{X} = R$, $B_1(\Omega, \mathfrak{X}) = L_1(\Omega)$, the class of real-valued random variables for which $\mathscr{E}\{x(\omega)\} < \infty$.

Now, let $B_p = B_p(\Omega, \mathfrak{X})$, $1 < p < \infty$, denote the class of all $\mathfrak{X}$-valued random variables $x(\omega)$ such that

$$\lim_{n\to\infty} (\mathrm{L})\int_\Omega \|x_n(\omega) - x(\omega)\|^p \, d\mu = 0$$

for some sequence $\{x_n(\omega)\}$ of simple random variables. The spaces B_p are also Banach spaces, with norm

$$[x]_p = \left((\mathrm{L})\int_\Omega \|x(\omega)\|^p \, d\mu\right)^{1/p} = (\mathscr{E}\{\|x(\omega)\|^p\})^{1/p}.$$

Clearly, any $x(\omega) \in B_p$ is a strong random variable and $\|x(\omega)\| \in L_p(\Omega)$. The converse is also true. Finally, we define $B_\infty = B_\infty(\Omega, \mathfrak{X})$ to be the class of all strong random variables $x(\omega)$ such that $\|x(\omega)\| \in L_\infty(\Omega)$.

We now list some properties of the strong expectation. We refer to Hille and Phillips [46, Sect. 3.7] and Zaanen [131, Chap. 6] for details.

1. $\mathscr{E}_s\{x(\omega)\}$ is uniquely defined.
2. If $\mathscr{E}_s\{x_1(\omega)\}$ and $\mathscr{E}_s\{x_2(\omega)\}$ exist, and α_1, α_2 are constants, then $\mathscr{E}_s\{\alpha_1 x_1(\omega) + \alpha_2 x_2(\omega)\}$ exists; and

$$\mathscr{E}_s\{\alpha_1 x_1(\omega) + \alpha_2 x_2(\omega)\} = \alpha_1 \mathscr{E}_s\{x_1(\omega)\} + \alpha_2 \mathscr{E}_s\{x_2(\omega)\}.$$

3. If $\mathscr{E}_s\{x(\omega)\}$ exists, then $\|\mathscr{E}_s\{x(\omega)\}\| \leqslant \mathscr{E}\{\|x(\omega)\|\}$.
4. If $\{x_n(\omega)\} \subset B_1$ converges almost surely to some limit random variable $x(\omega)$ and there exists a fixed nonnegative random variable $y(\omega) \in L_1(\Omega)$ such that $\|x_n(\omega)\| \leqslant y(\omega)$ for all n and ω, then $x(\omega) \in B_1$ and

$$\lim_{n\to\infty} \mathscr{E}_s\{x_n(\omega)\} = \mathscr{E}_s\{x(\omega)\}.$$

An analogue of Theorem 1.7 for Pettis integrals is the following result (cf. Hille and Phillips [46, p. 83]).

THEOREM 1.9. *Let L be a closed linear operator*† *on $\mathfrak{X}$ to itself. If $\mathscr{E}_s\{x(\omega)\}$ and $\mathscr{E}_s\{L[x(\omega)]\}$ exist, then $L[\mathscr{E}_s\{x(\omega)\}] = \mathscr{E}_s\{L[x(\omega)]\}$.*

3. ***Higher moments of Banach space-valued random variables.*** We restrict our attention to separable Banach spaces, and define higher moments via the Bochner integral.

***Definition* 1.26.** Let $x(\omega) \in B_1$, and let $k \geqslant 1$ be a real number. The *kth absolute moment* of $x(\omega)$ with respect to $\mathscr{E}\{x(\omega)\}$, denoted by $\mathscr{E}^{(k)}\{x(\omega)\}$, is

† Closed linear operators are defined in Sect. 2.2.

defined by the integral

$$\int_{\Omega} \|x(\omega) - \mathscr{E}\{x(\omega)\}\|^k \, d\mu$$

if the integral exists. Hence $\mathscr{E}^{(k)}\{x(\omega)\}$ exists if $\|x(\omega) - \mathscr{E}\{x(\omega)\}\| \in L_k(\Omega)$.

In view of the above, we can define the *variance* of a Banach space-valued random variable as follows: $\operatorname{Var}\{x(\omega)\} = \mathscr{E}\{\|x(\omega) - \mathscr{E}\{x(\omega)\}\|^2\}$. Hence, second-order Banach space-valued random variables are elements of $B^2(\Omega, \mathfrak{X})$. For another definition of the variance we refer to Birnbaum [12].

4. ***Conditional expectation of Banach space-valued random variables.*** As in the definition of the conditional expectation of real-valued random variables, let $\mathfrak{A}_0$ denote a σ-subalgebra of $\mathfrak{A}$.

***Definition* 1.27.** A strong random variable $\mathscr{E}_s\{x|\mathfrak{A}_0\}(\omega)$ is said to be the (*strong*) *conditional expectation* of a random variable $x(\omega) \in L_1(\Omega, \mathfrak{X})$ relative to $\mathfrak{A}_0$ if and only if it satisfies the following conditions:

i. $\mathscr{E}_s\{x|\mathfrak{A}_0\}(\omega)$ is measurable with respect to $\mathfrak{A}_0$ and is an element of $L_1(\Omega, \mathfrak{X})$;

ii. (B) $\int_A \mathscr{E}_s\{x|\mathfrak{A}_0\}(\omega) \, d\mu =$ (B) $\int_A x(\omega) \, d\mu$ for every $A \in \mathfrak{A}_0$.

In order to develop martingale theory in Banach spaces, the above definition was introduced by Chatterji [15], Driml and Hanš [21], Moy [87], and Scalora [109]; various properties of the (strong) conditional expectation were studied by these authors. We remark that it is also possible to define the (weak) conditional expectation $\mathscr{E}_w\{x|\mathfrak{A}_0\}(\omega)$ of Banach space-valued random variables; and a formal representation of the conditional expectation has been obtained by Brooks [13] when $x(\omega)$ is Pettis integrable.

We now list some properties of the (strong) conditional expectation:

1. $\mathscr{E}_s\{x|\mathfrak{A}_0\}(\omega)$ is unique in the sense of equality a.s.
2. If $x(\omega) = x_0$ (a constant) on Ω, then $\mathscr{E}_s\{x|\mathfrak{A}_0\}(\omega) = x_0$ a.s.
3. $\mathscr{E}_s\{\sum_{i=1}^n \alpha_i x_i|\mathfrak{A}_0\}(\omega) = \sum_{i=1}^n \alpha_1 \mathscr{E}_s\{x_i|\mathfrak{A}_0\}(\omega)$ a.s., where the α_i are scalars.
4. $[\mathscr{E}_s\{x|\mathfrak{A}_0\}(\omega)]_1 \leqslant \mathscr{E}\{\|x\||\mathfrak{A}_0\}$ a.s.
5. If L is a bounded linear operator on $\mathfrak{X}$ to itself, then

$$L[\mathscr{E}_s\{x|\mathfrak{A}_0\}(\omega)] = \mathscr{E}_s\{L[x]|\mathfrak{A}_0\}(\omega) \quad \text{a.s.}$$

6. If $x_n(\omega) \to x(\omega)$ strongly a.s., and there is a nonnegative real-valued random variable $y(\omega) \in L_1(\Omega)$ such that $\|x_n(\omega)\| \leqslant y(\omega)$ a.s., then

$$\lim_{n\to\infty} \mathscr{E}_s\{x_n|\mathfrak{A}_0\}(\omega) = \mathscr{E}_s\{x|\mathfrak{A}_0\}(\omega) \quad \text{a.s.}$$

We refer to Kappos [66, Chap. VIII] for another formulation of the conditional expectation of Banach space-valued random variables; and to Umegaki and Bharucha-Reid [120] for a proof of the existence and uniqueness of the strong conditional expectation using tensor product methods.

F. Banach algebra-valued random variables

1. *Introduction.* In this section we consider random variables with values in a Banach algebra. The study of Banach algebra-valued random variables is of great importance in the theory of random equations since many of the Banach spaces encountered are also algebras; and, as we shall see in Chaps. 2 and 3, Banach algebras of operators are of fundamental importance in the study of random operators and solutions of random equations.

***Definition* 1.28.** A *Banach algebra* is a (complex) Banach space $\mathfrak{X}$ which is also an *algebra* (that is, the operations of addition, scalar multiplication, and multiplication are defined for elements of $\mathfrak{X}$) in which the multiplicative structure is related to the norm by the inequality $\|xy\| \leqslant \|x\| \cdot \|y\|$. We will assume that $\mathfrak{X}$ has an *identity* (or *unit*) *element* e, such that $xe = ex = x$ for all $x \in \mathfrak{X}$, and $\|e\| = 1$.

It follows from the above multiplicative inequality that the operation of multiplication is jointly continuous in any Banach algebra; that is, if $x_n \to x$ and $y_n \to y$, then $x_n y_n \to xy$.

In Sect. 1.3D we showed that if $x(\omega)$ and $y(\omega)$ are random variables with values in a separable Banach space $\mathfrak{X}$, then αx (α a scalar) and $x + y$ are Banach space-valued random variables. If $\mathfrak{X}$ is also a Banach algebra, we must show that the product $x(\omega)y(\omega)$ of two random variables with values in $\mathfrak{X}$ is an $\mathfrak{X}$-valued random variable. Since $\mathfrak{X}$ is separable, we will assume that $x(\omega)$ and $y(\omega)$ are strong random variables. Hence $x(\omega)$ and $y(\omega)$ can be approximated (in norm) by simple random variables $x_n(\omega)$ and $y_n(\omega)$. Hence $x_n(\omega)y_n(\omega)$ is well defined. Since, in a Banach algebra, the operation of multiplication is continuous, we have $x_n(\omega)y_n(\omega) \to x(\omega)y(\omega)$; hence $z(\omega) = x(\omega)y(\omega)$ is a well-defined $\mathfrak{X}$-valued random variable.

***Definition* 1.29.** A Banach algebra $\mathfrak{X}$ is said to be *commutative* in case $xy = yx$ for all $x, y \in \mathfrak{X}$.

We will also need the following notion.

***Definition* 1.30.** An element $x \in \mathfrak{X}$ is said to be a *regular element* (or *invertible*) if x has an inverse in $\mathfrak{X}$; that is if there exists an element $x^{-1} \in \mathfrak{X}$

such that $x^{-1}x = xx^{-1} = e$. Elements of $\mathfrak{X}$ which are not regular are called *singular*.

We remark that the regular elements form an open set in $\mathfrak{X}$, and the set of singular elements in $\mathfrak{X}$ is closed and contains at least the null element θ.

Examples of Banach algebras can be found in any of the standard texts devoted to the theory of Banach algebras: for example, Hille and Phillips [46, Chaps. IV and V], Loomis [78, Chaps. IV and V], Naimark [88], and Rickart [105]. We list several examples of Banach algebras which will be encountered in this book.

1. The Banach space $\mathfrak{X} = C[a,b]$ of all complex-valued continuous functions on $[a,b]$ with the uniform norm is a Banach algebra.
2. The Banach space $L_\infty(S,\mathfrak{F},m)$ of all essentially bounded complex-valued measurable functions on a measure space $(S,\mathfrak{F},m)$ with the essential uniform norm is a Banach algebra.

Examples 1 and 2 are examples of *function algebras*, and multiplication is defined in the pointwise fashion, with $(xy)(s) = x(s)y(s)$.

3. The set $\mathfrak{L}(\mathfrak{X})$ of all bounded linear operators on a Banach space $\mathfrak{X}$ to itself is a Banach algebra.

Example 3 is an example of an *operator algebra*, and multiplication is defined by composition, that is, $L_1 L_2[x] = L_1[L_2 x]$.

An important class of Banach algebras is the class of so-called $*$-algebras.

***Definition* 1.31.** Let $\mathfrak{X}$ be a Banach algebra. A mapping $x \to x^*$ of $\mathfrak{X}$ onto itself is called an *involution* provided the following conditions are satisfied: (i) $(x^*)^* = x$, (ii) $(\alpha_1 x_1 + \alpha_2 x_2)^* = \bar{\alpha}_1 x_1^* + \bar{\alpha}_2 x_2^*$, and (iii) $(xy)^* = y^* x^*$. A Banach algebra with an involution is called a *$*$-algebra*.

The element x^* in a $*$-algebra is called the *adjoint* of x. An element x is said to be *self-adjoint* if $x = x^*$; and x is said to be *normal* if $xx^* = x^*x$. We remark that in a commutative $*$-algebra every element is normal; and for normal elements $\|x^2\| = \|x\|^2$, and $\|x\| = \|x^*\|$.

Of the examples of Banach algebras listed above, $C[a,b]$ and $L_\infty(S,\mathfrak{F},m)$ are $*$-algebras with the involution defined by $x^* = \bar{x}$; and $\mathfrak{L}(H)$ is a $*$-algebra, where L^* is the adjoint of L.

***Definition* 1.32.** A $*$-algebra is said to be a *B^*-algebra* if $\|xx^*\| = \|x\| \cdot \|x^*\|$ for all elements in the algebra. An *H^*-algebra* is a $*$-algebra $\mathfrak{H}$ which is also a

Hilbert space, and which has the property $(xy, z) = (y, x^* z)$. It is also assumed that $\|x^*\| = \|x\|$, and that $x \neq \theta$ implies $x^* x \neq \theta$.

An important class of B^*-algebras is the class of C^*-algebras. A *C^*-algebra* is a uniformly closed self-adjoint Banach subalgebra of bounded linear operators on a Hilbert space H. For a C^*-algebra the involution is defined via the inner product: $(Lx, y) = (x, L^* y)$; and the involution operation $L \to L^*$ has the following properties: (i) $L^{**} = L$, (ii) $(\alpha_1 L_1 + \alpha_2 L_2)^* = \bar{\alpha}_1 L_1^* + \bar{\alpha}_2 L_2^*$, (iii) $(L_1 L_2)^* = L_2^* L_1^*$, (iv) $\|L^* L\| = \|L\|^2$, and (v) $(I + L^* L)^{-1} \in \mathfrak{L}(H)$, where I is the identity operator. Gel'fand and Naimark (cf. any standard reference) have shown that every C^*-algebra is isometrically isomorphic to an algebra of bounded linear operators on a suitable Hilbert space. Furthermore, it is known that if a C^*-algebra is commutative, then it is isometrically isomorphic to the algebra of continuous complex-valued functions on a suitable compact Hausdorff space.

In this section we consider B^*-algebra-valued and H^*-algebra-valued random variables. In Chap. 2 we will consider operator-valued random variables, and the spectral theory of random operators; and in Sect. 1.6 we will consider some limit theorems for Banach algebra-valued random variables.

2. *Random variables with values in a commutative B^*-algebra.* Let $x(\omega)$ be a random variable with values in a separable commutative B^*-algebra $\mathfrak{X}$. Since $\mathfrak{X}$ is separable, we know that $\mathscr{E}\{x(\omega)\}$ exists if $\|x(\omega)\| \in L_1(\Omega)$ (cf. Mourier [84]). Following Theodorescu [117], we consider the second moment $\mathscr{E}^{(2)}\{x(\omega)\}$ and study its properties. By definition, $\mathscr{E}^{(2)}\{x(\omega)\} = \mathscr{E}\{x^2(\omega)\}$ (if it exists) is the *second moment* of $x(\omega)$.

THEOREM 1.10. *Let $\mathfrak{X}$ be a separable commutative B^*-algebra. If $\mathscr{E}\{\|x\|^2\} < \infty$, then $\mathscr{E}\{x^2(\omega)\}$ exists.*

***Definition* 1.33.** Let $x(\omega)$ and $y(\omega)$ be two $\mathfrak{X}$-valued random variables. $x(\omega)$ and $y(\omega)$ are said to be *uncorrelated* if $\mathscr{E}\{x(\omega) y(\omega)\} = \mathscr{E}\{x(\omega)\} \mathscr{E}\{y(\omega)\}$.

We now list some properties of the second moment $\mathscr{E}^{(2)}\{x(\omega)\}$, several of which are analogues of the properties of the weak expectation $\mathscr{E}_w\{x(\omega)\}$.

1. *Let α be a complex number, and suppose that $\mathscr{E}^{(2)}\{x(\omega)\}$ exists. Then $\mathscr{E}^{(2)}\{\alpha x(\omega)\}$ exists, and $\mathscr{E}^{(2)}\{\alpha x(\omega)\} = \alpha^2 \mathscr{E}^{(2)}\{x(\omega)\}$.*
2. *Let $x(\omega) = x_0$ a.s., and let $\xi(\omega) \in L_2(\Omega)$. Then $\mathscr{E}^{(2)}\{\xi(\omega) x(\omega)\}$ exists, and $\mathscr{E}^{(2)}\{\xi(\omega) x(\omega)\} = \mathscr{E}^{(2)}\{\xi(\omega)\} x_0^2$.*

3. *Let $x(\omega)$ and $y(\omega)$ be uncorrelated. If $\mathscr{E}^{(2)}\{x(\omega)\}$ and $\mathscr{E}^{(2)}\{y(\omega)\}$ exist and $\mathscr{E}\{x(\omega)\} = \mathscr{E}\{y(\omega)\} = \theta$, then $\mathscr{E}^{(2)}\{x(\omega)+y(\omega)\}$ exists and*

$$\mathscr{E}^{(2)}\{x(\omega)+y(\omega)\} = \mathscr{E}^{(2)}\{x(\omega)\} + \mathscr{E}^{(2)}\{y(\omega)\}.$$

4. Since $\mathfrak{X}$ is separable, $\|x(\omega)\|$ is a nonnegative real-valued random variable. *If $\|x(\omega)\| \in L_2(\Omega)$ and $\mathscr{E}^{(2)}\{x(\omega)\}$ exists, then $\|\mathscr{E}^{(2)}\{x(\omega)\}\| \leqslant \mathscr{E}\{\|x(\omega)\|^2\}$.* This result follows from properties of the Pettis integral.

For other results on random variables with values in a separable commutative B^*-algebra, we refer to the paper of Haïnis [42].

In closing, we refer to a result of Srinivasan [115] on Gaussian random variables with values in a separable commutative B^*-algebra $\mathfrak{X}$. Let $x_1(\omega), x_2(\omega), \ldots, x_n(\omega)$ be n independent observations on a Gaussian random variable with values in $\mathfrak{X}$. Then the function

$$\Phi_n = (1/n) \sum_{i=1}^{n} (x_i(\omega) - \mathscr{E}\{x(\omega)\})^2$$

is a sufficient estimator of the variance of $x(\omega)$.

3. *Random variables with values in an H^*-algebra.* Let $\mathfrak{H}$ be a separable H^*-algebra. Since $\mathfrak{H}$ is a Hilbert space, a mapping $x: \Omega \to \mathfrak{H}$ is an $\mathfrak{H}$-valued random variable if $(x(\omega), y)$ is a real-valued random variable for every $y \in \mathfrak{H}$. Let $B_2(\Omega, \mathfrak{H})$ denote the class of equivalent $\mathfrak{H}$-valued random variables such that $\int_\Omega \|x(\omega)\|^2 \, d\mu < \infty$.

Following Haïnis [41], we now state several results for random variables with values in the H^*-algebra $\mathfrak{H}$.

1. *If $x(\omega)$ is an $\mathfrak{H}$-valued random variable, then $(x(\omega))^*$ is also.* The proof follows from the relation $(x^*, y) = (x, y^*)$.

2. *If $x(\omega)$ and $y(\omega)$ are $\mathfrak{H}$-valued random variables, then $x(\omega)+y(\omega)$ is also an $\mathfrak{H}$-valued random variable.*

3. *If $x(\omega)$ is an $\mathfrak{H}$-valued random variable, and ξ is an element of $\mathfrak{H}$, then $\xi x(\omega)$ is also an $\mathfrak{H}$-valued random variable.* Since $\mathfrak{H}$ is an H^*-algebra, we have $(\xi x(\omega), y) = (x(\omega), \xi^* y)$ for every $y \in \mathfrak{H}$.

4. *If $x(\omega)$ and $y(\omega)$ are $\mathfrak{H}$-valued random variables, then their product $x(\omega)y(\omega)$ is defined, and is also an $\mathfrak{H}$-valued random variable.* The proof is based on the fact that if $x(\omega)$ is an $\mathfrak{H}$-valued random variable, then $\|x(\omega)\|$ is a nonnegative real-valued random variable, and that

$$\begin{aligned} 4(x(\omega)y(\omega), z) &= 4(y(\omega), x^*(\omega)z) \\ &= \|y(\omega) + x^*(\omega)z\| + \|y(\omega) - x^*(\omega)z\| \\ &\quad + i\|y(\omega) + ix^*(\omega)z\| - i\|y(\omega) - ix^*(\omega)z\|. \end{aligned}$$

Finally, we state the following result concerning the expectation of an $\mathfrak{H}$-valued random variable.

5. *If* $\mathscr{E}\{x(\omega)\}$ *exists, then* $\mathscr{E}\{x^*(\omega)\}$ *exists and* $\mathscr{E}\{x^*(\omega)\} = (\mathscr{E}\{x(\omega)\})^*$. For

$$\begin{aligned}(y, \mathscr{E}\{x^*(\omega)\}) &= \mathscr{E}\{(y, x^*(\omega))\} = \mathscr{E}\{(x(\omega), y^*)\} \\ &= (\mathscr{E}\{x(\omega)\}, y^*) = (y, (\mathscr{E}\{x(\omega)\})^*).\end{aligned}$$

We refer to the paper of Haïnis for other results on H^*-algebra-valued random variables.

1.4 Banach Space-Valued Random Functions

A. Introduction

Let $(\mathfrak{X}, \mathfrak{B})$ be a measurable space where $\mathfrak{X}$ is a Banach space and $\mathfrak{B}$ is the σ-algebra of Borel subsets of $\mathfrak{X}$; and let $(T, \mathfrak{T})$ be a measurable space where T is a subset of the extended real line $\bar{R}$ and $\mathfrak{T}$ is the σ-algebra of Borel subsets of T, that is, $\mathfrak{T} = T \cap \bar{\mathfrak{R}}$, where $\bar{\mathfrak{R}}$ is the σ-algebra of Borel subsets of $\bar{R}$.

***Definition* 1.34.** An *$\mathfrak{X}$-valued random function* on T is a mapping $x(t, \omega): T \times \Omega \to \mathfrak{X}$ such that for every $t \in T$, x is an $\mathfrak{X}$-valued random variable. For every fixed $\omega \in \Omega$, the function x defined on T is called a *realization*, *trajectory*, or *sample function* of $x(t, \omega)$.

***Definition* 1.35.** Let $x(t, \omega)$ and $y(t, \omega)$ be two $\mathfrak{X}$-valued random functions defined on the same probability space and the same interval T. $x(t, \omega)$ and $y(t, \omega)$ are said to be *equivalent* if $x(t, \omega) = y(t, \omega)$ a.s. for every $t \in T$.

We now introduce the important notions of separability and measurability of $\mathfrak{X}$-valued random functions.

***Definition* 1.36.** An $\mathfrak{X}$-valued random function $x(t, \omega)$ is said to be *separable* if there exists a countable set $S \subset T$ and an event A_0, with $\mu(A_0) = 0$, such that for all events of the form $\{\omega : x(t, \omega) \in F,\ t \in I \cap T\}$, where F is a closed set in $\mathfrak{X}$ and I is an open interval, the symmetric difference

$$\{\omega : x(t, \omega) \in F, \quad t \in I \cap S\} \bigtriangleup \{\omega : x(t, \omega) \in F, \quad t \in I \cap T\} \subset A_0.$$

The set S is called a *separant*, or a *set of separability*, for $x(t, \omega)$.

Consider the product measure space $(T \times \Omega, \mathfrak{T} \otimes \mathfrak{A}, \lambda \times \mu)$ and the measurable space $(\mathfrak{X}, \mathfrak{B})$.

***Definition* 1.37.** A measurable mapping $x(t, \omega): (T \times \Omega, \mathfrak{T} \otimes \mathfrak{A}) \to (\mathfrak{X}, \mathfrak{B})$ is called a *measurable $\mathfrak{X}$-valued random function.*

We state the following analogue of Fubini's theorem for Banach space-valued functions (cf. Hille and Phillips [46, p. 84]).

THEOREM 1.11. *If $x(t,\omega)$ is Bochner integrable on $T \times \Omega$, then the functions $y(\omega) = (\mathrm{B}) \int_T x(t,\omega)\,d\lambda$ and $z(t) = (\mathrm{B}) \int_\Omega x(t,\omega)\,d\mu$ are defined almost everywhere in Ω and T, respectively, and*

$$(\mathrm{B})\int_{T\times\Omega} x(t,\omega)\,d(\lambda \times \mu) = (\mathrm{B})\int_T z(t)\,d\lambda = (\mathrm{B})\int_\Omega y(\omega)\,d\mu.$$

As a consequence of the above theorem, we have

$$\mathscr{E}_s\{y(\omega)\} = \int_T \mathscr{E}_s\{x(t,\omega)\}\,d\lambda.$$

Finally, we introduce some concepts of continuity for $\mathfrak{X}$-valued random functions.

***Definition* 1.38.** An $\mathfrak{X}$-valued random function $x(t,\omega)$ is said to be (1) *weakly continuous in probability at a point $t_0 \in T$* if

$$\lim_{t\to t_0} \mu(\{\omega : |x^*(x(t_0,\omega) - x(t,\omega))| > \epsilon\}) = 0$$

for every $\epsilon > 0$ and every $x^* \in \mathfrak{X}^*$, (2) *strongly continuous in probability at a point $t_0 \in T$* if

$$\lim_{t\to t_0} \mu(\{\omega : \|x(t_0,\omega) - x(t,\omega)\| > \epsilon\}) = 0$$

for every $\epsilon > 0$. $x(t,\omega)$ is said to be *weakly* (resp. *strongly*) *continuous in probability on T* if it is weakly (resp. strongly) continuous in probability at every point of T. (3) An $\mathfrak{X}$-valued random function $x(t,\omega)$ such that $\|x(t,\omega)\| \in L_p(\Omega)$ for every $t \in T$ is said to be *continuous in mean of order p at $t_0 \in T$* if

$$\lim_{t\to t_0} [x(t_0,\omega) - x(t,\omega)]_p = 0;$$

and is *continuous in mean of order p on T* if the above holds at every $t_0 \in T$.

Let Ω_{t_0} denote the complement of the set

$$\{\omega : x(t_0,\omega) = \lim_{t\to t_0} x(t,\omega)\},$$

where the limit is taken in the weak (resp. strong) sense.

***Definition* 1.39.** A separable $\mathfrak{X}$-valued random function $x(t,\omega)$ is said to be *weakly* (resp. *strongly*) *continuous a.s. at a point $t_0 \in T$* if $\mu(\Omega_{t0}) = 0$. $x(t,\omega)$ is said to be *weakly* (resp. *strongly*) *continuous a.s. on T* if it is weakly (resp. strongly) continuous a.s. at every point of T.

In the remaining subsections we define and discuss briefly some of the basic classes of random functions with values in a Banach space which are encountered in the theory of random equations; namely, stationary random functions, Gaussian random functions, Markov processes, Wiener processes, and martingales. Some authors have defined Banach space-valued random functions in a weak sense; that is, a Banach space-valued random function $x(t, \omega)$ is said to be of a given type if the scalar-valued process $x^*(x(t, \omega))$ is of the same type for all $x^* \in \mathfrak{X}^*$. In several cases, however, it is possible to give direct definitions of Banach space-valued random functions that belong to the basic classes of random functions.

B. Stationary random functions

Let $x(t, \omega)$ be a real- (or complex-) valued random function on a parameter set $T \subset R$ with finite second moments, that is, $\mathscr{E}\{|x(t, \omega)|^2\} < \infty$ for all $t \in T$. $x(t, \omega)$ is said to be *stationary in the wide sense*, or *weakly stationary*, if (i) its expectation $\mathscr{E}\{x(t, \omega)\}$ is a constant, not depending on t (which we can take to be zero); and (ii) its *covariance function* $\mathscr{E}\{x(s, \omega)\overline{x(t, \omega)}\} = R(s, t)$ depends only on the difference $s - t$. Let $L_2(\Omega)$ denote the Hilbert space of second-order real (or complex) random variables with inner product $(x, y) = \mathscr{E}\{x\bar{y}\}$. Then we can also define a (wide-sense) stationary random function as a mapping $x(t)\colon T \to L_2(\Omega)$ such that (i) above holds, and (ii)

$$R(s, t) = R(s - t) = (x(s, \omega), x(t, \omega)).$$

Condition (ii) is equivalent to the following: There exists in the closed linear subspace H spanned by $x(t)$, $t \in T$, in $L_2(\Omega)$, a group of unitary operators $\{U(t), t \in T\}$ such that $U(t)[x(s, \omega)] = x(s + t, \omega)$. Hence, any wide-sense stationary random function can be represented as an H-valued function given by the relation $x(t, \omega) = U(t)[x(0, \omega)]$. For a detailed discussion of stationary random functions and their properties we refer to the books of Cramér and Leadbetter [16], Hida [45], Rozanov [106], and Yaglom [129].

Generalizations of the theory of real-valued stationary random functions have been considered by a number of probabilists. We refer, in particular, to the papers of Kallianpur and Mandrekar [57], Payen [94], and Vo-Khac [128] on Hilbert space-valued stationary random functions, Loynes [80] on stationary random functions with values in the class of so-called LVH-spaces; Vo-Khac [127] on Banach space-valued stationary random functions; Haïnis [42] and Theodorescu [117] on Banach algebra-valued stationary random functions; and Gel'fand and Vilenkin [33, Chap. III], Hida [45], Itô [49], and Urbanik [121] on generalized stationary random functions. We will restrict our attention to Hilbert space-, Banach space-, and Banach algebra-valued stationary random functions.

Let H be a separable Hilbert space with inner product $(\cdot,\cdot)$, and with reference to Sect. 1.3E, let $B_2(\Omega, H)$ denote the Banach space of H-valued random variables such that (B) $\int_\Omega \|x(\omega)\|^2 \, d\mu < \infty$. It is easy to show that $B_2(\Omega, H)$ is a Hilbert space with inner product

$$\langle x, y\rangle = \int_\Omega (x(\omega), y(\omega)) \, d\mu$$

and norm $[x]_2 = \langle x, x\rangle^{1/2}$. We remark that $B_2(\Omega, H)$ is the tensor product Hilbert space $L_2(\Omega) \hat{\otimes} H$ (cf. Umegaki and Bharucha-Reid [120]); hence a *second-order H-valued random function* $x(t, \omega)$ can be defined as a mapping $x(t): T \to L_2(\Omega) \hat{\otimes} H$. In this case the covariance function is defined by $R(s, t) = \mathscr{E}\{x(s, \omega)\overline{x(t, \omega)}\} = \langle x(s, \omega), x(t, \omega)\rangle$; and $x(t, \omega)$ is said to be weakly stationary if $R(s, t) = R(s - t)$. Since $B_2(\Omega, H)$ is a Hilbert space, the theory of stationary H-valued random functions can be developed in a manner analogous to the classical case.

Vo-Khac [127] has introduced the notion of a stationary random function of order p with values in a complex Banach space $\mathfrak{X}$. Let T be a locally compact abelian topological group, and let $x(t, \omega): T \times \Omega \to \mathfrak{X}$ be such that $\|x(t, \omega)\| \in L_p(\Omega)$ for every $t \in T$. We denote by $\tau_h x$ the function $t \to x(t + h)$, and let $S = \{\tau_h : h \in T\}$. Let $V(Sx)$ denote the linear hull of Sx.

***Definition* 1.40.** A Banach space-valued random function $x(t, \omega)$ is said to be a *stationary random function of order p* if (i) $x(t, \omega)$ is continuous in mean of order p, (ii) for every $y \in V(Sx)$, the set $y[T]$ (the image of T under y) is (a) relatively compact in the weak topology of $B_p(\Omega, \mathfrak{X})$ and (b) a subset of a sphere with center θ (that is, $\mathscr{E}\{\|y(t, \omega)\|^p\}$ is independent of t).

We remark that the collection of stationary random functions defined above is a subset of $C(T, B_p(\Omega, \mathfrak{X}))$.

It is of interest to consider the case where $\mathfrak{X} = H$ (a Hilbert space) with inner product $(\cdot,\cdot)$, and $p = 2$ (cf. [128]). Let $x(t, \omega)$ and $y(t, \omega)$ be two second-order H-valued random functions defined on T; and let A and B be elements of $\mathfrak{L}(H)$. In this case a *covariance tensor* can be defined as

$$\begin{aligned} R_{xy}(s, t)(A, B) &= \langle R_{xy}(s, t) A, B\rangle \\ &= \mathscr{E}\{(Ax(s, \omega), By(t, \omega))\}, \end{aligned}$$

where $\langle u, v\rangle = v(u)$ denotes the duality between $\mathfrak{L}(H)$ and $\mathfrak{L}^*(H)$. Using Hilbert space methods, Vo-Khac obtained the spectral representation of a second-order H-valued stationary random function (cf. also Payen [94]).

Let $\mathfrak{X}$ be a separable commutative B^*-algebra. As before, a *second-order $\mathfrak{X}$-valued random function* can be defined as a mapping $x(t): T \to B_2(\Omega, \mathfrak{X})$. A *covariance function* can be defined as follows:

$$R(s, t) = \mathscr{E}\{x(s, \omega) x^*(t, \omega)\}, \qquad s, t \in T,$$

where * denotes the operation of involution. Stationarity is defined in the usual manner. We note that if $x(t,\omega)$ is a second-order random function with values in a separable commutative B^*-algebra with unit element, the the covariance $R(s,t)$ exists. For

$$\begin{aligned}\mathscr{E}\{\|x(s,\omega)\,x^*(t,\omega)\|\} &= \mathscr{E}\{\|x(s,\omega)\|\,\|x^*(t,\omega)\|\} \\ &= \mathscr{E}\{\|x(s,\omega)\|\,\|x(t,\omega)\|\} \\ &\leqslant (\mathscr{E}\{\|x(s,\omega)\|^2\}\,\mathscr{E}\{\|x(t,\omega)\|^2)^{1/2} < \infty.\end{aligned}$$

C. Gaussian random functions

A real-valued random function $x(t,\omega)$ is said to be a *Gaussian* or *normal random function* if its finite-dimensional distributions are Gaussian; that is, if the characteristic functions of the joint distributions of the values $x(t_1,\omega),\ldots,x(t_n,\omega)$ of the random function are of the form

$$\varphi_{t_1,\ldots,t_n}(\lambda_1,\ldots,\lambda_n) = \exp\left\{i\sum_{k=1}^{n} M(t_k)\lambda_k - \tfrac{1}{2}\sum_{k,j=1}^{n} R(t_k,t_j)\lambda_k\lambda_j\right\},$$

where $M(t)=\mathscr{E}\{x(t,\omega)\}$ and $R(s,t)=\mathscr{E}\{(x(s,\omega)-M(s))(x(t,\omega)-M(t))\}$. For Banach space-valued random functions the following definition is often employed.

***Definition* 1.41.** A random function with values in a Banach space $\mathfrak{X}$ is said to be *Gaussian*, or *normal*, if for every $x^*\in\mathfrak{X}^*$ all n-dimensional random vectors $(x^*(x(t_1,\omega)),\ldots,x^*(x(t_n,\omega)))$ have a Gaussian distribution.

In the next section, Sect. 1.5, Banach space-valued Gaussian random functions will be considered from a measure-theoretic viewpoint; that is Gaussian random functions will be defined via Gaussian measures on Banach spaces.

D. Markov processes

Let $(\Omega,\mathfrak{A},\mu)$ be a probability measure space, let $T\subset R$, and let $(\mathfrak{X},\mathfrak{B})$ be a measurable space where $\mathfrak{X}$ is a separable Banach space and $\mathfrak{B}$ is the σ-algebra of Borel sets of $\mathfrak{X}$. Let $x(t,\omega)$ be an $\mathfrak{X}$-valued random function. Further, let $\mathfrak{A}_{(t_1,t_2)}$ denote the σ-algebra of all subsets $\{\omega:x(t,\omega)\in B\}$, $t\in[t_1,t_2]\cap T$, $B\in\mathfrak{B}$. A Banach space-valued Markov process can be defined using conditional probabilities as in the case of real-valued Markov processes.†

† For detailed treatments of real-valued Markov processes we refer to the books of Doob [20], Dynkin [25], and Loève [77].

***Definition* 1.42.** A random function $x(t,\omega): T \times \Omega \to \mathfrak{X}$ is said to be a *Markov process* or *random function of Markov type* if the conditional probabilities $\mathscr{P}\{A|\mathfrak{A}_{(-\infty,s)}\}$ of events $A \in \mathfrak{A}_{(t,\infty)}$ with respect to the σ-algebra $\mathfrak{A}_{(-\infty,s)}$ satisfy for $s \leqslant t$ the relation

$$\mathscr{P}\{A|\mathfrak{A}_{(-\infty,s)}\} = \mathscr{P}\{A|x(s,\omega)\}$$

almost surely. More precisely, for any event $A \in \mathfrak{A}_{(-\infty,t_1)}$ and $B \in \mathfrak{A}_{(t_2,\infty)}$ and any $t \in [t_1, t_2]$,

$$\mathscr{P}\{A \cap B|x(t,\omega)\} = \mathscr{P}\{A|x(t,\omega)\}\mathscr{P}\{B|x(t,\omega)\}$$

almost surely.

A function $P(s,\xi,t,B)$, where $s,t \in T$ $(s \leqslant t)$, $\xi \in \mathfrak{X}$, $B \in \mathfrak{B}$, is called a *transition function* of a Banach space-valued Markov process if it is (1) a probability measure on $\mathfrak{B}$ for fixed $s,t \in T$ and $\xi \in \mathfrak{X}$, and (2) a measurable function of $\xi \in \mathfrak{X}$ for fixed $s,t \in T$, $B \in \mathfrak{B}$, such that

$$P\{s, x(s,\omega), t, B\} = \mathscr{P}\{x(t,\omega) \in B|x(s,\omega)\}$$

almost surely, and $P(s,\xi,s,B) = 1$ for $\xi \in B$, and 0 for $\xi \notin B$. Since $(\mathfrak{X},\mathfrak{B})$ is separable, $P(s,\xi,t,B)$ will exist if the probability measures

$$\nu_t(B) = \mu(\{\omega : x(t,\omega) \in B\}),$$

$B \in \mathfrak{B}$, are perfect.† For any $s,t \in T$, $B \in \mathfrak{B}$ and almost all $\xi \in \mathfrak{X}$ (with respect to the corresponding measures ν_t), the transition probabilities satisfy the *Chapman–Kolmogorov equation*

$$P(s,\xi,t,B) = \int_{\mathfrak{X}} P(s,\xi,\tau,d\lambda)P(\tau,\lambda,t,B), \qquad \tau \in [s,t].$$

Banach space-valued Markov processes have not been studied in any systematic manner; however, Hilbert space-valued Markov processes have been studied by a number of mathematicians in connection with solutions of random differential and integral equations in Hilbert spaces (cf. Baklan [4], Daletskiĭ [17], Kandelaki [60], and papers referred to in Chap. 7).

E. Wiener processes

A real-valued Gaussian process $x(t,\omega)$ with independent increments on an interval $T = [a,b], -\infty \leqslant a < b \leqslant \infty$, such that

$$\mathscr{E}\{x(t,\omega) - x(s,\omega)\} = 0, \qquad \operatorname{Var}\{x(t,\omega) - x(s,\omega)\} = t - s,$$

for any $s,t \in T$, $s \leqslant t$, is called a *real-valued Wiener process*. As is well known, the Wiener process is the mathematical model for Brownian motion. The

† See Sect. 1.6 for the definition of a perfect measure.

Wiener process $x(t,\omega)$ on $T=[0,1]$ can be defined as a Gaussian process with expectation zero and correlation function $R(s,t)=\min(s,t)$. In Sect. 1.5C we consider Wiener measure, the probability measure induced by a Wiener process.

As in the case of Markov processes, Wiener processes in Banach spaces have not been studied in any systematic manner; however, Wiener processes in Hilbert spaces have been introduced in connection with the study of stochastic integrals (cf. Falb [26], Kandelaki [60], Kannan and Bharucha-Reid [64]) and random integral equations in Hilbert spaces (cf. Chap. 7).

***Definition* 1.43.** Let H be a Hilbert space with inner product $(\cdot,\cdot)$. An H-valued random function $\{w(t,\omega),\ t\in[a,b]\}$ is said to be a *Wiener process* in H if (i) $\mathscr{E}\{w(t,\omega)\}=\theta$ for all $t\in[a,b]$, (ii) the increments of $w(t,\omega)$ over disjoint intervals are independent, (iii) $w(t,\omega)$ is almost surely continuous as a function of t, and (iv)

$$\mathscr{E}\{\|w(t,\omega)-w(s,\omega)\|^2\}=\mathscr{E}\{(w(t,\omega)-w(s,\omega),w(t,\omega)-w(s,\omega))\}=|t-s|.$$

F. Martingales

Let $(\Omega,\mathfrak{A},\mu)$ be a probability measure space, and let T be an interval of the extended line $\bar{I}$ of integers (discrete parameter case) or an interval of the extended line $\bar{R}$ (continuous parameter case). Let $\{\mathfrak{A}_t, t\in T\}$ be an increasing family of sub-σ-algebras of $\mathfrak{A}$. A real-valued random function $x(t,\omega)$, which is adapted to the family $\{\mathfrak{A}_t\}$ (that is, $x(t,\omega)$ is $\mathfrak{A}_t$-measurable), is said to be a *martingale* if (i) $\mathscr{E}\{|x(t,\omega)|\}<\infty$ for every $t\in T$ and (ii) for every pair $s,t\in T\ (s<t)$, $x(s,\omega)=\mathscr{E}\{x(t,\omega)|\mathfrak{A}_s\}$ almost surely. For detailed expositions of the theory of real-valued martingales we refer to the books of Doob [20], Meyer [83], and Neveu [92].

Utilizing the definition of the conditional expectation of a Banach space-valued random variable (Definition 1.27) it is possible to introduce the notion of a Banach space-valued martingale. Let $x(t,\omega)$ be a random function with values in a separable Banach space $\mathfrak{X}$, and let the family of sub-σ-algebras $\{\mathfrak{A}_t, t\in T\}$ be as before.

***Definition* 1.44.** An $\mathfrak{X}$-valued random function $x(t,\omega)$ which is adapted to the family $\{\mathfrak{A}_t\}$ is said to be a *Banach space-valued martingale* if (i) $\mathscr{E}\{\|x(t,\omega)\|\}<\infty$ for all $t\in T$, and (ii) for every pair $s,t\in T\ (s<t)$, $x(s,\omega)=\mathscr{E}\{x(t,\omega)|\mathfrak{A}_s\}$ almost surely.

The theory of Banach space-valued martingales was initiated by Chatterji [14], Driml and Hanš [21], Moy [87], and Scalora [109]. They extended the basic results for real-valued martingales to Banach space-valued martingales;

in particular, they were concerned with the formulation and study of martingale convergence theorems in Banach spaces. For some recent results we refer to Ip [47, 48], Kannan [62], Kunita [74], and Uhl [118]. Some convergence theorems for Banach space-valued martingales will be given in Sect. 1.6.

1.5 Probability Measures on Banach Spaces

A. Introduction

Let $x(\omega)$ be a real-valued random variable defined on a probability measure space $(\Omega, \mathfrak{A}, \mu)$; and let $B \in \mathfrak{B}$, where $\mathfrak{B}$ is the Borel algebra of R. Put

$$\nu_x(B) = \mu \circ x^{-1}(B) = \mu(\{\omega : x(\omega) \in B\}).$$

Then, ν_x defines the *probability distribution*, or simply the *distribution*, of the random variable x. It is easy to see that ν_x is a probability measure on $\mathfrak{B}$; hence ν_x is called the probability measure induced on $\mathfrak{B}$ by μ and x, and $(R, \mathfrak{B}, \nu_x)$ is called the *induced probability measure space.*

The *distribution function* $F_x(\xi)$ of x is defined as follows:

$$F_x(\xi) = \nu_x((-\infty, \xi)) = \mu(\{\omega : x(\omega) < \xi\}), \qquad \xi \in R.$$

It can be shown that $F_x(\xi)$ is nondecreasing and continuous from the left on R, with $F_x(-\infty) = 0$ and $F_x(+\infty) = 1$. Conversely, every real function with the above properties is the distribution function of a real-valued random variable on some probability measure space.

An important tool in the study of real-valued random variables and their probability measures (or distribution functions) is the *characteristic function* (or *Fourier–Stieltjes transform*) of ν_x (or F_x). For any real-valued random variable x with probability measure ν_x (or distribution function F_x) the characteristic function $\varphi_x(\lambda)$, for all $\lambda \in R$, is defined as follows:

$$\begin{aligned} \varphi_x(\lambda) &= \mathscr{E}\{e^{i\lambda x(\omega)}\} = \int_\Omega e^{i\lambda x(\omega)}\, d\mu(\omega) \\ &= \int_R e^{i\lambda\xi}\, d\nu_x(\xi) = \int_{-\infty}^{\infty} e^{i\lambda\xi}\, dF_x(\xi). \end{aligned}$$

Necessary and sufficient conditions that a complex-valued function $\varphi(\lambda)$ on R be the characteristic function of a probability measure ν are given by the following result due to Bochner.

THEOREM 1.12. $\varphi(\lambda)$ *is a characteristic function if and only if it is positive definite and continuous at* 0 *with* $\varphi(0) = 1$.

All of the above notions can be extended to the case where the range of the random variable is $\bar{R}$, R_n, or $\bar{R}_n$.

We now consider the probability measure corresponding to, or associated with, a stochastic process. Let $\mathfrak{X} = R$; then, from Definition 1.34, a real-valued random function is a mapping $x(t,\omega): T \times \Omega \to R$ such that for every $t \in T$, x is a real-valued random variable. Let R^T denote the space of all functions $x(t)$ defined on T with values in R. Now, a mapping of Ω into R^T, defining a stochastic process $y(t)$ with values in R, takes $\mathfrak{A}$-measurable subsets of Ω into some σ-algebra of subsets of R^T. Since for every $t \in T$ the set $\{\omega: y(t,\omega) \in B, B \in \mathfrak{B}\} \in \mathfrak{A}$, sets of the form $\Gamma_t(B) = \{x(\cdot): x(t) \in B\}$, $t \in T$ (that is, the *one-dimensional cylinder sets*), clearly belong to this σ-algebra. Sets of the form $\bigcap_{i=1}^n \Gamma_{t_i}(B_i)$ are called *cylinder sets over* $\{t_1, t_2, \ldots, t_n\}$. Let $\mathfrak{B}^T$ denote the minimal σ-algebra of R^T containing all cylinder sets. Under the mapping $y(\cdot,\omega): \Omega \to R^T$, the inverse image of a set in $\mathfrak{B}^T$ is clearly $\mathfrak{A}$-measurable; and this mapping defines a probability measure, say ν, on $\mathfrak{B}^T$. The measure ν is called the *probability measure corresponding to*, or *associated with, the stochastic process* $y(t,\omega)$. Frequently, the probability measure ν is referred to as the stochastic process, since the probability measure space $(R^T, \mathfrak{B}^T, \nu)$ can be associated with this process; and the natural mapping $y(t, x(\cdot)) = x(t)$ gives the stochastic process corresponding to the probability measure ν.

In order to specify a stochastic process in the above measure-theoretic sense it is sufficient to know the values of the induced measure on the cylinder sets; that is, it is sufficient to know the *finite-dimensional distribution functions*

$$F_{t_1, t_2, \ldots, t_n}(B_1, B_2, \ldots, B_n) = \nu\left(\bigcap_{i=1}^n \Gamma_{t_i}(B_i)\right).$$

Now, the celebrated Kolmogorov consistency or extension theorem (cf., for example, Billingsley [10, pp. 228–230]) asserts that to every system of finite-dimensional distribution function which satisfies certain consistency conditions there corresponds a stochastic process, that is a probability measure on $(R^T, \mathfrak{B}^T)$.

In this section we consider probability measures on Banach spaces. In the development of probability theory in Banach spaces it is clear that the study of probability measures on Banach spaces plays a fundamental role, since the proper measure-theoretic framework is required in order to discuss the distributions of Banach space-valued random variables, and to define Banach space-valued random functions. When the range of a random variable is a Banach space, we are obliged to work with either its probability distribution (or measure) or its Fourier–Stieltjes transform (or characteristic functional), since we can no longer define a distribution function as in the case of scalar-valued random variables.

The remainder of this section is divided into four subsections. Section 1.5B is devoted to some basic definitions and concepts; and in Sect. 1.5C we consider the construction of probability measures on some concrete Banach spaces. In Sect. 1.5D we consider the important case of Gaussian measures on Banach spaces; and in Sect. 1.5E we consider briefly the absolute continuity of probability measures on Banach spaces.

The study of probability measures on Banach spaces belongs to the theory of probability measures on linear topological spaces (cf. Badrikian [3], Gel'fand and Vilenkin [33, Chap. IV], and Mourier [86]); however, in this book we do not need to consider spaces more general than Banach spaces.

B. Basic definitions and concepts

Let $\mathfrak{X}$ be a separable Banach space, and let $\mathfrak{B}$ be the Borel algebra of $\mathfrak{X}$. Since $\mathfrak{X}$ is separable, $\mathfrak{B}$ is generated by spheres. When we speak of a probability measure on the Banach space $\mathfrak{X}$ we mean, of course, a measure on the measurable space $(\mathfrak{X}, \mathfrak{B})$. A detailed exposition of the topics considered in this subsection can be found in Billingsley [10] or Parthasarathy [93].

1. Regular and tight measures

***Definition* 1.45.** A probability measure ν on a measurable space $(\mathfrak{Y}, \mathfrak{B})$, where $\mathfrak{Y}$ is a metric space, is said to be *regular* if for every $B \in \mathfrak{B}$ and $\epsilon > 0$ there exist a closed set F and an open set G such that $F \subseteq B \subseteq G$ and $\nu(G - F) < \epsilon$.

A fundamental result is the following: *Every probability measure on* $(\mathfrak{X}, \mathfrak{B})$ *is regular.*

A smaller class of probability measures is the class of tight measures. The concept of tightness is important in the study of weak convergence of probability measures and its applications (cf., de Acosta [19], Billingsley [10], Parthasarathy [93]).

***Definition* 1.46.** A probability measure ν on a measurable space $(\mathfrak{Y}, \mathfrak{B})$, where $\mathfrak{Y}$ is a metric space, is said to be *tight* if for every $\epsilon > 0$ there exists a compact set $K_\epsilon \subseteq \mathfrak{Y}$ such that $\nu(\mathfrak{Y} - K_\epsilon) < \epsilon$.

In view of the above definition, tight measures have the property of being determined by their values for compact sets. For separable Banach spaces we have the following result: *Every probability measure on* $(\mathfrak{X}, \mathfrak{B})$ *is tight.*

Another concept of interest is that of a perfect measure.

***Definition* 1.47.** A measure ν on $(\mathfrak{Y}, \mathfrak{B})$, where $\mathfrak{Y}$ is a metric space, is said to be *perfect* if for any $\mathfrak{B}$-measurable real-valued function f and any set E on the real line such that $f^{-1}(E) \in \mathfrak{B}$, there are Borel sets F_1 and F_2 on the real line such that (i) $F_1 \subseteq E \subseteq F_2$ and (ii) $\nu(f^{-1}(F_2 - F_1)) = 0$. If a probability measure ν is perfect, then $(\mathfrak{Y}, \mathfrak{B}, \nu)$ is called a perfect measure space.

We remark that (1) every probability measure space $(\mathfrak{X}, \mathfrak{B}, \nu)$, where $\mathfrak{X}$ is a separable Banach space, is perfect; and (2) any mapping $\psi = \psi(x)$ of $\mathfrak{X}$ into an arbitrary space $\mathfrak{Y}$ induces on $\mathfrak{Y}$ a perfect measure $\tilde{\nu}$; that is $\tilde{\nu}$ is the probability measure induced on the σ-algebra of all sets E whose inverse images $\{\psi \in E\} \in \mathfrak{B}$.

2. ***The space of probability measures on a separable Banach space.*** Let $M(\mathfrak{X})$ denote the space of probability measures on $(\mathfrak{X}, \mathfrak{B})$; and let $\{\nu_n\}$ be a sequence in $M(\mathfrak{X})$. Three notions of convergence can be defined for sequences in $M(\mathfrak{X})$, namely, uniform, strong, and weak. We will restrict our attention to weak convergence.

***Definition* 1.48.** A sequence $\{\nu_n\}$ of probability measures is said to *converge weakly* to a probability measure ν if

$$\int_{\mathfrak{X}} f(x)\, d\nu_n(x) \to \int_{\mathfrak{X}} f(x)\, d\nu(x)$$

for every bounded and continuous function f on $\mathfrak{X}$ (that is, for every $f \in C(\mathfrak{X})$).

We now introduce a metric in $M(\mathfrak{X})$. Let $\nu_1, \nu_2 \in M(\mathfrak{X})$; and let F be an arbitrary closed set in $\mathfrak{X}$. Consider the open set

$$E = \{x : \inf_{y \in F} \|x - y\| < \epsilon\}.$$

Define ϵ_1 as the greatest lower bound of the ϵ's satisfying $\nu_2(F) < \nu_1(E) + \epsilon$, and ϵ_2 as the greatest lower bound of the ϵ's satisfying $\nu_1(F) < \nu_2(E) + \epsilon$. The *Prohorov metric* is defined by

$$P(\nu_1, \nu_2) = \max(\epsilon_1, \epsilon_2).$$

Prohorov [97] (cf. also, Billingsley [10] and Parthasarathy [93]) has shown that $(M(\mathfrak{X}), P)$ is a complete, separable metric space; and that convergence in this metric space is equivalent to weak convergence.

A necessary and sufficient condition for weak convergence is given by the following result.

THEOREM 1.13. *Let ν_n be a sequence of measures in $M(\mathfrak{X})$. Then ν_n converges weakly to ν if and only if*

$$\lim_{n \to \infty} \sup_{f \in E} \left| \int f\, d\nu_n - \int f\, d\nu \right| = 0$$

for every family $E \subseteq C(\mathfrak{X})$ which is equicontinuous for all $x \in \mathfrak{X}$ and uniformly bounded.

3. *Characteristic functionals.* The concept of the characteristic function of a real-valued random variable was extended to Banach space-valued random variables by Kolmogorov [67]. Let $x(\omega)$ be a random variable with values in an arbitrary Banach space $\mathfrak{X}$, and let x^* be any real, bounded linear functional on $\mathfrak{X}$.

***Definition* 1.49.** The *characteristic functional of* $x(\omega)$ (or $\nu = \nu_x \in M(\mathfrak{X})$) is the function

$$\varphi(x^*) = \mathscr{E}\{\exp[ix^*(x(\omega))]\} = \int_{\mathfrak{X}} \exp[ix^*(x)]\, d\nu(x).$$

defined for all real $x^* \in \mathfrak{X}^*$.

Some of the basic properties of characteristic functionals are given below.

1. The characteristic functional determines uniquely a probability measure ν on the minimal σ-algebra $\mathfrak{B}_{\mathrm{m}}$ relative to which all linear functionals are measurable. When $\mathfrak{X}$ is separable, $\mathfrak{B}_{\mathrm{m}}$ and $\mathfrak{B}$ (the σ-algebra of all Borel sets of $\mathfrak{X}$) coincide.
2. $\varphi(x^*)$ is a uniformly continuous function of x^* in the strong topology of $\mathfrak{X}^*$, and is a continuous function of x^* in the weak topology of $\mathfrak{X}^*$.
3. Characteristic functionals are positive-definite; that is, for every positive integer n, every n elements $x_1^*, x_2^*, \ldots, x_n^* \in \mathfrak{X}^*$, and every n complex numbers $\lambda_1, \lambda_2, \ldots, \lambda_n$ the quantity $\sum_{j,k} \varphi(x_j^* - x_k^*)\lambda_j \bar{\lambda}_k$ is a nonnegative real number.
4. $\varphi(\theta^*) = 1$, where θ^* is the null element of $\mathfrak{X}^*$.
5. If $\varphi_{x_1}(x^*)$ and $\varphi_{x_2}(x^*)$ are the characteristic functionals of two independent $\mathfrak{X}$-valued random variables $x_1(\omega)$ and $x_2(\omega)$, respectively (where $\mathfrak{X}$ is separable), then the characteristic functional of $x_1(\omega) + x_2(\omega)$ is $\varphi_{x_1+x_2}(x^*) = \varphi_{x_1}(x^*)\varphi_{x_2}(x^*)$.
6. If $\mathscr{E}\{x\}$ and $\mathscr{E}\{\|x\|^2\}$ exist, then $\varphi(x^*)$ admits the representation

$$\varphi(x^*) = 1 + ix^*(\mathscr{E}\{x\}) - \tfrac{1}{2}\mathscr{E}\{|x^*(x)|^2\} + \|x^*\|^2 h(x^*),$$

where $h(x^*) \to 0$ if $\|x^*\| \to 0$.

7. Let $\{\nu_n\}$ be a sequence of probability measures in $M(\mathfrak{X})$ with associated characteristic functionals $\varphi_n(x^*)$. If $\{\nu_n\}$ converges weakly to a probability measure ν with characteristic functional $\varphi(x^*)$, then $\varphi_n(x^*) \to \varphi(x^*)$ for all $x^* \in \mathfrak{X}^*$.

For real-valued random variables, Bochner's theorem (Theorem 1.12) asserts that the characteristic function is a positive-definite function, and conversely, any positive-definite function φ such that $\varphi(0)=1$ and is continuous at 0 is the characteristic function of a real-valued random variable (or a probability measure $\nu \in M(R)$). If $x(\omega)$ is a random variable with values in a Banach space $\mathfrak{X}$, properties (3) and (4) show that the characteristic functional of x is a positive-definite function with $\varphi(\theta^*)=1$. However, for infinite-dimensional Banach spaces it is not true that any positive-definite function with $\varphi(\theta^*)=1$ is the characteristic functional of an $\mathfrak{X}$-valued random variable (or a probability measure $\nu \in M(\mathfrak{X})$).

In the case where $\mathfrak{X}$ is separable and reflexive, Mourier [84] has given a necessary and sufficient condition (referred to as "condition C") for a positive-definite function $\varphi(x^*)$ to be the characteristic functional of an $\mathfrak{X}$-valued random variable. "Condition C" is complicated and difficult to apply; however, Mourier has given the following sufficient condition which can be obtained from "condition C."

THEOREM 1.14. *If $\mathfrak{X}$ is a separable and reflexive Banach space, then $\varphi(x^*)$ is a characteristic functional provided there exist three positive numbers k, M, and N, and a sequence of $\mathfrak{X}$-valued random variables $x_n(\omega)$ with characteristic functionals $\varphi_n(x^*)$ such that* (i) $\mathscr{E}\{\|x_n(\omega)\|^k\} \leqslant M$ *and* (ii) $\lim_{n\to\infty} \varphi_n(x^*) = \varphi(x^*)$ *uniformly in x^* for all x^* such that $\|x^*\| \leqslant N$. If $x(\omega)$ is the $\mathfrak{X}$-valued random variable whose characteristic functional is $\varphi(x^*)$, then $\mathscr{E}\{\|x(\omega)\|^k\} \leqslant 3M$.*

Getoor [34] has obtained the following result in the case where $\mathfrak{X}$ is a reflexive Banach space, but not necessarily separable (cf. also, Badrikian [2]).

THEOREM 1.15. *Let $\mathfrak{X}$ be a reflexive Banach space, and let $\varphi(x^*)$ be a positive-definite function of $\mathfrak{X}^*$. A necessary and sufficient condition that $\varphi(x^*)$ be the characteristic functional of an $\mathfrak{X}$-valued random variable is that*: (i) $\varphi(\theta)=1$ *and $\varphi(x^*)$ is continuous on all finite-dimensional subspaces of $\mathfrak{X}^*$.* (ii) *For every separable subspace $\mathfrak{X}_0^*$ of $\mathfrak{X}^*$ and for all $\epsilon, \lambda > 0$ there exists a $\delta = \delta(\mathfrak{X}_0^*, \epsilon, \lambda)$ such that for any finite collection $\{x_i^*, \ldots, x_n^*\} \subset \mathfrak{X}^*$, with $\|x_i^*\| \leqslant \delta$,*

$$\int_{-\epsilon}^{\epsilon} \cdots \int_{-\epsilon}^{\epsilon} dF_n(x_1^*, \xi_1; \ldots, x_n^*, \xi_n) \geqslant 1 - \lambda,$$

where $F_n(x_1^*, \xi_1; \ldots, x_n^*, \xi_n) = \mu(\{\omega : x_i^*(x(\omega)) \leqslant \xi_i, i = 1, \ldots, n\})$.

For other results on characteristic functionals and their properties we refer to Ahmad [1], Dudley [22], Grenander [38], LeCam [75], Mourier [84, 86], and Prohorov [97, 98]. In the next section we will consider the characteristic functionals of probability measures on some concrete Banach spaces.

4. *L-measures, Gaussian, and infinitely divisible measures.* We now define three classes of probability measures on Banach spaces which are of theoretical importance and wide applicability.

***Definition* 1.50.** A probability measure ν on a Banach space $\mathfrak{X}$ is said to be an *L-measure* if every bounded linear functional x^* is ν-measurable.

***Definition* 1.51.** A probability measure $\nu_x \in M(\mathfrak{X})$, induced by an $\mathfrak{X}$-valued random variable x, is said to be *Gaussian* if for every linear functional x^* on $\mathfrak{X}$, $x^*(x)$ has a Gaussian distribution and there exists an element $m_{\nu_x} \in \mathfrak{X}$ such that $x^*(m_{\nu_x}) = \mathscr{E}\{x^*(x)\}$.

***Definition* 1.52.** A probability measure $\nu \in M(\mathfrak{X})$ is said to be *infinitely divisible* if for every positive integer n

$$\nu = \underbrace{\nu_n * \nu_n * \cdots * \nu_n}_{n \text{ times}},$$

where ν_n is a probability measure on $\mathfrak{X}$ and $*$ denotes convolution.

C. Probability measures on some concrete Banach spaces

1. *Introduction.* As in the classical case (cf. Sect. 1.5A), if $x(\omega)$ is an $\mathfrak{X}$-valued random variable defined on $(\Omega, \mathfrak{A}, \mu)$, then x and μ induce a probability measure $\nu_x = \mu \circ x^{-1}$ on $(\mathfrak{X}, \mathfrak{B})$. In the development of the theory of probability measures on Banach spaces, and in the study of solutions of random equations in Banach spaces, the construction of induced probability measures on concrete Banach spaces and the study of their properties is of fundamental importance. In this section we consider three main topics: probability measures on Banach spaces with Schauder bases, Wiener measure, and probability measures on Hilbert spaces.

2. *Probability measures on Banach spaces with Schauder bases.* Let $\mathfrak{X}$ be a real Banach space. $\mathfrak{X}$ is said to have a *Schauder basis* if there exists a sequence† $\{e_n\} \subset \mathfrak{X}$ such that for each element $x \in \mathfrak{X}$ there corresponds a unique sequence $\{\eta_n\}$ of real numbers with the property that the sequence $\{x_n\}$, where $x_n = \sum_{i=1}^{n} \eta_i e_i$, converges strongly to x; that is, $\lim_{n\to\infty} \|x - x_n\|_{\mathfrak{X}} = 0$. Spaces with Schauder bases constitute an important class of Banach spaces, one reason being that every known example of a separable Banach space has a Schauder basis. Hence, in view of our earlier remarks concerning the role of separable Banach spaces in probability theory, spaces with Schauder bases

† We will assume $\|e_n\|_{\mathfrak{X}} = 1$.

are important in the development of probability theory in Banach spaces. For detailed expositions of Banach spaces with Schauder bases we refer to the books of Day [18], Marti [81], and Singer [112].

In this subsection we first give Kampé de Fériet's method of constructing probability measures on Banach spaces with Schauder bases [58], and then give a theorem of Bochner type due to de Acosta [19] for characteristic functionals of probability measures on Banach spaces with Schauder bases.

Let R_∞ denote the space of all sequences of real numbers. If $\mathfrak{X}$ has a Schauder basis, then, by definition, $\mathfrak{X}$ is isomorphic to a subspace $\tilde{\mathfrak{X}} \subset R_\infty$: $x = A\tilde{x}$, $\tilde{x} = A^{-1}x$, $x \in \mathfrak{X}$, $\tilde{x} \in \tilde{\mathfrak{X}}$. Define a sequence $\{\psi_n\}$ of functions as follows:

$$\psi_n(\eta_1, \ldots, \eta_n) = \|x_n\|_{\mathfrak{X}} = \|\sum_{i=1}^{n} \eta_i e_i\|_{\mathfrak{X}} .$$

The following properties of ψ_n are immediate:

(i) $\psi_n(\eta_1, \ldots, \eta_n) > 0$ if $\sum_{i=1}^n |\eta_i| > 0$,

(ii) $\psi_n(0, \ldots, 0) = 0$,

(iii) $\psi_n(\eta_1' + \eta_1'', \ldots, \eta_n' + \eta_n'') \leqslant \psi_n(\eta_1', \ldots, \eta_n') + \psi_n(\eta_1'', \ldots, \eta_n'')$,

(iv) $\psi_n(\lambda\eta_1, \ldots, \lambda\eta_n) = |\lambda|\, \psi_n(\eta_1, \ldots, \eta_n)$,

(v) $\psi_{n+1}(\eta_1, \ldots, \eta_n, 0) = \psi_n(\eta_1, \ldots, \eta_n)$,

(vi) $\psi_n(\eta_1, \ldots, \eta_{n+1}) \leqslant \psi_n(\eta_1, \ldots, \eta_n) + |\eta_{n+1}|$,

(vii) $0 \leqslant \psi_n(\eta_1, \ldots, \eta_n) \leqslant \sum_{i=1}^n |\eta_i|$,

(viii) $|\psi_n(\eta_1'', \ldots, \eta_n'') - \psi_n(\eta_1', \ldots, \eta_n')| \leqslant \psi_n(\eta_1'' - \eta_1', \ldots, \eta_n'' - \eta_n')$,

(ix) $|\psi_n(\eta_1'', \ldots, \eta_n'') - \psi_n(\eta_1', \ldots, \eta_n')| \leqslant \sum_{i=1}^n |\eta_i'' - \eta_i'|$,

(x) ψ_n is a continuous function on R_n.

We define a function $\varphi_{n,k}$ as follows:

$$\begin{aligned}\varphi_{n,k}(\eta_{n+1}, \ldots, \eta_{n+k}) &= \|x_{n+k} - x_n\| \\ &= \psi_{n+k}(0, \ldots, 0, \eta_{n+1}, \ldots, \eta_{n+k}).\end{aligned}$$

Now, the sequence $\{x_n\}$ is strongly convergent if and only if

$$\inf_{n\geqslant 1} \sup_{k\geqslant 1} \|x_{n+k} - x_n\| = 0;$$

hence the space $\tilde{\mathfrak{X}} \subset R_\infty$ which is isomorphic to $\mathfrak{X}$ can be defined as follows:

$$\tilde{\mathfrak{X}} = \{\{\eta_n\} \subset R_\infty : \inf_{n\geqslant 1} \sup_{k\geqslant 1} \varphi_{n,k}(\eta_{n+1}, \ldots, \eta_{n+k}) = 0\}.$$

$\tilde{\mathfrak{X}}$ becomes a Banach space under the norm

$$\|\tilde{x}\|_{\tilde{\mathfrak{X}}} = \sup_{n\geqslant 1} \|x_n\| = \sup_{n\geqslant 1} \psi_n(\eta_1, \ldots, \eta_n),$$

where $\tilde{x} = \{\eta_n\}$; and we have

$$\|\tilde{x}\|_{\tilde{\mathfrak{X}}}/\|A^{-1}\| \leqslant \|x\|_{\mathfrak{X}} \leqslant \|\tilde{x}\|_{\tilde{\mathfrak{X}}}.$$

Now, since $\tilde{\mathfrak{X}}$ is isomorphic to $\mathfrak{X}$ it possesses a basis $\{\epsilon_n\}$, where $\epsilon_n = A^{-1}e_n$, with $\|\epsilon_n\|_{\tilde{\mathfrak{X}}} = \|e_n\|_{\mathfrak{X}} = 1$; and $\epsilon_n = \{\delta_{nj}\}$, where δ_{ij} is the Kronecker delta.

Let $\mathfrak{X}^*$ and $\tilde{\mathfrak{X}}^*$ be the duals of $\mathfrak{X}$ and $\tilde{\mathfrak{X}}$, respectively. $\mathfrak{X}^*$ and $\tilde{\mathfrak{X}}^*$ are isomorphic under the isomorphism $x^*(A\tilde{x}) = \tilde{x}^*(x)$, $\tilde{x}^*(A^{-1}x) = x^*(x)$. Put $e_n^*(x) = \eta_n(A^{-1}x)$. Hence it follows that the coordinates η_n are linear functionals on $\tilde{\mathfrak{X}}$. We remark that the sequences $\{e_n\}$ and $\{e_n^*\}$ are biorthogonal; that is, $e_i^*(e_j) = \delta_{ij}$.

Because $\mathfrak{X}$ and $\tilde{\mathfrak{X}}$ are isomorphic, the construction of a probability measure ν on $\mathfrak{X}$ reduces to the construction of a probability measure $\tilde{\nu}$ on $\tilde{\mathfrak{X}}$. The construction of $\tilde{\nu}$ can be carried out as follows. Let

$$\begin{aligned} \tilde{x}_1 &= (\eta_1) \in R, \\ \tilde{x}_2 &= (\eta_1, \eta_2) \in R_2, \\ &\vdots \\ \tilde{x}_n &= (\eta_1, \ldots, \eta_n) \in R_n; \end{aligned}$$

and, for every n, let $\tilde{\nu}_n$ be a probability measure on $(R_n, \mathfrak{B}_n)$, where $\mathfrak{B}_n$ is the σ-algebra of all Borel sets of R_n. Let the sequence $\{\tilde{\nu}_n\}$ satisfy the consistency conditions

$$\tilde{\nu}_n(R_n) = 1, \qquad \tilde{\nu}_{n+1}(\Gamma_{n+1}(B_n)) = \nu_n(B_n),$$

where $\Gamma_{n+1}(B_n)$ is a cylinder set in R_{n+1} with base B_n ($\in \mathfrak{B}_n$) in R_n; that is,

$$\Gamma_{n+1}(B_n) = \{\tilde{x}_{n+1} : \tilde{x}_n \in B_n, \quad -\infty < \eta_{n+1} < \infty\}.$$

Let $\mathfrak{B}_\infty$ be the σ-algebra of subsets of R_∞ generated by the cylinder sets

$$\Gamma_\infty(B_n) = \{\tilde{x} \in R_\infty : \tilde{x}_n \in B_n, \quad -\infty < \eta_{n+k} < \infty, \quad k = 1, 2, \ldots\}.$$

Now, by Kolmogorov's extension theorem, there exists a unique probability measure $\tilde{\nu}$ on R_∞ such that

$$\tilde{\nu}(R_\infty) = 1, \qquad \tilde{\nu}(\Gamma_\infty(B_n)) = \tilde{\nu}_n(B_n).$$

We remark that the set $\tilde{\mathfrak{X}}$ is $\tilde{\nu}$-measurable. This follows from the continuity of the functions $\varphi_{n,k}$ on R_{n+k}, and the fact that $\tilde{\mathfrak{X}}$ is defined by countable operations on the $\varphi_{n,k}$.

Finally, let $\tilde{\mathfrak{B}}$ be the restriction of $\mathfrak{B}_\infty$ to $\tilde{\mathfrak{X}}$. For $B \in \tilde{\mathfrak{B}}$, let $H = A[B]$, $B = A^{-1}[H]$; and define $\tilde{\nu}$ so that $\nu(H) = \tilde{\nu}(B)$. Hence, through the isomorphism between $\tilde{\mathfrak{X}}$ and $\mathfrak{X}$, we obtain a measure space $(\mathfrak{X}, \mathfrak{H}, \nu) = A(\tilde{\mathfrak{X}}, \tilde{\mathfrak{B}}, \tilde{\nu})$, where $\mathfrak{H}$ is the σ-algebra of subsets of $\mathfrak{X}$ defined by $H = A[B]$.

It is clear that $0 \leqslant \nu(\mathfrak{X}) = \tilde{\nu}(\tilde{\mathfrak{X}}) \leqslant 1$. We now give conditions in order that $\nu(\mathfrak{X}) = 1$; that is, that $(\mathfrak{X}, \mathfrak{H}, \nu)$ be a probability measure space. For a given $\lambda > 0$, consider the cylinder set

$$\Gamma_{n,k}(\lambda) = \{\tilde{x} : \tilde{x}_{n+k} \in F_{n,k}(\lambda), \quad -\infty < \eta_{n+k+j} < \infty, \quad j = 1, 2, \ldots\}$$

with base

$$F_{n,k}(\lambda) = \bigcap_{j=1}^{k} \{\tilde{x}_{n+k} : \varphi_{n,j}(\eta_{n+1}, \ldots, \eta_{n+j}) < \lambda\}.$$

We have

$$\tilde{\mathfrak{X}} = \bigcap_{\lambda>0} \bigcup_{n\geq 1} \bigcap_{k\geq 1} \Gamma_{n,k}(\lambda) \qquad \text{and} \qquad \tilde{\nu}(\tilde{\mathfrak{X}}) = \inf_{\lambda>0} \sup_{n\geq 1} \inf_{k\geq 1} \nu(\Gamma_{n,k}(\lambda)).$$

Since $\tilde{\nu}(\Gamma_{n,k}(\lambda)) = \tilde{\nu}_{n+k}(F_{n,k}(\lambda))$, we obtain the following result.

THEOREM 1.16. *The measure space $(\mathfrak{X}, \mathfrak{H}, \nu)$ induced by the isomorphism A of $(\tilde{\mathfrak{X}}, \tilde{\mathfrak{B}}, \tilde{\nu})$ is a probability measure space if and only if the sequence $\{\tilde{\nu}_n\}$ satisfies, in addition to the consistency conditions $\tilde{\nu}(R_\infty) = 1$ and $\tilde{\nu}_{n+1}(\Gamma_{n+1}(B_n)) = \tilde{\nu}_n(B_n)$, the condition*

$$\inf_{\lambda>0} \sup_{n\geq 1} \inf_{k\geq 1} \tilde{\nu}_{n+k}(F_{n,k}(\lambda)) = 1.$$

It is of interest to note that Kampé de Fériet has shown that ν is an L-measure (Definition 1.50) if and only if the $\eta_n(\tilde{x})$ are $\tilde{\nu}$-measurable.

Kampé de Fériet used the method outlined above to construct probability measures on separable Hilbert spaces, l_p and L_p $(1 \leqslant p < \infty)$, $C[0,1]$, and $C_0[0,1]$, where $C_0[0,1]$ is the space of real functions in $C[0,1]$ which assume the value 0 at 0 and 1. Kannan and Bharucha-Reid [65] extended Kampé de Fériet's method to complex Banach spaces with Schauder bases, and utilized the results to construct probability measures on H_p spaces $(1 < p < \infty)$, that is, the Hardy spaces of analytic functions).

Since the method of Kampé de Fériet gives us an effective method of constructing probability measures on separable Banach spaces, we remark that given a probability measure on a separable Banach space the Banach–Mazur theorem (Theorem 1.1) enables us to obtain a probability measure on a closed linear subspace of $C[0,1]$. To be more precise, consider the probability measure space $(\mathfrak{X}, \mathfrak{B}, \nu)$, where $\mathfrak{X}$ is a separable Banach space. Then, according to the Banach–Mazur theorem, $\mathfrak{X}$ is isometrically isomorphic, or congruent, to a closed subspace, say $\tilde{C}$ of $C[0,1]$. Let γ denote the congruence from $\mathfrak{X}$ onto $\tilde{C}$. It is clear that $\mathfrak{B} = \gamma^{-1}(\tilde{\mathcal{C}})$, where $\tilde{\mathcal{C}}$ is the σ-algebra of Borel sets of $\tilde{C}$, and $\tilde{\nu}$ defined by $\tilde{\nu} = \nu \circ \gamma^{-1}$ is a probability measure on $\tilde{C}$.

In Sect. 1.5B we stated some theorems of Bochner type on characteristic functionals due to Mourier and Getoor. We now state the following result, due to de Acosta, concerning characteristic functionals of probability measures on Banach spaces with bases.

THEOREM 1.17. *Let $\mathfrak{X}$ be a Banach space with Schauder basis $\{e_i\}$, and let $\mathfrak{X}_n$ denote the subspace of $\mathfrak{X}$ generated by $\{e_1, \ldots, e_n\}$. Let φ be a complex-valued function defined on $\mathfrak{X}_0^*$, where $\mathfrak{X}_0^*$ is the linear subspace of $\mathfrak{X}^*$ spanned by the* *coordinate functions* $\{f_i\}$ *for the basis* $\{e_i\}$, *which is positive-definite with* $\varphi(\theta^*)=1$. *If the restriction of* φ *to every finite-dimensional subspace of* $\mathfrak{X}_0^*$ *is continuous, then for each n there exists a probability measure* ν_n *on* $(\mathfrak{X},\mathfrak{B})$ *such that* $\nu_n(\mathfrak{X}_n)=1$ *and the characteristic functional of* ν_n *is* $\varphi(\sum_{i=1}^n x^*(e_i)f_i)$.

3. ***Probability measures on*** $C[0,1]$. The importance of the space of continuous functions in probability theory has been pointed out several times in this text; in particular, the study of probability measures on spaces of continuous functions is tantamount to the study of random functions with continuous trajectories. Also, the existence of a probability measure on the space of continuous functions enables us to study problems in analysis concerning continuous functions using probabilistic methods.

Consider the measurable space $(C,\mathfrak{C})$, where $C=C[0,1]$ and $\mathfrak{C}$ is the σ-algebra of Borel sets of C. We first state the following result:

THEOREM 1.18. *The* σ*-algebra* $\mathfrak{C}$ *coincides with the minimal* σ*-algebra of subsets of C with respect to which the projections* $\varphi_t\colon x\to x(t)$ *are measurable for all* $t\in[0,1]$. *Let* ν_1 *and* ν_2 *be two probability measures on C. Then, a necessary and sufficient condition that* $\nu_1=\nu_2$ *is that* $\nu_1^{t_1,\ldots,t_n}=\nu_2^{t_1,\ldots,t_n}$ *for all n and* $t_1,\ldots,t_n\in[0,1]$, *where* $\nu_n^{t_1,\ldots,t_n}$ $(i=1,2)$ *are probability measures on* R_n *induced by* ν_i $(i=1,2)$, *respectively, through the projections*

$$\varphi_{t_1,\ldots,t_n}\colon x\to(x(t_1),\ldots,x(t_n)).$$

A fundamental problem is the characterization of random functions $x(t,\omega)$, $t\in[0,1]$ whose associated probability measures are defined on $C[0,1]$. More precisely, let $x(t,\omega)$, $t\in[0,1]$ be a random function and let $\lambda^{t_1,\ldots,t_n}$ be its finite-dimensional distributions. Then, what additional condition must be satisfied in order that there exist a probability measure ν on C such that $\lambda^{t_1,\ldots,t_n}=\nu^{t_1,\ldots,t_n}$ for all n and $t_1,\ldots,t_n\in[0,1]$? If there exists such a ν then, by Theorem 1.18, it is unique. The following result, due to Kolmogorov, gives a sufficient condition.

THEOREM 1.19. *Let* $x(t,\omega)$, $t\in[0,1]$ *be a random function and let* $\lambda^{t_1,\ldots,t_n}$ *be the probability distribution on* $(R_n,\mathfrak{B}_n)$ *of the random vector* $(x(t_1,\omega),\ldots,x(t_n,\omega))$. *If there are positive constants* α, β *and M such that*

$$\mathscr{E}\{|x(t,\omega)-x(s,\omega)|^\alpha\}\leqslant M|t-s|^{1+\beta}$$

for all $s,t\in[0,1]$, *then there exists a unique probability measure* ν *on C such that* $\lambda^{t_1,\ldots,t_n}=\nu^{t_1,\ldots,t_n}$ *for all n and* $t_1,\ldots,t_n\in[0,1]$.

We now consider Wiener measure on $C[0,1]$. *Wiener measure* is the probability measure on $C[0,1]$ associated with the Wiener process or the Brownian motion process. An application of Theorem 1.19 yields the following result on the existence of Wiener measure.

THEOREM 1.20. *There exists a unique probability measure w on $(C, \mathfrak{C})$, called Wiener measure, with the following properties:*

i. $w(\{x : x(0,\omega) = 0\}) = 1$;
ii. *if* $0 \leqslant t_1 < t_2 < \cdots < t_n \leqslant 1$, *the C-valued random variables* $x(t_1,\omega)$, $x(t_k,\omega) - x(t_{k-1},\omega)$, $1 < k \leqslant n$, *are independent;*
iii. *if* $0 \leqslant s \leqslant t \leqslant 1$, *the C-valued random variable* $x(t,\omega) - x(s,\omega)$ *is Gaussian with expectation* 0 *and variance* $t - s$.

We refer to the books of Billingsley [10] and Parthasarathy [93] for detailed treatments of probability measures on $C[0,1]$; and we refer to books of Itô and McKean [50], Kac [55], Kampé de Fériet [59], Nelson [91], and Rankin [101] for other treatments of Wiener measure and related problems.

4. *Probability measures on Hilbert spaces.* Consider the measurable space $(H, \mathfrak{B})$, where H is a real separable Hilbert space with inner product $(\cdot,\cdot)$ and $\mathfrak{B}$ is the σ-algebra of Borel sets of H. Let $x(\omega)$ be an H-valued random variable, and let $\nu = \nu_x$ be the probability measure on H induced by x. We denote by $M(H)$ the space of all probability measures on H. We first define the characteristic functional of a probability measure on H, and then introduce the notions of the mean and covariance operator of a probability measure on H.

***Definition* 1.53.** The *characteristic functional* of a probability measure $\nu \in M(H)$ is defined for all $y \in H$ by

$$\varphi(y) = \int_H \exp\{i(x,y)\}\, d\nu(x).$$

The basic properties of the characteristic functional are as follows:

i. $\varphi(y)$ is uniformly continuous in the strong topology.
ii. Let $\nu_1, \nu_2 \in M(H)$ with characteristic functionals $\varphi_1(y)$ and $\varphi_2(y)$, respectively. If $\varphi_1(y) = \varphi_2(y)$ for all $y \in H$, then $\nu_1 = \nu_2$.
iii. $\varphi_{\nu_1 * \nu_2}(y) = \varphi_1(y)\varphi_2(y)$ for all $y \in H$ and $\nu_1, \nu_2 \in M(H)$, where $\nu_1 * \nu_2$ denotes the convolution of ν_1 and ν_2.
iv. $\varphi(-y) = \bar{\varphi}(y)$.

***Definition* 1.54.** The *mean* (or *expectation*) of a probability measure $\nu \in M(H)$ is an element $m \in H$ defined for all $y \in H$ by

$$(m, y) = \int_H (x, y)\, d\nu(x).$$

The mean of a measure in $M(H)$ need not exist; however, if $\int_H \|x\| \, d\nu(x) < \infty$, then $m = \int_H x \, d\nu(x)$ exists and $\|m\| \leqslant \int_H \|x\| \, d\nu(x)$.

We now consider a class of operators, introduced by Prohorov [97], which play a fundamental role in the study of probability theory in Hilbert spaces.

***Definition* 1.55.** Consider a probability measure $\nu \in M(H)$ with the property that $\int \|x\|^2 \, d\nu < \infty$. Then, the *covariance operator* S of ν is defined by the equation

$$(Sy, y) = \int_H \|y\|^2 \, d\nu(x).$$

***Definition* 1.56.** A linear operator L on H is said to be an *S-operator* if it is a positive, self-adjoint operator with finite trace, that is, for some orthonormal basis $\{e_i\} \subset H$, $\sum_{i=1}^{\infty} (Le_i, e_i) < \infty$. S-operators are Hilbert–Schmidt operators,† and hence compact. The covariance operator defined above is an S-operator.

The following remarks concerning S-operators are of interest:

1. S-operators are the infinite-dimensional analogues of the central second-moment matrices associated with a distribution on R_n.
2. Since S-operators are Hilbert–Schmidt operators, they admit the following representation: $Sx = \sum_k \lambda_k (x, \epsilon_k) \epsilon_k$, where the λ_k are the eigenvalues of S (with $\sum_k \lambda_k^2 < \infty$) and $\{\epsilon_k\}$ is a complete orthogonal system of functions. For other representations of S-operators, obtained using tensor product methods, we refer to Kannan and Bharucha-Reid [63].
3. Let ν_1 and ν_2 be two probability measures in $M(H)$ induced by random variables x_1 and x_2. If $\int \|x_i\|^2 \, d\nu_i < \infty$, $i = 1, 2$, then $S_{\nu_1 * \nu_2} = S_{\nu_1} + S_{\nu_2}$.

Let $\mathfrak{S}$ denote the collection of all S-operators. The class of sets $\{\{x : (Sx, x) < 1\}, S \in \mathfrak{S}\}$ defines a system of neighborhoods at the origin for a topology, which is called the *S-topology*.

We now state the following theorem concerning characteristic functionals and the S-topology.

THEOREM 1.21. *A function $\varphi(y)$, $y \in H$, is the characteristic functional of a probability measure ν on H if and only if* (i) *$\varphi(y)$ is positive-definite in y, with $\varphi(\theta) = 1$, and* (ii) *$\varphi(y)$ is continuous at θ in the S-topology.*

The above result is due to Sazonov [108]. We also have the following result of Prohorov [97].

† See Dunford and Schwartz [24, Chapter XI.6].

THEOREM 1.22. *Let $\varphi(y)$ be a positive-definite function on H with $\varphi(\theta)=1$. Then $\varphi(y)$ is the characteristic functional of a probability measure on H if and only if for every $\epsilon>0$ there is a positive-definite, self-adjoint operator of trace class S_ϵ such that $1-\mathcal{R}(\varphi(y)) \leqslant (S_\epsilon y, y)+\epsilon$.*

An important corollary of Theorem 1.22 is as follows: *Let S be a positive-definite, self-adjoint operator on H. The function $\varphi(y)=\exp\{-\frac{1}{2}(Sy,y)\}$ is the characteristic functional of a probability measure ν on H, with $\int \|x\|^2 d\nu<\infty$, if and only if S is an S-operator. S is the covariance operator of ν.*

For other results we refer to Kolmogorov [68], Parthasarathy [93], and Prohorov and Sazonov [100]. Of particular interest is the result of Kuelbs and Mandrekar [73], who generalized Bochner's theorem to l_p spaces. For the case $H=l_2$, their theory coincides with some results of Gross [39], and is related to the works referred to above.

The important class of Gaussian (or normal) probability measures on Hilbert spaces will be considered in Sect. 1.5D. Other classes of probability measures on Hilbert spaces have been considered by several authors; for examples, stable measures have been considered by Jajte [53], and infinitely divisible measures have been studied by Jajte [54], Prakasa Rao [95], and Varadhan [126] (cf. also Parthasarathy [93]). We refer also to the work of Daletskiĭ [17] on probability measures on scales of Hilbert spaces, studied in connection with random integral equation of Itô type (cf. Sect. 7.3C).

D. Gaussian probability measures on Banach spaces

Gaussian (or normal) distribution play an important and fundamental role in classical probability and mathematical statistics; hence it should come as no surprise that the notion of a Gaussian probability measure on a Banach space is of considerable interest in the development of probability theory in Banach spaces, and applications of this theory in many applied fields. We first consider Gaussian probability measures on separable Hilbert spaces, and then consider the case of separable Banach spaces.

Let H be a real separable Hilbert space with inner product $(\cdot,\cdot)$. The following theorem is due to Prohorov [97] (cf. also Varadhan [126]).

THEOREM 1.23. *A probability measure ν on H is Gaussian if and only if its characteristic functional $\varphi(y)$ is of the form*

$$\varphi(y)=\exp\{i(x_0,y)-\tfrac{1}{2}(Sy,y)\},$$

where x_0 is a fixed element of H and S is an S-operator.

It is clear that $x_0 = m$, the mean of ν; and that S is the covariance operator of ν. Hence a Gaussian measure on a Hilbert space is uniquely determined by its mean and covariance operator.

If $x(\omega)$ is an H-valued random variable, then we denote by ν_x the measure on H induced by x and μ. Hence, we can refer to m_x as the mean or expectation of x and S_x as the covariance operator of x. We now state several results which are of interest in applications.

THEOREM 1.24. *Let $x_1(\omega)$ and $x_2(\omega)$ be two independent, Gaussian random variables with values in H, and let m_i and S_i, $i = 1, 2$, denote their expectations and covariance operators, respectively. The sum $x_1(\omega) + x_2(\omega) = y(\omega)$ is a Gaussian H-valued random variable with expectation $m_y = m_1 + m_2$ and covariance operator $S_y = S_1 + S_2$.*

THEOREM 1.25. *If* (i) $y(\omega) = x_1(\omega) + x_2(\omega)$, *where $x_1(\omega)$ and $x_2(\omega)$ are independent H-valued random variables and* (ii) $y(\omega)$ *is Gaussian, then $x_1(\omega)$ and $x_2(\omega)$ are Gaussian.*

Two results of particular interest in the theory of random equations are the following (cf. Grenander [38]):

THEOREM 1.26. *Let $x(\omega)$ be an H-valued Gaussian random variable with expectation m_x and covariance operator S_x, and let $L \in \mathfrak{L}(H)$. Then $y(\omega) = L[x(\omega)]$ is an H-valued Gaussian random variable with expectation $m_y = L[m_x]$ and covariance operator $S_y = LS_xL^*$.*

THEOREM 1.27. *Let $\{\nu_n\}$ be a sequence of Gaussian probability measures on H with means m_n and covariance operators S_n. If $\lim_{n\to\infty} \|m_n - m\| = 0$ and $\lim_{n\to\infty} (S_n x, x) = (Sx, x)$ for all $x \in H$ and $S_n \leqslant T$, where T is some S-operator, then ν_n converges weakly to a Gaussian measure ν with mean m and covariance operator S.*

For other results concerning Gaussian measures on Hilbert spaces we refer to the papers of Birkhoff [11], Garsia *et al.* [32], Jajte [52], Prohorov and Fisz [99], Skorohod [113, 114], and Vahanija [122–125]; in particular, Vahanija has shown that for the Hilbert space l_2, the function

$$\varphi(x) = \exp\left\{-\tfrac{1}{2} \sum_{i,j=1}^{\infty} s_{ij} x_i x_j\right\}$$

for which $\sum_{i=1}^{\infty} s_{ii} < \infty$ is the characteristic functional of a unique Gaussian probability measure ν on l_2. (The mean of ν is assumed to be 0.) Conversely, the characteristic functional of any Gaussian measure on l_2 can be written in the above form if the covariance matrix operator $S=(s_{ij})$ satisfies the condition stated above. Since l_2 is isometrically isomorphic to any separable Hilbert space, the above representation gives the general form of any characteristic functional of a Gaussian probability measure on an arbitrary separable Hilbert space H (where the x_i are the coordinates of an element $x \in H$ with respect to an arbitrary basis).

We now consider Gaussian probability measures on real separable Banach spaces. Let $\mathfrak{X}$ be a real separable Banach space with norm $\|\cdot\|_{\mathfrak{X}}$, and let $\mathfrak{B}_{\mathfrak{X}}$ denote the σ-algebra of Borel sets of $\mathfrak{X}$. Following Kuelbs [71], we will show that given a Gaussian probability measure ν on $\mathfrak{X}$, then it is possible to extend ν to a Gaussian probability measure on a certain separable Hilbert space. We first state the following result.

LEMMA 1.1. *There exists an inner product $(\cdot,\cdot)_{\mathfrak{X}}$ on $\mathfrak{X}$ such that the norm generated by $(\cdot,\cdot)_{\mathfrak{X}}$ is weaker than $\|\cdot\|_{\mathfrak{X}}$, and if $\tilde{H}$ is the separable Hilbert space obtained by completing $\mathfrak{X}$ with respect to the inner product norm, then $\mathfrak{B}_{\mathfrak{X}} \subset \mathfrak{B}_{\tilde{H}}$ so that $\mathfrak{B}_{\mathfrak{X}} = \mathfrak{X} \cap \mathfrak{B}_{\tilde{H}}$.*

The next lemma follows immediately from the above result.

LEMMA 1.2. *If ν is a Gaussian probability measure on $(\mathfrak{X}, \mathfrak{B}_{\mathfrak{X}})$, and $(\tilde{H}, \mathfrak{B}_{\tilde{H}})$ is defined as in Lemma 1.1, then $\nu(B) = \nu(B \cap \mathfrak{X})$, $B \in \mathfrak{B}_{\tilde{H}}$, extends ν to a Gaussian probability measure on $(\tilde{H}, \mathfrak{B}_{\tilde{H}})$.*

In view of the above lemma any Gaussian probability measure defined on a real separable Banach space is also defined on the separable Hilbert space $\tilde{H}$; hence all of the results for Gaussian probability measures on separable Hilbert spaces are applicable. Let m and S denote the mean and covariance operators of ν on $\tilde{H}$. Given S we can define another Hilbert space H as follows:

$$H = \{x \in \tilde{H} : x \in \operatorname{span}\{\epsilon_1, \epsilon_2, \ldots\}, \ \sum_i (x, \epsilon_i)_{\tilde{H}}^2 / \lambda_i < \infty\},$$

where the $\lambda_i > 0$ are the eigenvalues of S and the ϵ_i are the eigenfunctions of S. For $x, y \in H$, the inner product of x and y is defined as

$$(x, y)_H = \sum_i \frac{(x, \epsilon_i)_{\tilde{H}} (y, \epsilon_i)_{\tilde{H}}}{\lambda_i}.$$

We now state the following theorem which gives necessary and sufficient conditions that two Gaussian probability measures on a separable Banach space be equivalent.

THEOREM 1.28. *If ν_1 and ν_2 are Gaussian probability measures on $\mathfrak{X}$, then ν_1 is equivalent to ν_2 if and only if* (i) $m_1 - m_2 \in H$ *and*

$$\text{(ii)} \quad (S_2 x, y)_{\tilde{H}} = (S_1 x, S_1 y)_H - (KS_1 x, S_1 y)_H$$

for all $x, y \in \tilde{H}$, where K is a symmetric Hilbert–Schmidt operator on H which does not have one as an eigenvalue.

The above theorem can be applied to the study of the equivalence of Gaussian and Wiener probability measures (cf. Shepp [110]), and to the study of Gaussian probability measures on l_p spaces (cf. the papers of Vahanija).

In closing this subsection we refer to the work of Kallianpur [56], Sato [107], and de Acosta [19]. Kallianpur has studied Gaussian probability measures on a separable Banach space $\mathfrak{X}$ by considering a Hilbert space which acts as a reproducing kernel Hilbert space for the Gaussian probability measure on $\mathfrak{X}$. Sato has shown that any Gaussian probability measure on a separable or reflexive Banach space is an abstract Wiener measure in the sense of Gross (cf. Gross [40]), and can be extended to a Radon probability measure on the same space. de Acosta has constructed Gaussian probability measures on Banach spaces with a Schauder basis, and has obtained the representation of the characteristic functional of such measures.

For some results on Gaussian random functions and Gaussian probability measures on Banach spaces we refer to Jain and Kallianpur [51], Kuelbs [72], and LePage [76].

E. Absolute continuity of probability measures

Consider the measurable space $(X, \mathfrak{B})$, where X is an arbitrary set and $\mathfrak{B}$ is a σ-algebra of subsets of X. Let ν_1 and ν_2 be two measures on X. The measure ν_2 is said to be *absolutely continuous* with respect to ν_1 (written $\nu_2 \ll \nu_1$) if $\nu_2(B) = 0$ for all $B \in \mathfrak{B}$ for which $\nu_1(B) = 0$. If $\nu_2 \ll \nu_1$ and $\nu_1 \ll \nu_2$, then ν_1 and ν_2 are said to be *equivalent* (written $\nu_1 \sim \nu_2$). The measure ν_2 is said to be *singular* with respect to ν_1 if there is a set B such that $\nu_1(B) = 0$ and $\nu_2(X - B) = 0$. If ν_2 is singular with respect to ν_1, then ν_1 is singular with respect to ν_2; and in this case ν_1 and ν_2 are said to be *orthogonal* (written $\nu_1 \perp \nu_2$).

The well-known Radon–Nikodym theorem states that for finite measures, $\nu_2 \ll \nu_1$ if and only if there exists a $\mathfrak{B}$-measurable function $\rho(x)$ such that for all $B \in \mathfrak{B}$

$$\nu_2(B) = \int_B \rho(x)\, d\nu_1(x).$$

The function $\rho(x)$ is called the *Radon–Nikodym derivative*, or *density*, of ν_2 with respect to ν_1. $\rho(x)$, which is written $\rho(x) = [d\nu_2/d\nu_1](x)$, is uniquely defined within equivalence with respect to ν_1. We refer to Halmos [43, Chap. VI] for a discussion of absolute continuity and Radon–Nikodym derivatives.

Let $x_1(t,\omega)$ and $x_2(t,\omega)$ be two stochastic processes with associated probability measures ν_1 and ν_2, respectively. If $\nu_2 \ll \nu_1$, then it is possible to find sets B for which $\nu_2(B) = 1$, provided, of course, such sets are known for ν_1. That is, the process $x_2(t,\omega)$ will satisfy almost surely all those properties that the process $x_1(t,\omega)$ satisfies almost surely. In view of the above, the study of the absolute continuity of probability measures associated with stochastic processes is of great interest in the theory of random equations. For example, if $x_1(t,\omega)$ represents a (known) random input to a differential or integral equation the solution of which is $x_2(t,\omega)$, it is of particular importance to determine conditions under which $\nu_2 \ll \nu_1$, for in this case the solution process $x_2(t,\omega)$ will satisfy almost surely all those properties which the input process $x_1(t,\omega)$ satisfies almost surely.

In this section we state a few results on the absolute continuity of probability measures which we feel are of interest in the study of random equations. We refer to the fundamental paper of Gihman and Skorohod [35] for a detailed discussion of the densities of probability measures on function spaces, and for the proofs of the theorems stated in this section.

We first consider the absolute continuity of Gaussian probability measures on Hilbert spaces. Let ν_1 and ν_2 be Gaussian probability measures on a separable Hilbert space H with means $m_1 = m_2 = 0$ and covariance operators S_1 and S_2, respectively. The following result gives a simple necessary condition for the absolute continuity of ν_2 with respect to ν_1.

THEOREM 1.29. *If $\nu_2 \ll \nu_1$, then there exists a positive constant λ such that*

$$\lambda^{-1} \leqslant (S_2 x, x)/(S_1 x, x) \leqslant \lambda.$$

Let $\{f_i\}$ be an orthonormal sequence of eigenfunctions of S_1, and $\{\lambda_i\}$ the corresponding sequence of eigenvalues.

THEOREM 1.30. *If $\nu_2 \ll \nu_1$, then there exist numbers α and β_{ij} (with $\sum \beta_{ij}^2 < \infty$) such that*

$$\frac{d\nu_2}{d\nu_1}(x) = \exp\{\alpha + \sum \beta_{ij}[(x,f_i)(x,f_j)(\lambda_i\lambda_j)^{1/2} - \delta_{ij}]\}.$$

and $\beta_{ij} = ((I - LL^*)f_i, f_j)$, where L is a bounded operator such that $S_1^{1/2} = LS_2^{1/2}$.

We now state a theorem which gives necessary and sufficient conditions for the equivalence† of two Gaussian probability measures ν_1 and ν_2 with means m_1, m_2 and covariance operators S_1, S_2, respectively.

† See also Theorem 1.28.

THEOREM 1.31. *Two Gaussian probability measures ν_1 and ν_2 on a Hilbert space are either equivalent or orthogonal. Let $\{f_i\}$ denote the eigenfunctions of S_1 and $\{\lambda_i\}$ the associated eigenvalues. Then, $\nu_1 \sim \nu_2$ if and only if* (i) $\sum_{i=1}^{\infty}(m_2 - m_1, f_i)^2 \lambda_i^{-1} < \infty$, *and* (ii) *there exists a bounded operator L, with bounded inverse, such that $S_1^{1/2} = LS_2^{1/2}$ and* $\mathrm{Tr}(I - LL^*)^2 < \infty$.

For detailed treatments of the Radon–Nikodym derivatives of Gaussian probability measures we refer to Rao and Varadarajan [102] and Shepp [110]. In view of Kuelb's result (Lemma 1.2), the above theorems can be utilized, with appropriate modifications, to study the absolute continuity of Gaussian probability measures on separable Banach spaces.

Let ν_1 be a probability measure on the measurable space $(X, \mathfrak{B})$, and let L denote a transformation of X onto itself. Put $\nu_2(B) = \nu_1(L^{-1}(B))$ for all $B \in \mathfrak{B}$. A problem of fundamental importance in the theory of random equations is the determination of conditions under which $\nu_2 \ll \nu_1$. The first problem of this type was investigated by Cameron and Martin (cf. [35]), who calculated the density, with respect to Wiener measure $w = \nu_1$ on the space of continuous functions, of the probability measure ν_2 induced by a measurable transformation of a Wiener process. In recent years a number of results have been obtained for Gaussian probability measures on Hilbert spaces for certain classes of transformations (cf. Gihman and Skorohod [35], Baklan and Šatašvili [5, 6]); however, a considerable amount of research needs to be done for classes of transformations which rise in the study of random differential and integral equations. In Sect. 6.6 we consider a problem of this type for a nonlinear random integrodifferential equation.

1.6 Limit Theorems

A. Introduction

The study of probability theory in any type of algebraic or topological structure is not complete without an investigation of limit theorems, for as Gnedenko and Kolmogorov have remarked, "...the epistemological value of the theory of probability is revealed only be limit theorems." In this section we consider some limit theorems for Banach space-valued random variables. Thus far limit theorems have not been employed extensively in the study of random equations; but it is clear that as the theory of random equations develops limit theorems will undoubtedly play a major role, especially in the study of approximate solutions of random equations.

The remaining subsections of this section are devoted to strong laws of large numbers, central limit theorems, and martingale convergence theorems.

B. Strong laws of large numbers

The well-known strong law of large numbers due to Kolmogorov can be stated as follows (cf. Loève [77, p. 239]): *Let* $\{x_n(\omega)\}$ *be a sequence of independent and identically distributed real-valued random variables. Put* $S_n(\omega) = (1/n)\sum_{i=1}^{n} x_i(\omega)$. *Then* $S_n \to y$ *almost surely if and only if* $\mathscr{E}\{|x(\omega)|\} < \infty$; *and then* $y = \mathscr{E}\{x(\omega)\}$.

The first strong law of large numbers for Banach space-valued random variables is due to Mourier [84].

THEOREM 1.32. *Let* $\mathfrak{X}$ *be a separable Banach space, and let* $\{x_n(\omega)\}$ *be a sequence of independent and identically distributed* $\mathfrak{X}$*-valued random variables such that* $\mathscr{E}\{\|x\|\} < \infty$. *Put* $S_n(\omega) = (1/n)\sum_{i=1}^{n} x_i(\omega)$. *Then* $\lim_{n\to\infty} S_n(\omega) = \mathscr{E}\{\|x\|\}$ *strongly almost surely.*

Other strong laws of Kolmogorov type have been obtained by Hanš [44] for Banach space-valued random variables, and by Bharucha-Reid [9] for Orlicz space-valued random variables. In Banach spaces it is possible to formulate many versions of the strong law of large numbers. For example, Beck (cf. [7]) considered a sequence $\{x_n(\omega)\}$ of independent Banach space-valued random variables with $\mathscr{E}\{x_n(\omega)\} = 0$, $\mathscr{E}\{\|x_n(\omega)\|^2\} \leqslant M < \infty$, and showed that a strong law holds under the assumption that $\mathfrak{X}$ is a uniformly convex Banach space; and Beck and Warren [8] have proved a strong law for weakly orthogonal† sequences of Banach space-valued random variables. For other results we refer to Fortet and Mourier [29], Grenander [38, Sect. 6.4], Parthasarathy [93, pp. 208–210], and Révész [104, Chap. 9].

C. Central limit theorems

The classical central limit theorem in its most simple form can be stated as follows: *Let* $\{x_n(\omega)\}$ *be a sequence of independent and indentically distributed real-valued random variables for which* $\sigma_n = (\mathrm{Var}\{x_n(\omega)\})^{1/2} > 0$ *exist,* $n = 1, 2, \ldots$. *Put* $S_n(\omega) = \sum_{i=1}^{n} x_i(\omega)$ *and* $\tilde{S}_n(\omega) = [S_n(\omega) - \mathscr{E}\{x_n(\omega)\}]/\sigma_n$. *If* $F_n(\xi)$ *is the distribution function of* $\tilde{S}_n(\omega)$, *then* $\lim_{n\to\infty} F_n(\xi) = \Phi(\xi)$ *uniformly for* $\xi \in (-\infty, \infty)$, *where* $\Phi(\xi)$ *is the standard normal distribution.* The well-known Lindeberg–Feller version of the central limit theorem can be stated as follows: *Let* $\{x_n(\omega)\}$ *be a sequence of independent real-valued random variables for which* $m_n = \mathscr{E}\{x_n(\omega)\}$ *and* $\sigma_n = (\mathrm{Var}\{x_n(\omega)\})^{1/2}$ *exist,* $n = 1, 2, \ldots$. *Put* $s_n = (\sum_{i=1}^{n} \sigma_i^2)^{1/2}$, *and let* $F_n(\xi)$ *be the distribution function of* $x_n(\omega) - m_k$; *and put*

$$\tilde{x}_n(\omega) = \sum_{i=1}^{n} (x_i(\omega) - m_i)/s_n.$$

† A sequence of $\mathfrak{X}$-valued random variables is said to be *weakly orthogonal* if, for all $x^* \in \mathfrak{X}^*$, $\int_\Omega x^*(x_i(\omega))x^*(x_j(\omega))\,d\mu = 0$ for all $i \neq j$.

Then $\lim_{n\to\infty}\mathscr{P}\{\tilde{x}_n(\omega)<\xi\}=\Phi(\xi)$ *if and only if*

$$\lim_{n\to\infty}(1/s_n^2)\sum_{i=1}^{n}\int_{|\xi|>\epsilon s_n}\xi^2\,dF_i(\xi)=0$$

for every $\epsilon>0$.

For a detailed exposition of the central limit problem for real-valued random variables we refer to Loève [77, Chap. VI].

The study of Banach space analogues of the central limit theorem was initiated by Mourier [84]. We have

THEOREM 1.33. *Let H be a separable Hilbert space, and let* $\{x_n(\omega)\}$ *be a sequence of independent and identically distributed H-valued random variables. Suppose that* $\mathscr{E}\{\|x\|^2\}<\infty$; *and let m and S denote the expectation and covariance operator of* $x(\omega)$, *respectively. Put* $\tilde{x}_n(\omega)=n^{-1/2}\sum_{i=1}^{n}(x_i(\omega)-m)$. *Then the sequence* $\{\nu_n(\omega)\}$ *of probability measures in* $M(H)$ *associated with* $\{\tilde{x}_n(\omega)\}$ *converges weakly to a Gaussian probability measure in* $M(H)$ *with expectation zero and covariance operator S.*

Kandelaki and Sazonov [61] (cf. also Reinschke [103]) have proved a central limit theorem of Lindeberg–Feller type for H-valued random variables.

The central limit theorem for a certain class of Banach spaces has been considered by Fortet and Mourier [29] (cf. also Mourier [85]). A Banach space $\mathfrak{X}$ is said to be a *G-space* if there is a constant K and a mapping $\psi:\mathfrak{X}\to\mathfrak{X}^*$ such that (i) $\|\psi(x)\|=\|x\|$, (ii) $(\psi(x),x)=\|x\|^2$, (iii) $\|\psi(x_1)-\psi(x_2)\|\leqslant K\|x_1-x_2\|$. A general characterization of G-spaces is not known; however, it is known that a wide class of uniformly convex spaces are G-spaces, including Hilbert spaces and all L_p spaces ($p\geqslant 2$). A central limit theorem for random variables with values in a Banach space $\mathfrak{X}$ which is a G-space can be stated as follows:

THEOREM 1.34. *Let* $\{x_n(\omega)\}$ *be a sequence of independent and identically distributed* $\mathfrak{X}$*-valued random variables, where* $\mathfrak{X}$ *is a reflexive G-space with basis. Let* $m=\mathscr{E}\{x\}$, *and suppose* $\mathscr{E}\{\|x\|^2\}<\infty$. *Then the sequence* $\{\nu_n(\omega)\}$ *of probability measures in* $M(\mathfrak{X})$ *associated with* $\{\tilde{x}_n(\omega)\}$ (cf. Theorem 1.33) *converges weakly to a Gaussian probability measure in* $M(\mathfrak{X})$.

For other results on central limit theorems in Banach spaces we refer to Dudley [22a], LeCam [75a], and Strassen and Dudley [115a].

D. Martingale convergence theorems

The classical (discrete) martingale convergence theorem, due to Doob (cf. [20, Chap. VII]) can be stated as follows: *Let* $\{x_n\}$ *be a discrete parameter martingale. Then* $\mathscr{E}\{|x_1(\omega)|\} \leqslant \mathscr{E}\{|x_2(\omega)|\} \leqslant \cdots$. *If* $\lim_{n\to\infty} \mathscr{E}\{|x_n(\omega)|\} = M < \infty$, *then* $x_n(\omega) = \tilde{x}(\omega)$ *almost surely, and* $\mathscr{E}\{\tilde{x}(\omega)\} \leqslant M$. *In particular,* $M < \infty$, *if the* $x_n(\omega)$ *are all real nonnegative random variables, or all real and nonpositive.* Martingale convergence theorems have been utilized in many applications of martingale theory in analysis and other areas of probability theory; hence the investigation of convergence theorems for Banach space-valued martingales is of interest.

In proving convergence theorems for Banach space-valued martingales,† Chatterji [14] and others observed that unless some additional conditions are imposed on the range space of the process, martingale convergence theorems of the classical (scalar) type could not be obtained. Counterexamples are given, for example, by Chatterji [14], and Driml and Hanš [21].

We first state a convergence theorem due to Chatterji [14].

THEOREM 1.35. *Let* $\mathfrak{X}$ *be a reflexive Banach space, and let* $\{x_n(\omega), n \geqslant 1\}$ *be an* $\mathfrak{X}$*-valued martingale such that* (i) $x_n(\omega) \in B_p(\Omega, \mathfrak{X})$, $1 < p < \infty$, *and* (ii) $[x_n(\omega)]_p < \infty$. *Then there exists an* $\tilde{x}(\omega) \in B_p(\Omega, \mathfrak{X})$ *such that*

$$\lim_{n\to\infty} [x_n(\omega) - \tilde{x}(\omega)]_p = 0 \quad \textit{and} \quad x_n(\omega) = \mathscr{E}\{\tilde{x}(\omega) | \mathfrak{A}_n\}.$$

Scalora [109] also assumed $\mathfrak{X}$ to be reflexive; and Driml and Hanš assumed the weak compactness (almost surely) of a particular subset of Ω.

For Hilbert space-valued martingales, we state the following result due to Scalora:

THEOREM 1.36. *Let* $\{x_n(\omega), n \geqslant 1\}$ *be a Hilbert space-valued martingale; and suppose that* $\lim_{n\to\infty} \mathscr{E}\{\|x_n(\omega)\|\} < \infty$. *Then, there exists a strong Hilbert space-valued random variable* $\tilde{x}(\omega)$ *such that* $\lim_{n\to\infty} x_n(\omega) = \tilde{x}(\omega)$ *weakly almost surely.*

For other results concerning the convergence of Banach space-valued martingales we refer to Chatterji [14, 15], Driml and Hanš [21], Ip [48], Kannan [62], Metivier [82], Scalora [109], and Uhl [119].

References

1. Ahmad, S., Éléments aléatoires dans les espaces vectoriels topologiques. *Ann. Inst. H. Poincaré Sect. B* **2** (1965), 97–135.
2. Badrikian, A., Résultats relatifs aux éléments aléatoires prenant leurs valeurs dans un espace de Banach réflexif non séparable. *C. R. Acad. Sci. Paris* **246** (1958), 882–884.
3. Badrikian, A., "Séminaire sur les fonctions aléatoires linéaires et les mesures cylindriques." Springer-Verlag, Berlin and New York, 1970.

† We refer to Sect. 1.4F for the definition of a Banach space-valued martingale.

4. Baklan, V. V., Variational differential equations and Markov processes in Hilbert space (Russian). *Dokl. Akad. Nauk SSSR* **159** (1964), 707–710.
5. Baklan, V. V., and Šatašvili, A. D., Conditions for the absolute continuity of probability measures corresponding to Gaussian random variables in a Hilbert space (Ukrainian). *Dopovidi Akad. Nauk Ukrain. RSR* (1965), pp. 23–26.
6. Baklan, V. V., and Šatašvili, A. D., Transformation of Gaussian measures by nonlinear mappings in Hilbert space (Ukrainian). *Dopovidi Akad. Nauk Ukrain RSR* (1965), pp. 1115–1117.
7. Beck, A., On the strong law of large numbers. *In* "Ergodic Theory" (F. Wright, ed.), pp. 21–53. Academic Press, New York, 1963.
8. Beck, A., and Warren, P., A strong law of large numbers for weakly orthogonal sequences of Banach space-valued random variables. Tech. Rep. 848. Math. Res. Center, Univ. of Wisconsin, Madison, Wisconsin, 1969.
9. Bharucha-Reid, A. T., On random elements in Orlicz spaces. *Bull. Acad. Polon. Sci. Cl. Troisieme* **4** (1956), 655–657.
10. Billingsley, P., "Convergence of Probability Measures." Wiley, New York, 1968.
11. Birkhoff, G., Fourier synthesis of homogeneous turbulence. *Comm. Pure Appl. Math.* **7** (1954), 19–44.
12. Birnbaum, S., The Tschebytscheff inequality for Banach space-valued random variables (Abstract). *Notices Amer. Math. Soc.* **18** (1971), 179.
13. Brooks, J. K., Representations of weak and strong integrals in Banach spaces. *Proc. Nat. Acad. Sci. U.S.A.* **63** (1969), 266–270.
14. Chatterji, S. D., Martingales of Banach-valued random variables. *Bull. Amer. Math. Soc.* **66** (1960), 395–398.
15. Chatterji, S. D., Martingale convergence and the Radon-Nikodym theorem in Banach spaces. *Math. Scand.* **22** (1968), 21–41.
16. Cramér, H., and Leadbetter, M. R., "Stationary and Related Stochastic Processes." Wiley, New York, 1967.
17. Daletskiĭ, Yu. L., Infinite-dimensional elliptic operators and parabolic equations associated with them (Russian). *Uspehi Mat. Nauk* **22** (4) (1967), 3–54.
18. Day, M. M., "Normed Linear Spaces." Springer-Verlag, Berlin and New York, 1958.
19. de Acosta, A. D., Existence and convergence of probability measures in Banach spaces. *Trans. Amer. Math. Soc.* **152** (1970), 273–298.
20. Doob, J. L., "Stochastic Processes." Wiley, New York, 1953.
21. Driml, M., and Hanš, O., Conditional expectations for generalized random variables. *Trans. 2nd Prague Conf. on Information Theory, Statist. Decision Functions, and Random Processes, (1959)*, pp. 123–143, 1960.
22. Dudley, R. M., Random linear functionals. *Trans. Amer. Math. Soc.* **136** (1969), 1–24.

22a. Dudley, R. M., Some uses of ϵ-entropy in probability theory. "Les probabilités sur les structures algébriques," pp. 113–122. CNRS, Paris, 1970.

23. Dunford, N., and Schwartz, J. T., "Linear Operators. Part I: General Theory." Wiley (Interscience), New York, 1958.
24. Dunford, N., and Schwartz, J. T., "Linear Operators. Part II: Spectral Theory." Wiley (Interscience), New York, 1963.
25. Dynkin, E. B., "Markov Processes," Vols. I and II, translated from the Russian. Academic Press, New York, 1965.
26. Falb, P. L., Infinite-dimensional filtering: The Kalman-Bucy filter in Hilbert space. *Information and Control* **11** (1967), 102–137.
27. Fortet, R., Normalverteilte Zufallselemente in Banachschen Räumen. Anwendungen auf Zufällige Funktionen. *In* "Bericht über die Tagung Wahrscheinlichkeitsrechnung

und Mathematische Statistik in Berlin." Deutscher Verlag der Wissenschaften, Berlin, 1956.

28. Fortet, R., and Mourier, E., Résultats complémentaires sur les éléments aléatoires prenant leur valeurs dans un espace de Banach. *Bull. Sci. Math.* (2), **78** (1954), 14–30.
29. Fortet, R., and Mourier, E., Les fonctions aléatoires comme éléments aléatoires dans des espaces de Banach. *Studia Math.* **15** (1955), 62–79.
30. Fréchet, M., Les éléments aléatoires de nature quelconque dans un espace distancié. *Ann. Inst. H. Poincaré* **10** (1947), 215–310.
31. Fréchet, M., Abstrakte Zufallselemente. *In* "Bericht über die Tagung Wahrscheinlichkeitsrechnung und Mathematische Statistik in Berlin." Deutscher Verlag der Wissenschaften, Berlin, 1956.
32. Garsia, A. M., Posner, E. C., and Rodemich, E. R., Some properties of the measures on function spaces induced by Gaussian processes. *J. Math. Anal. Appl.* **21** (1968), 150–161.
33. Gel'fand, I. M., and Vilenkin, N. Ya., "Generalized Functions, Vol. 4: Applications of Harmonic Analysis," translated from the Russian. Academic Press, New York, 1964.
34. Getoor, R. K., On characteristic functions of Banach space valued random variables. *Pacific J. Math.* **7** (1957), 885–896.
35. Gihman, I. I., and Skorohod, A. V., On the densities of probability measures in function spaces (Russian). *Uspehi Mat. Nauk* **21** (1966), 83–152.
36. Glivenko, V., Sur la loi des grands nombres dans les espaces fonctionnels. *Rend. Lincei* **8** (1928), 673–676.
37. Goffman, C., and Pedrick, G., "First Course in Functional Analysis." Prentice-Hall, Englewood Cliffs, New Jersey, 1965.
38. Grenander, U., "Probabilities on Algebraic Structures." Wiley, New York, 1963.
39. Gross, L., Harmonic analysis on Hilbert space. *Mem. Amer. Math. Soc.* **No. 46** (1963).
40. Gross, L., Abstract Wiener spaces. *Proc. 5th Berkeley Symp. Math. Statist. and Probability (1965)*, Vol. II, Pt. I, pp. 31–42, 1967.
41. Haïnis, J., Éléments aléatoires dans un H^*-algèbre de Banach séparable. *Bull. Soc. Roy. Sci. Liège* **33** (1964), 170–177.
42. Haïnis, J., Random variables with values in Banach algebras and random transformations in Hilbert spaces (Greek, French summary). *Bull. Soc. Math. Grèce* (N. S.) **7** (1966), 179–223.
43. Halmos, P. R., "Measure Theory." Van Nostrand-Reinhold, Princeton, New Jersey, 1950.
44. Hanš, O., Generalized random variables. *Trans. 1st Prague Conf. on Information Theory, Statist. Decision Functions, and Random Processes (1956)*, pp. 61–103, 1957.
45. Hida, T., "Stationary Stochastic Processes." Princeton Univ. Press, Princeton, New Jersey, 1970.
46. Hille, E., and Phillips, R. S., "Functional Analysis and Semi-Groups." Amer. Math. Soc., Providence, Rhode Island, 1957.
47. Ip, B.-F., Riesz decomposition of positive Banach space-valued supermartingales (Abstract). *Notices Amer. Math. Soc.* **17** (1970), 838.
48. Ip, B.-F., Banach space-valued martingales. Ph.D. Dissertation, Wayne State Univ., Detroit, Michigan, 1970.
49. Itô, K., Stationary random distributions. *Mem. Coll. Sci. Univ. Kyoto Ser. A* **28** (1953), 209–223.
50. Itô, K., and McKean, H. P., "Diffusion Processes and Their Sample Paths." Springer-Verlag, Berlin and New York, 1965.

51. Jain, N. C., and Kallianpur, G., Norm convergent expansions for Gaussian processes in Banach spaces. *Proc. Amer. Math. Soc.* **25** (1970), 890–895.
52. Jajte, R., On the probability measures in Hilbert spaces. *Studia Math.* **29** (1968), 221–241.
53. Jajte, R., On stable distributions in Hilbert space. *Studia Math.* **30** (1968), 63–71.
54. Jajte, R., On convergence of infinitely divisible distributions in Hilbert space. *Colloq. Math.* **19** (1968), 327–332.
55. Kac, M., "Probability and Related Topics in Physical Sciences." Wiley (Interscience), New York, 1959.
56. Kallianpur, G., Abstract Wiener processes and their reproducing kernel Hilbert spaces. *Z. Wahrscheinlichkeitstheorie und Verw. Gebiete* **17** (1971), 113–123.
57. Kallianpur, G., and Mandrekar, V., Multiplicity and representation theory of purely non-deterministic stochastic processes. *Teor. Verojatnost. i Primenen.* **10** (1965), 614–644.
58. Kampé de Fériet, J., Mesures de probabilité sur les espaces de Banach possédant une base dénombrable. *J. Math. Pures Appl.* **39** (1960), 119–163.
59. Kampé de Fériet, J., "Models for Random Functions." Academic Press, New York, to be published.
60. Kandelaki, N. P., Wiener process with values in a Hilbert space (Russian). *Sakharth. SSR Mecn. Akad. Gamothvl. Centr. Šrom.* **7** (1966), No. 1, 5–7.
61. Kandelaki, N. P., and Sazonov, V. V., On the central limit theorem for random elements with values in a Hilbert space (Russian). *Teor. Verojatnost. i Primenen.* **9** (1964), 43–52.
62. Kannan, D., Martingales in Banach spaces with Schauder bases. To be published.
63. Kannan, D., and Bharucha-Reid, A. T., Note on covariance operators of probability measures on a Hilbert space. *Proc. Japan Acad.* **46** (1970), 124–129.
64. Kannan, D., and Bharucha-Reid, A. T., An operator-valued stochastic integral. *Proc. Japan Acad.* **47** (1971), 472–476.
65. Kannan, D., and Bharucha-Reid, A. T., Probability measures on H_p spaces. *Ann. Inst. H. Poincaré Sect. B* **7** (1971), 205–217.
66. Kappos, D. A., "Probability Algebras and Stochastic Spaces." Academic Press, New York, 1969.
67. Kolmogorov, A. N., La transformation de Laplace dans des espaces linéaires. *C. R. Acad. Sci. Paris* **200** (1935), 1717–1718.
68. Kolmogorov, A. N., A note on the papers of R. A. Milnos and V. Sazonov (Russian). *Teor. Verojatnost. i Primenen.* **4** (1959), 237–239.
69. Kolmogorov, A. N., and Fomin, S. V., "Introductory Real Analysis," rev. Engl. ed., translated from the Russian. Prentice-Hall, Englewood Cliffs, New Jersey, 1970.
70. Krasnosel'skiĭ, M. A., and Rutickiĭ, Ya. B., "Convex Functions and Orlicz Spaces" (Russian). Gosudarstv. Izdat. Fiz.-Mat. Lit., Moscow, 1958.
71. Kuelbs, J., Gaussian measures on a Banach space. *J. Functional Analysis* **5** (1970), 354–367.
72. Kuelbs, J., Expansions of vectors in a Banach space related to Gaussian measures. *Proc. Amer. Math. Soc.* **27** (1971), 364–370.
73. Kuelbs, J., and Mandrekar, V., Harmonic analysis on certain vector spaces. *Trans. Amer. Math. Soc.* **149** (1970), 213–231.
74. Kunita, H., Stochastic integrals based on martingales taking values in Hilbert space. *Nagoya Math. J.* **39** (1970), 41–52.
75. LeCam, L., Convergence in distribution of stochastic processes. *Univ. California Publ. Statist.* **2** (11) (1957), 207–236.

75a. LeCam, L., Remarques sur le théorème limite central dans les espaces localement convexes. "Les probabilités sur les structures algébriques," pp. 233–249. CNRS, Paris, 1970.
76. LePage, R. D., Note relating Bochner integrals and reproducing kernels to series expansions on a Gaussian Banach space. *Proc. Amer. Math. Soc.* **32** (1972), 285–288.
77. Loève, M., "Probability Theory," 3rd ed. Van Nostrand-Reinhold, Princeton, New Jersey, 1963.
78. Loomis, L. H., "An Introduction to Abstract Harmonic Analysis." Van Nostrand-Reinhold, Princeton, New Jersey, 1953.
79. Lorch, E. R., "Spectral Theory." Oxford Univ. Press, London and New York, 1962.
80. Loynes, R. M., On a generalization of second-order stationarity. *Proc. London Math. Soc.* (3) **15** (1965), 385–398.
81. Marti, J. T., "Introduction to the Theory of Bases." Springer-Verlag, Berlin and New York, 1969.
82. Metivier, M., Limites projective de mesures; martingales; applications. *Ann. Mat. Pura Appl.* **63** (1963), 225–352.
83. Meyer, P. A., "Probability and Potentials." Ginn (Blaisdell), Boston, Massachusetts, 1966.
84. Mourier, E., Éléments aléatoires dans un espace de Banach. *Ann. Inst. H. Poincaré* **13** (1953), 161–244.
85. Mourier, E., L-random elements and L^*-random elements in Banach spaces. *Proc. 3rd Berkeley Symp. Math. Statist. and Probability (1955)*, Vol. II, pp. 231–242, 1956.
86. Mourier, E., Random elements in linear spaces. *Proc. 5th Berkeley Symp. Math. Statist. and Probability (1965)*, Vol. II, Pt. I, pp. 43–53, 1967.
87. Moy, S.-T. C., Conditional expectations of random variables values in a Banach space and their properties. Tech. Rep. Dept. of Math., Wayne State Univ., Detroit, Michigan, 1956.
88. Naimark, M. A., "Normed Rings." Nordhoff, Groningen, 1959.
89. Nasr, S. K., Determination of the Mourier mean of random variables situated in some Banach spaces. *Proc. Math. Phys. Soc. U.A.R. (Egypt)* **5** (1956), 79–85.
90. Nedoma, J., Note on generalized random variables. *Trans. 1st Prague Conf. on Information Theory, Statist. Decision Functions, and Random Processes (1956)*, pp. 139–141, 1957.
91. Nelson, E., "Dynamical Theories of Brownian Motion." Princeton Univ. Press, Princeton, New Jersey, 1967.
92. Neveu, J., "Mathematical Foundations of the Calculus of Probability." Holden-Day, San Francisco, California, 1965.
93. Parthasarathy, K. R., "Probability Measures on Metric Spaces." Academic Press, New York, 1967.
94. Payen, R., Fonctions aléatoires du second ordre à valeurs dans un espace de Hilbert. *Ann. Inst. H. Poincaré Sect. B* **3** (1967), 323–396.
95. Prakasa Rao, B. L. S., On a characterization of infinitely divisible characteristic functionals on a Hilbert space. *Z. Wahrscheinlichkeitstheorie und Verw. Gebiete* **14** (1970), 254–256.
96. Price, G. B., The theory of integration. *Trans. Amer. Math. Soc.* **47** (1940), 1–50.
97. Prohorov, Yu. V., Convergence of random processes and limit theorems in probability theory (Russian). *Teor. Verojatnost. i Primenen.* **1** (1956), 177–238.
98. Prohorov, Yu. V., The methods of characteristic functionals. *Proc. 4th Berkeley Symp. Math. Statist. and Probability (1960)*, Vol. II, pp 403–419, 1961.
99. Prohorov, Yu. V., and Fisz, M., A characterization of normal distributions in Hilbert space (Russian). *Teor. Verojatnost. i Primenen.* **2** (1957), 475–477.

100. Prohorov, Yu. V., and Sazonov, V. V., Some results associated with Bochner's theorem (Russian). *Teor. Verojatnost. i Primenen.* **6** (1961), 87–93.
101. Rankin, B., ed., "Differential Space, Quantum Systems, and Prediction." MIT Press, Cambridge, Massachusetts, 1966.
102. Rao, C. R., and Varadarajan, V. S., Discrimination of Gaussian processes. *Sankhyā, Ser. A* **25** (1963), 303–330.
103. Reinschke, K., Zum zentralen Grenzwertsatz für zufallige Elemente mit Werten aus einem Hilbertraum. *Z. Wahrscheinlichkeitstheorie und Verw. Gebiete* **6** (1966), 161–169.
104. Révész, P., "The Laws of Large Numbers." Academic Press, New York, 1968.
105. Rickart, C. E., "General Theory of Banach Algebras." Van Nostrand-Reinhold, Princeton, New Jersey, 1960.
106. Rozanov, Yu. V., "Stationary Random Functions," translated from the Russian. Holden-Day, San Francisco, California, 1967.
107. Sato, H., Gaussian measure on a Banach space and abstract Wiener measure. *Nagoya Math. J.* **36** (1969), 65–81.
108. Sazonov, V. V., A remark on characteristic functionals (Russian). *Teor. Verojatnost. i Primenen.* **3** (1958), 201–205.
109. Scalora, F., Abstract martingale convergence theorems. *Pacific J. Math.* **11** (1961), 347–374.
110. Shepp, L. A., Radon–Nikodym derivatives of Gaussian measures. *Ann. Math. Statist.* **37** (1966), 321–354.
111. Simmons, G. F., "Introduction to Topology and Modern Analysis." McGraw-Hill, New York, 1963.
112. Singer, I., "Bases in Banach Spaces," Vol. I. Springer-Verlag, Berlin and New York, 1970.
113. Skorohod, A. V., Note on Gaussian measures in a Banach space (Russian). *Teor. Verojatnost. i Primenen.* **15** (1970), 519–520.
114. Skorohod, A. V., On admissible translations of measures in Hilbert space (Russian). *Teor. Verojatnost. i Primenen.* **15** (1970), 577–598.
115. Srinivasan, R., A remark on Gaussian random elements with values in a Banach algebra. *Rev. Roumaine Math. Pures. Appl.* **7** (1962), 285–286.
115a. Strassen, V., and Dudley, R. M., The central limit theorem and ϵ-entropy. *In* "Lecture Notes in Mathematics, Vol. 89: Probability and Information Theory," pp. 224–231. Springer-Verlag, Berlin and New York, 1969.
116. Taylor, A. E., "Introduction to Functional Analysis." Wiley, New York, 1958.
117. Theodorescu, R., Some remarks on abstract random variables and functions. *Bull. Math. Soc. Sci. Math. Phys. R. P. Roumaine* **2** (1958), 343–351.
118. Uhl, J. J., Martingales of vector valued set functions. *Pacific J. Math.* **30** (1969), 533–548.
119. Uhl, J. J., Applications of Radon–Nikodym theorems to martingale convergence. *Trans. Amer. Math. Soc.* **145** (1969), 271–285.
120. Umegaki, H., and Bharucha-Reid, A. T., Banach space-valued random variables and tensor products of Banach spaces. *J. Math. Anal. Appl.* **31** (1970), 49–67.
121. Urbanik, K., Generalized stochastic processes. *Studia Math.* **16** (1958), 268–334.
122. Vahanija, N. N., On normal distributions in l_p spaces (Russian). *Teor. Verojatnost. i Primenen.* **9** (1964), 737–738.
123. Vahanija, N. N., Sur une propiété des répartitions normales de probabilités dans les espaces l_p $(1 \leqslant p < \infty)$ et *H. C. R. Acad. Sci. Paris* **260** (1965), 1334–1336.
124. Vahanija, N. N., "Sur les répartition de probabilités dans les espaces de suites numériques. *C. R. Acad. Sci. Paris* **260** (1965), 1560–1562.

125. Vahanija, N. N., On non-degenerate probability distributions in the space l_p $(1 \leqslant p < \infty)$ (Russian). *Teor. Verojatnost. i Primenen.* **11** (1966), 524–528.
126. Varadhan, S. R. S., Limit theorems for sums of independent random variables with values in a Hilbert space. *Sankhyā* **24** (1962), 213–238.
127. Vo-Khac, K., Fonctions et distributions vectorielles aléatoires d'ordre *p*. *C. R. Acad. Sci. Paris Ser. A* **264** (1967), 1012–1015.
128. Vo-Khac, K., Processus stochastiques stationnaires du second ordre à valeurs vectorielles. *C. R. Acad. Sci. Paris Ser. A* **264** (1967), 1069–1072.
129. Yaglom, A. M., "An Introduction to the Theory of Stationary Random Functions," translated from the Russian. Prentice-Hall, Englewood Cliffs, New Jersey, 1962.
130. Zaanen, A. C., "Linear Analysis." North-Holland Publ., Amsterdam, 1953.
131. Zaanen, A. C., "Integration." North-Holland Publ., Amsterdam, 1967.

CHAPTER 2

Operator-Valued Random Variables

2.1 Introduction

In Chap. 1 we introduced the notion of a Banach space-valued random variable and presented an introductory survey of probability theory in Banach spaces. In this chapter we introduce the notion of a random operator and study the properties of such operators. As we will see, a random operator is a measurable mapping defined on a probability measure space with values in some collection of operators; that is, a random operator is an operator-valued random variable. If the collection of operators is the Banach algebra $\mathfrak{L}(\mathfrak{X})$ of bounded linear operators on a Banach space $\mathfrak{X}$, then the results outlined in Chap. 1 are applicable. However, applications require the study of random variables with values in other collections of operators, for example the class of closed operators; hence the study of random operators does not in general reduce to the study of Banach space-valued random variables.

The development of a theory of random operators is of importance in its own right as a generalization of deterministic operator theory; and just as operator theory is of fundamental importance in the study of deterministic operator equations based on the methods of functional analysis, a theory of random operators is required for the study of random operator equations.

In Sect. 2.2 we consider some basic definitions, concepts and results from the theory of operators on Banach spaces. In Sect. 2.3 the notion of a random operator on a Banach space is introduced, and the basic properties of random operators are considered. Section 2.4 is devoted to some results from the developing spectral theory of random operators. In Sect. 2.5 we consider some classes of operator-valued random functions; and in Sect. 2.6 we study some limit theorems for random operators.

2.2 Operators on Banach Spaces

A. Introduction

Let $\mathfrak{X}$ and $\mathfrak{Y}$ be two Banach spaces of the same scalar type, that is, the scalar fields over which $\mathfrak{X}$ and $\mathfrak{Y}$ are defined are either both real or both complex. Let $\mathfrak{D}$ be a subset of $\mathfrak{X}$. If to every element $x \in \mathfrak{D}$ there corresponds an element $y \in \mathfrak{Y}$ according to some rule $y = Tx$, then T is called an *operator*, *transformation*, or *mapping* on $\mathfrak{D}$ with values in the *range space* $\mathfrak{Y}$. $\mathfrak{D}$ is called the *domain* of T, and we write $\mathfrak{D} = \mathfrak{D}(T)$. The range of T is denoted by $\mathfrak{R}(T)$. We will assume that $\mathfrak{D}(T)$ is a linear subspace of $\mathfrak{X}$. In addition to the correspondence $Tx = y$, an operator is often denoted in several other ways, for example (i) $T\colon \mathfrak{X} \to \mathfrak{Y}$, (ii) $\mathfrak{X} \xrightarrow{T} \mathfrak{Y}$, (iii) $T\colon x \to y$, $x \in \mathfrak{X}$, $y \in \mathfrak{Y}$.

In this section we present a survey of some basic definitions and theorems from the theory of operators on Banach spaces which will be utilized in the study of random operators in this chapter and the study of random integral equations in subsequent chapters. The reader is referred to the following books for detailed expositions of operator theory: Dunford and Schwartz [17, 18], Hille and Phillips [28], Kato [30], Taylor [49], and Zaanen [55].

B. Bounded linear operators: Some general properties

***Definition* 2.1.** An operator T on a Banach space $\mathfrak{X}$ is said to be *linear* if

$$T[\alpha_1 x_1 + \alpha_2 x_2] = \alpha_1 Tx_1 + \alpha_2 Tx_2 \qquad \text{for all} \quad x_1, x_2 \in \mathfrak{X}, \quad \alpha_1, \alpha_2 \text{ scalars.}$$

***Definition* 2.2.** An operator T on one Banach space to another is said to be *bounded* if it takes bounded sets into bounded sets. If T is a bounded linear operator, then

$$\|T\| = \sup_{\|x\| \leqslant 1} \|Tx\|, \qquad x \in \mathfrak{D}(T),$$

is called the *bound* or *norm* of T.

***Definition* 2.3.** An operator T from $\mathfrak{X}$ into $\mathfrak{Y}$ is *continuous at* x_0 if

$$\lim_{n\to\infty} \|x_n - x_0\|_{\mathfrak{X}} = 0 \qquad \text{implies that} \qquad \lim_{n\to\infty} \|Tx_n - Tx_0\|_{\mathfrak{Y}} = 0.$$

The relationship between continuity and boundedness of linear operators is given by the following result.

THEOREM 2.1. *A continuous linear operator T on $\mathfrak{X}$ into $\mathfrak{Y}$ is bounded on $\mathfrak{X}$.*

Let $\mathfrak{L}(\mathfrak{X},\mathfrak{Y})$ denote the collection of all bounded linear operators on $\mathfrak{X}$ into $\mathfrak{Y}$. Let T, T_1 and T_2 be elements of $\mathfrak{L}(\mathfrak{X},\mathfrak{Y})$, and let α be a scalar. Addition and scalar multiplication of elements of $\mathfrak{L}(\mathfrak{X},\mathfrak{Y})$ are defined as follows: (i) $(T_1+T_2)x=T_1x+T_2x$, where $\mathfrak{D}(T_1+T_2)=\mathfrak{D}(T_1)\cap\mathfrak{D}(T_2)$, (ii) $(\alpha T)x=\alpha Tx$. The *null operator* Θ on $\mathfrak{X}$ to $\mathfrak{Y}$ is defined by $\Theta x=\theta$; and the *identity operator* I on $\mathfrak{X}$ to $\mathfrak{Y}$ is such that $Ix=x$. Clearly, Θ and I map $\mathfrak{X}$ into itself.

THEOREM 2.2. *$\mathfrak{L}(\mathfrak{X},\mathfrak{Y})$ is a Banach space, the norm of an element $T\in\mathfrak{L}(\mathfrak{X},\mathfrak{Y})$ being given by*

$$\|T\| = \sup_{\|x\|\leqslant 1} \|Tx\|, \qquad x\in\mathfrak{D}(T).$$

C. Closed linear operators

Bounded linear operators constitute an important class of operators on Banach spaces; but many applications involve linear operators which are unbounded (or discontinuous). We now consider a class of linear operators which are called closed operators. Many linear operators encountered in analysis and applied mathematics are discontinuous (for example, ordinary and partial differential operators), but in concrete applications it is often possible to employ a set-up in which the operators are closed.

***Definition* 2.4.** Let $\mathfrak{X}\times\mathfrak{Y}$ denote the product of the Banach spaces $\mathfrak{X}$ and $\mathfrak{Y}$. The *graph* of a linear operator T on $\mathfrak{X}$ is a set of pairs of elements $\{(x,Tx),\ x\in\mathfrak{D}(T)\}\subset\mathfrak{X}\times\mathfrak{Y}$.

***Definition* 2.5.** A linear operator T on $\mathfrak{X}$ into $\mathfrak{Y}$ is said to be *closed* if its graph is a closed subset of $\mathfrak{X}\times\mathfrak{Y}$.

In other words, T is closed if whenever $x_n\to x$ ($\{x_n\}\subset\mathfrak{D}(T)$) and $\lim_{n\to\infty} y_n=\lim_{n\to\infty} Tx_n=y$, then $x\in\mathfrak{D}(T)$ and $y=Tx$.

We now state a theorem, referred to as the *closed graph theorem*, which has many applications.

THEOREM 2.3. *If T is a linear operator on $\mathfrak{X}$ into $\mathfrak{Y}$, with $\mathfrak{D}(T)=\mathfrak{X}$, then T is continuous if and only if its graph is closed.*

Let $\mathfrak{C}(\mathfrak{X},\mathfrak{Y})$ denote the class of all closed linear operators on $\mathfrak{X}$ into $\mathfrak{Y}$. Since every $T\in\mathfrak{L}(\mathfrak{X},\mathfrak{Y})$ is closed, we have $\mathfrak{L}(\mathfrak{X},\mathfrak{Y})\subset\mathfrak{C}(\mathfrak{X},\mathfrak{Y})$. The closed graph theorem can also be stated as follows: *If $T\in\mathfrak{C}(\mathfrak{X},\mathfrak{Y})$ and $\mathfrak{D}(T)=\mathfrak{X}$, then $T\in\mathfrak{L}(\mathfrak{X},\mathfrak{Y})$.* It is of interest to remark that a metric can be introduced in $\mathfrak{C}(\mathfrak{X},)\mathfrak{Y}$ so as to make it into a metric space (cf. Kato [30, Chap. IV]).

D. Some examples of operators

The literature of analysis and applied mathematics is replete with examples of linear operators. In this subsection we give a few examples of linear operators. The probabilistic analogues of some of these operators will be considered in this and other chapters of this book.

1. Let $\mathfrak{X} = \mathfrak{Y} = R_n$. In this case a linear operator A has a unique representation as an $n \times n$ matrix,

$$A = \begin{pmatrix} a_{11} & a_{12} & \cdots & a_{1n} \\ a_{21} & a_{22} & \cdots & a_{2n} \\ \vdots & \vdots & & \vdots \\ a_{n1} & a_{n2} & \cdots & a_{nn} \end{pmatrix}. \tag{2.1}$$

Here the correspondence $Ax = y$ has the form

$$y_i = \sum_{j=1}^{n} a_{ij} x_i, \qquad i = 1, 2, \ldots, n, \tag{2.2}$$

for $x = (x_1, x_2, \ldots, x_n)$, $y = (y_1, y_2, \ldots, y_n)$. The *matrix operator* A is bounded, and has norm

$$\|A\| = \left(\sum_{i,j=1}^{n} a_{ij}^2 \right)^{1/2}.$$

Finite-dimensional matrix operators also arise as transformations on the l_p^n spaces; and infinite-dimensional matrix operators represent transformations on the l_p spaces.

2. Let $\mathfrak{X} = \mathfrak{Y} = C[0,1]$, and let $C'[0,1]$ denote the subspace of $C[0,1]$ consisting of all functions with continuous first derivatives. Define the *linear differential operator* T: $C'[0,1] \to C[0,1]$ by $Tx(t) = x'(t)$, $t \in [0,1]$. T is a closed operator; however it is unbounded (cf. Taylor [49, p. 175]).

3. Let $\mathfrak{X} = \mathfrak{Y} = l_2$, and let $x = \{x_n\} \in l_2$. A *linear difference operator* T on l_2 to itself can be defined as follows: $T[x_n] = \alpha x_{n+1} + \beta x_{n-1}$, where α, β are constants.

4. Let $\mathfrak{X} = \mathfrak{Y} = C[a,b]$. The the *linear integral operator* T defined

$$T[x(t)] = \int_a^b K(t,s)\, x(s)\, ds \tag{2.3}$$

maps $C[a,b]$ into itself provided the function $K(t,s)$ is continuous on $[a,b] \times [a,b]$. The integral operator T is bounded; and if $|K(t,s)| \leqslant M$ on $[a,b] \times [a,b]$, then $\|T\| \leqslant M(b-a)$.

Since this book is concerned with integral equations, many examples of linear and nonlinear integral operators will be given in subsequent chapters.

E. Inverse and adjoint operators

Let $\mathfrak{X}$ and $\mathfrak{Y}$ be two Banach spaces.

***Definition* 2.6.** An operator $T \in \mathfrak{L}(\mathfrak{X}, \mathfrak{Y})$ is said to have an *inverse*, denoted by T^{-1}, if for every $y \in \mathfrak{Y}$ the "equation" $Tx = y$ has a unique solution.

Some useful results are summarized as

THEOREM 2.4. (*a*) *If* $T \in \mathfrak{L}(\mathfrak{X},\mathfrak{Y})$ *has an inverse* T^{-1}, *then* $T^{-1} \in \mathfrak{L}(\mathfrak{Y}, \mathfrak{X})$. (*b*) *If* $T \in \mathfrak{C}(\mathfrak{X},\mathfrak{Y})$ *has an inverse* T^{-1}, *then* $T^{-1} \in \mathfrak{C}(\mathfrak{Y}, \mathfrak{X})$.

Let T be a linear operator on $\mathfrak{X}$ to $\mathfrak{Y}$ with $\mathfrak{D}(T)$ dense in $\mathfrak{X}$, and let $\mathfrak{X}^*$ and $\mathfrak{Y}^*$ denote the adjoint spaces of $\mathfrak{X}$ and $\mathfrak{Y}$, respectively.

***Definition* 2.7.** The *adjoint operator* T^* of T is defined as follows: $\mathfrak{D}(T^*)$ is the set of all $y^* \in \mathfrak{Y}^*$ for which there exists an $x^* \in \mathfrak{X}^*$ such that $y^*(Tx) = x^*(x)$ for all $x \in \mathfrak{D}(T)$. In this case we define $T^* y^* = x^*$.

THEOREM 2.5. *Let T be a linear operator on $\mathfrak{X}$ to $\mathfrak{Y}$ with $\mathfrak{D}(T)$ dense in $\mathfrak{X}$. Then T^* is a closed linear operator with $\mathfrak{D}(T^*) \subset \mathfrak{Y}^*$ and range $\mathfrak{X}^*$. If, in addition, $T \in \mathfrak{L}(\mathfrak{X},\mathfrak{Y})$, then $T^* \in \mathfrak{L}(\mathfrak{Y}^*, \mathfrak{X}^*)$ and $\|T^*\| = \|T\|$.*

THEOREM 2.6. *Let T be a linear operator with inverse T^{-1} and such that $\overline{\mathfrak{D}(T)} = \mathfrak{X}$ and $\overline{\mathfrak{R}(T)} = \mathfrak{Y}$. Then $(T^*)^{-1} = (T^{-1})^*$; and T^{-1} is bounded if and only if $(T^*)^{-1}$ is bounded on $\mathfrak{X}^*$.*

F. Compact linear operators

We now consider a class of linear operators, called compact (or completely continuous) operators, which are in many respects analogous to operators on finite-dimensional Banach spaces. Compact operators are of particular importance in the theory of integral equations.

***Definition* 2.8.** A linear operator T on $\mathfrak{X}$ into $\mathfrak{Y}$ is said to be *compact* if, for every bounded sequence $\{x_n\}$ in $\mathfrak{X}$, the sequence $\{Tx_n\}$ contains a subsequence converging to some limit in $\mathfrak{Y}$.

Some useful facts about compact operators are presented in the following theorems.

THEOREM 2.7. (i) *Every compact linear operator is bounded.* (ii) *Every linear operator whose domain is a finite-dimensional Banach space is compact.*

THEOREM 2.8. *If T is a compact linear operator, then T^* is also compact.*

THEOREM 2.9. *If T_1 and T_2 are compact linear operators, and α, β are arbitrary complex numbers, then $\alpha T_1 + \beta T_2$ is compact. If T is compact and $S \in \mathfrak{L}(\mathfrak{X}, \mathfrak{Y})$, then ST and TS are compact.*

THEOREM 2.10. *If $\{T_n\}$ is a sequence of compact linear operators and $\lim_{n\to\infty} \|T_n - T\| = 0$, then T is compact.*

G. The operator algebra $\mathfrak{L}(\mathfrak{X})$

In Sect. 2.2B we introduced the Banach space $\mathfrak{L}(\mathfrak{X}, \mathfrak{Y})$ of all bounded linear operators on a Banach space $\mathfrak{X}$ into a Banach space $\mathfrak{Y}$. We now consider the important case when $\mathfrak{X} = \mathfrak{Y}$.

***Definition* 2.9.** A bounded linear operator on a Banach space $\mathfrak{X}$ to itself is called an *endomorphism* of $\mathfrak{X}$.

For a given Banach space $\mathfrak{X}$ we denote the set of all endomorphisms of $\mathfrak{X}$ by $\mathfrak{L}(\mathfrak{X})$, rather than $\mathfrak{L}(\mathfrak{X}, \mathfrak{X})$. For endomorphisms the operation of multiplication is defined; that is, if $T_1, T_2 \in \mathfrak{L}(\mathfrak{X})$, then $(T_1 T_2)x = T_1 T_2 x$. Hence multiplication is defined by composition. We also have $\|T_1 T_2\| \leqslant \|T_1\| \cdot \|T_2\|$; and $\|I\| = 1$, where I is the *identity operator*.

THEOREM 2.11. *$\mathfrak{L}(\mathfrak{X})$ is a Banach algebra with unit element the identity operator. $\mathfrak{L}(\mathfrak{X})$ is noncommutative if the dimension of $\mathfrak{X}$ is greater than one.*

***Definition* 2.10.** An operator $T \in \mathfrak{L}(\mathfrak{X})$ with domain $\mathfrak{D}(T)$ is said to satisfy a *Lipschitz condition* if there is a constant $k > 0$ such that

$$\|Tx_1 - Tx_2\| \leqslant k\|x_1 - x_2\| \qquad \text{for all} \quad x_1, x_2 \in \mathfrak{D}(T)$$

(or, equivalently, $d(Tx_1, Tx_2) \leqslant kd(x_1, x_2)$). T is said to be a *contraction operator if $k < 1$.*

In Sect. 1.2D we considered the weak and strong topologies in Banach spaces and the associated concepts of convergence. We now consider some topologies in $\mathfrak{L}(\mathfrak{X})$ and convergence in these topologies.

***Definition* 2.11.** The *uniform operator topology* in $\mathfrak{L}(\mathfrak{X})$ is the metric topology in $\mathfrak{L}(\mathfrak{X})$ induced by its norm $\|T\| = \sup_{\|x\| \leqslant 1} \|Tx\|$, $T \in \mathfrak{L}(\mathfrak{X})$, $x \in \mathfrak{X}$.

Convergence of a sequence $\{T_n\}$ to T in $\mathfrak{L}(\mathfrak{X})$ in the uniform operator topology is called *uniform convergence*, and is denoted by $\lim_{n\to\infty} \|T_n - T\| = 0$.

***Definition* 2.12.** The *strong operator topology* in $\mathfrak{L}(\mathfrak{X})$ is the topology defined by the set of neighborhoods

$$\begin{aligned} N(T) &= N(T, \mathfrak{X}_0, \epsilon) \\ &= \{S: S \in \mathfrak{L}(\mathfrak{X}), \quad \|Tx - Sx\| < \epsilon, \quad x \in \mathfrak{X}_0\}, \end{aligned}$$

where $\mathfrak{X}_0$ is an arbitrary finite subset of $\mathfrak{X}$, and ϵ is an arbitrary positive number.

***Definition* 2.13.** The *weak operator topology* in $\mathfrak{L}(\mathfrak{X})$ is the topology defined by the set of neighborhoods

$$\begin{aligned} N(T) &= N(T, \mathfrak{X}_0, \mathfrak{X}_0^*, \epsilon) \\ &= \{S: S \in \mathfrak{L}(\mathfrak{X}), \quad |x^*(Tx) - x^*(Sx)| < \epsilon, \quad x \in \mathfrak{X}_0, \quad x^* \in \mathfrak{X}^*\}, \end{aligned}$$

where $\mathfrak{X}_0$ and $\mathfrak{X}_0^*$ are arbitrary finite subsets of $\mathfrak{X}$ and $\mathfrak{X}^*$ respectively, and ϵ is an arbitrary positive number.

The notions of weak and strong convergence in $\mathfrak{L}(\mathfrak{X})$ are given by the following definitions.

***Definition* 2.14.** Let $\{T_n\}$ be a sequence of operators in $\mathfrak{L}(\mathfrak{X})$, and let $T \in \mathfrak{L}(\mathfrak{X})$. If $\lim_{n\to\infty} \|T_n x - Tx\| = 0$ for every $x \in \mathfrak{X}$, then T_n is said to *converge strongly* to T, or *converge in the strong operator topology* of $\mathfrak{L}(\mathfrak{X})$.

***Definition* 2.15.** Let $\{T_n\}$ be a sequence of operators in $\mathfrak{L}(\mathfrak{X})$, and let $T \in \mathfrak{L}(\mathfrak{X})$. If $\lim_{n\to\infty} |x^*(T_n x) - x^*(Tx)| = 0$ for every $x \in \mathfrak{X}$, $x^* \in \mathfrak{X}^*$, then T_n is said to *converge weakly* to T, or *converge in the weak operator topology* of $\mathfrak{L}(\mathfrak{X})$.

The following relations obtain between the modes of convergence in $\mathfrak{L}(\mathfrak{X})$: *uniform convergence implies strong convergence*; *and strong convergence implies weak convergence.*

H. Operators on Hilbert spaces

In this subsection we consider various types of operators on Hilbert spaces. We first consider the matrix representation of an operator $T \in \mathfrak{L}(H)$, where $\mathfrak{L}(H)$ is the algebra of bounded linear operators on a Hilbert space H.

Let H be a separable Hilbert space, and let $\{\varphi_n\}$ be a complete orthonormal sequence in H. If $T \in \mathfrak{L}(H)$, we define a finite or infinite matrix $U = (u_{ij})$ by

$$u_{ij} = (T\varphi_j, \varphi_i). \tag{2.4}$$

In (2.4), $(\cdot,\cdot)$ denotes the inner product in H. The elements u_{ij} are scalars; and U is finite if and only if H is finite-dimensional. U is the *matrix representation* of $T \in \mathfrak{L}(H)$.

THEOREM 2.12. $u_{ij} = (T\varphi_j, \varphi_i)$ *defines a one-to-one correspondence between* $\mathfrak{L}(H)$ *and the collection of all matrices* U *whose elements satisfy the condition*

$$\sum_j \left| \sum_i u_{ji} \alpha_i \right|^2 \leqslant M \sum_i |\alpha_i|^2$$

for M *a positive constant (depending on* U*) and all sequences* $\{\alpha_i\} \in l_2$.

Let T be a linear operator on H with $\mathfrak{D}(T)$ dense in H. Consider the pairs $\{y,z\}$ of elements of H such that $(Tx,y) = (x,z)$ for all $x \in \mathfrak{D}(T)$. Let Y denote the set of all first elements y in the pair $\{y,z\}$ satisfying the above relation; and let T^* be the operator defined on $\mathfrak{D}(T^*)$ by $T^*y = z$.

***Definition* 2.16.** The operator T^* defined by $(Tx,y) = (x,T^*y)$, for $x \in \mathfrak{D}(T)$, $y \in \mathfrak{D}(T^*)$ is called the *Hilbert space adjoint* of T.

***Definition* 2.17.** Let T be a linear operator with $\mathfrak{D}(T)$ dense in H. T is said to be *self-adjoint* if (i) $\mathfrak{D}(T) = \mathfrak{D}(T^*)$, (ii) $Tx = T^*x$ for all $x \in \mathfrak{D}(T)$.

***Definition* 2.18.** An operator T on H is said to be *symmetric* if $(Tx,y) = (x,Ty)$ for all $x,y \in \mathfrak{D}(T)$.

***Definition* 2.19.** If a bounded self-adjoint operator T on H is such that $(Tx,x) \geqslant 0$ for all $x \in H$, then T is called a *positive* operator.

Let $\{\varphi_n\}$ be a complete orthonormal sequence in H.

***Definition* 2.20.** A bounded linear operator T on H is said to be a *Hilbert–Schmidt operator* if the quantity $\|T\|$ defined by

$$\|T\| = \left(\sum_{n=1}^{\infty} \|T\varphi_n\|^2 \right)^{1/2} < \infty.$$

The class of Hilbert–Schmidt operators will be denoted by $\mathfrak{S}_2(H)$. Every Hilbert–Schmidt operator is compact.

2.3 Random Operators

A. Introduction

In this section we introduce various definitions of a random operator and consider some of the basic properties of random operators. Throughout this section $(\Omega, \mathfrak{A}, \mu)$ will denote a complete probability measure space, $\mathfrak{X}$ and $\mathfrak{Y}$ will be Banach spaces (not necessarily separable), and $\mathfrak{L}(\mathfrak{X},\mathfrak{Y})$ and $\mathfrak{L}(\mathfrak{X})$ will denote the collections of bounded linear operators defined in Sect. 2.2.

B. Definitions of a random operator

As in the case of Banach space-valued random variables, there are a number of ways of defining a random operator; and all of these utilize, in one way or another, the notion of a Banach space-valued random variable.

***Definition* 2.21.** A mapping $T(\omega)$: $\Omega \times \mathfrak{X} \to \mathfrak{Y}$ is said to be a *random operator* if $\{\omega : T(\omega)x \in B\} \in \mathfrak{A}$ for all $x \in \mathfrak{X}$, $B \in \mathfrak{B}_{\mathfrak{Y}}$.

The above definition of a random operator is based on Def. 1.10, and simply says that $T(\omega)$ is a random operator if $T(\omega)x = y(\omega)$, say, is a $\mathfrak{Y}$-valued random variable for every $x \in \mathfrak{X}$.

***Definition* 2.22.** A mapping $T(\omega)$: $\Omega \times \mathfrak{X} \to \mathfrak{Y}$ is said to be a *weak random operator* if $y^*(T(\omega)x)$ is a real-valued random variable for all $x \in \mathfrak{X}$, $y^* \in \mathfrak{Y}^*$; that is, $T(\omega)x = y(\omega)$ is a weak $\mathfrak{Y}$-valued random variable for every $x \in \mathfrak{X}$.

***Definition* 2.23.** A mapping $T(\omega)$: $\Omega \times \mathfrak{X} \to \mathfrak{Y}$ is said to be a *strong random operator* if $T(\omega)x$ is a strong $\mathfrak{Y}$-valued random variable for every $x \in \mathfrak{X}$.

The above definitions are based on the definitions of weak and strong random variables respectively (cf. Defs. 1.13 and 1.14); and, as in Sect. 1.3B, we can define a random operator utilizing the notion of Price measurability (Def. 1.15). If $\mathfrak{Y}$ is separable, then all of the notions of a random operator introduced above are equivalent.

In many applications it is of interest to consider bounded linear random operators. In this connection we have

***Definition* 2.24.** A random operator $T(\omega)$ on $\mathfrak{X}$ is said to be (a) *linear* if $T(\omega)[\alpha x_1 + \beta x_2] = \alpha T(\omega)x_1 + \beta T(\omega)x_2$ almost surely for all $x_1, x_2 \in \mathfrak{X}$, α, β scalars, and (b) *bounded* if there exists a nonnegative real-valued random variable $M(\omega)$ such that for all $x_1, x_2 \in \mathfrak{X}$, $\|T(\omega)x_1 - T(\omega)x_2\| \leqslant M(\omega)\|x_1 - x_2\|$ almost surely.

We remark that if $T(\omega)$ is linear, then (b) can be replaced by (b′) $\|T(\omega)x\| \leqslant M(\omega)\|x\|$ almost surely.

We now introduce some definitions of bounded linear random operators utilizing the notion of an operator-valued function defined on a probability measure space. We refer to Dinculeanu [14, pp. 101–106] and Hille and Phillips [28, pp. 74–75] for discussions of operator-valued measurable functions.

***Definition* 2.25.** A mapping $T(\omega)\colon \Omega \to \mathfrak{L}(\mathfrak{X}, \mathfrak{Y})$ is said to be a *uniform random operator* if there exists a sequence of countably-valued operator-valued random variables† $\{T_n(\omega)\}$ in $\mathfrak{L}(\mathfrak{X}, \mathfrak{Y})$ converging almost surely to $T(\omega)$ in the uniform operator topology.

We remark that the notion of a uniform random operator is equivalent to that of a strong random operator if we consider $T(\omega)$ as a *strong random variable* with values in the Banach space $\mathfrak{L}(\mathfrak{X}, \mathfrak{Y})$. A *weak random operator* can also be defined as a weak $\mathfrak{L}(\mathfrak{X}, \mathfrak{Y})$-valued random variable.

We now state a theorem due to Dunford [16] (cf. also Hille and Phillips [28, pp. 74–75]) which establishes the connection between the different notions of operator-valued functions.

THEOREM 2.13. *A necessary and sufficient condition that* $T(\omega)$ *be* (1) *a strong random operator is that* $T(\omega)$ *be a weak random operator and that* $T(\omega)x$ *be almost separably-valued‡ in* $\mathfrak{Y}$ *for every* $x \in \mathfrak{X}$; (2) *a uniform random operator is that* $T(\omega)$ *be a weak random operator and almost separably-valued in* $\mathfrak{L}(\mathfrak{X}, \mathfrak{Y})$.

Of special interest in applications is the case when $\mathfrak{X} = \mathfrak{Y}$, and $\mathfrak{X}$ is a separable Banach space.

***Definition* 2.26.** A mapping $T(\omega)\colon \Omega \to \mathfrak{L}(\mathfrak{X})$ is said to be a *random endomorphism* of $\mathfrak{X}$ if $T(\omega)$ is an $\mathfrak{L}(\mathfrak{X})$-valued random variable.

Finally, we introduce the notion of a random contraction operator which is of fundamental importance in the study of random equations.

***Definition* 2.27.** A random endomorphism $T(\omega)$ on $\mathfrak{X}$ is said to be a *random contraction operator* if there exists a mapping $k(\omega)\colon \Omega \to R$ such that $k(\omega) < 1$ almost surely and such that $\|T(\omega)x_1 - T(\omega)x_2\| \leqslant k(\omega)\|x_1 - x_2\|$ almost surely (equivalently, $d(Tx_1, Tx_2) \leqslant k(\omega)d(x_1, x_2)$) for all $x_1, x_2 \in \mathfrak{X}$. If $k(\omega) = k$ (a constant) for all $\omega \in \Omega$, $T(\omega)$ is called a *uniform random contraction operator*.

† Compare Def. 1.11.

‡ Compare Def. 1.12.

We close this subsection by remarking that all of the definitions of random operators introduced have been based, in one way or another, on the notion of a Banach space-valued random variable; and in the important case of bounded linear random operators the definitions were those of Banach space-valued random variables. In view of the above, the results of Chap. 1 can be applied to most of the random operators which we will encounter. In particular, we can study probability measures on the Banach space $\mathfrak{L}(\mathfrak{X},\mathfrak{Y})$ and the Banach algebra $\mathfrak{L}(\mathfrak{X})$.

C. Examples of random operators

In this subsection we list a few examples of concrete random operators, some of which will be encountered in subsequent chapters of this book.

1. *Random matrices.* Let $\mathfrak{X} = R_n$. An $n \times n$ *random matrix* $M(\omega)$ is a matrix whose elements m_{ij}, $i,j = 1,2,\ldots,n$, are random variables. That is $M(\omega) = (m_{ij}(\omega))$, $\omega \in \Omega$. A random matrix can also be defined as a mapping $M: \Omega \to \mathfrak{L}(R_n)$, where $\mathfrak{L}(R_n)$ is the Banach algebra of $n \times n$ matrices.

Random matrices constitute a very important class of random operators which are encountered in mathematical statistics (cf. Anderson [1]) and physics (cf. Mehta [36]). Random matrices also arise in the study of systems of algebraic, difference, and differential equations with random coefficients.

2. *Random difference operators.* Let $\mathfrak{X} = l_2$. A *random difference operator* $T(\omega)$ on l_2 to itself can be defined as follows: $T(\omega)[x_k] = \sum_{i=1}^{n} a_i(\omega)\tau^i[x_k]$, where τ^i denotes the translation operator $\tau^i[x_k] = x_{k+i}$, $i = 0,1,\ldots,n$. The coefficients $a_i(\omega)$ are assumed to be real-valued random variables. Random difference operators arise in the study of many discrete parameter stochastic processes which arise in engineering and time series analysis. They also occur in the study of approximate solutions of random differential equations.

3. *Random ordinary differential operators.* Let $\mathfrak{X} = C[a,b]$, and let $C^{(n)}[a,b]$ denote the subspace of $C[a,b]$ consisting of all functions whose first n derivatives are continuous. We can define a *random differential operator* $T(\omega): \Omega \times C^{(n)}.[a,b] \to C[a,b]$ as follows:

$$T(\omega)[x(t)] = \sum_{k=0}^{n} a_k(\omega)\, d^k x/dt^k, \qquad t \in [a,b],$$

where the coefficients $a_k(\omega)$ are real-valued random variables. In some applications the coefficients are real-valued random functions $a_k(t,\omega)$, $t \in [a,b]$.

4. *Random partial differential operators.* An example of a random partial differential operator is the *random Helmholtz operator* which occurs in the study of wave propagation in random media:

$$(\Delta + k_0^2 n^2(r,\omega))\,[\psi(r)]$$

(cf. Frisch [20]). In the above $n^2(r,\omega)$ is the index of refraction, and is often assumed to be a real homogeneous and isotropic random function. Random partial differential operators also arise in the study of quantum-mechanical systems with random Hamiltonians, and diffusion or heat conduction problems with random parameters.

5. *Random integral operators.* Since this book is devoted to random integral equations, we refer the reader to Chaps. 4–7 where many examples of random linear and nonlinear integral operators which are probabilistic analogues of some of the classical integral operators are considered.

It is of interest to give an example of a random integral operator studied by Zadeh [56]. Consider the random linear integral operator

$$L(\omega)\,[f] = \int_{-\infty}^{\infty} K(t,\lambda,\omega)\, e^{it\lambda}\, dF(\lambda),$$

where $K(t,\lambda,\omega)$, $t \in (-\infty,\infty)$, $\lambda \in (-\infty,\infty)$, is a complex-valued weakly stationary random function, and $F(\lambda)$ is the Fourier–Stieltjes transform of $f = f(t)$. We remark that when $K(t,\lambda,\omega)$ is a polynomial in λ, the operator $L(\omega)$ reduces to a random differential operator. Also, the random function $K(t,\lambda,\omega)$ can be expressed in terms of $L(\omega)$ by the relation

$$K(t,\lambda,\omega) = e^{-it\lambda} L(\omega)\,[e^{it\lambda}].$$

D. A composition theorem

We now state and prove a theorem due to Hanš [25] which will be utilized in the study of solutions of random equations.

THEOREM 2.14. *Let $\mathfrak{X}$ be a separable Banach space and let $\mathfrak{B}$ denote the σ-algebra of Borel sets of $\mathfrak{X}$. Let $T(\omega)$ be a random endomorphism of $\mathfrak{X}$; and let $x(\omega)$ be an $\mathfrak{X}$-valued random variable. Put $y(\omega) = T(\omega)x(\omega)$. Then $y(\omega)$ is an $\mathfrak{X}$-valued random variable.*

Proof. Since $x(\omega)$ is an $\mathfrak{X}$-valued random variable, it can be approximated by a sequence of countably-valued $\mathfrak{X}$-valued random variables $\{x_n(\omega)\}$. For every $\omega \in \Omega$, put $y_n(\omega) = T(\omega)x_n(\omega)$. Then, for every $B \in \mathfrak{B}$

$$\{\omega : y_n(\omega) \in B\} = \bigcup_{i=1}^{\infty} \{\omega : T(\omega)\,\xi_i \in B\} \cap \{\omega : x_n(\omega) = \xi_i\}.$$

Therefore, $\{y_n(\omega)\}$ is a sequence of $\mathfrak{X}$-valued random variables. Since $T(\omega)$ is continuous, $\lim_{n\to\infty} y_n(\omega) = y(\omega)$ almost surely; and $y(\omega)$ is an $\mathfrak{X}$-valued random variable. The above theorem states that the class $\mathscr{V}(\Omega, \mathfrak{X})$ of random variables with values in a separable Banach space $\mathfrak{X}$ (cf. Sect. 1.3D) is closed under bounded linear random transformations.

E. Inverse and adjoint random operators

Let $T(\omega)$ be a random operator with values in $\mathfrak{L}(\mathfrak{X}, \mathfrak{Y})$, where $\mathfrak{X}$ and $\mathfrak{Y}$ are separable. The *inverse* $T^{-1}(\omega)$ of $T(\omega)$ from $\Omega \times \mathfrak{X} \to \mathfrak{Y}$ is defined if and only if $T(\omega)$ is one to one almost surely, which is the case if and only if $T(\omega)x = \theta$ almost surely implies $x = \theta$ almost surely.

***Definition* 2.28.** If $T(\omega)$ is a random operator with values in $\mathfrak{L}(\mathfrak{X}, \mathfrak{Y})$, then $T^{-1}(\omega)$ is the random operator with values in $\mathfrak{L}(\mathfrak{Y}, \mathfrak{X})$ which maps $T(\omega)x$ into x almost surely. Hence $T^{-1}(\omega)T(\omega)x = x$ almost surely, $x \in \mathfrak{D}(T(\omega))$, and $T(\omega)T^{-1}(\omega)y = y$ almost surely, $y \in \mathfrak{R}(T(\omega))$. $T(\omega)$ is said to be *invertible* if $T^{-1}(\omega)$ exists.

We now state a theorem due to Hanš [26] on the inverse of a bounded linear random operator which is the probabilistic analogue of Theorem 2.4(a). We omit the proof since it is only concerned with establishing the measurability of the inverse operator.

THEOREM 2.15. *Let $T(\omega)$ be an invertible random operator with values in $\mathfrak{L}(\mathfrak{X}, \mathfrak{Y})$, where $\mathfrak{X}$ and $\mathfrak{Y}$ are separable. Then $T^{-1}(\omega)$ is a random operator with values in $\mathfrak{L}(\mathfrak{Y}, \mathfrak{X})$.*

For other results on the inverse of a random operator we refer to Hanš [26] and Špaček [48]. Nashed and Salehi [41a] have recently obtained results on the inverse of a separable random operator (cf. Sect. 2.3H) and on the measurability of the generalized inverse of a random bounded linear operator between two Hilbert spaces. In Sects. 2.4B and 3.3C we consider the inverse of operators of the form $T(\omega) - \lambda I$.

Let $\mathfrak{X}$ and $\mathfrak{Y}$ be separable Banach spaces, and let $\mathfrak{X}^*$ and $\mathfrak{Y}^*$ denote the adjoint spaces of $\mathfrak{X}$ and $\mathfrak{Y}$, respectively.

***Definition* 2.29.** Let $T(\omega) \in \mathfrak{L}(\mathfrak{X}, \mathfrak{Y})$. The *adjoint operator* $T^*(\omega)$ of $T(\omega)$ is defined as follows: $T^*(\omega) \in \mathfrak{L}(\mathfrak{Y}^*, \mathfrak{X}^*)$ is the adjoint of $T(\omega)$ if, for almost every $\omega \in \Omega$, the equality $x^* = T^*(\omega)y^*$ is equivalent to the equality $y^*(T(\omega)x) = x^*(x)$ for all $x \in \mathfrak{X}$.

The fact that $T^*(\omega)$ is a random operator (that is, it is measurable) follows from the following result due to Hanš [26].

THEOREM 2.16. *Let $T(\omega) \in \mathfrak{L}(\mathfrak{X},\mathfrak{Y})$. The following conditions are equivalent:* (i) *the mapping $T(\omega)$ of $\mathfrak{X}$ onto $\mathfrak{Y}$ is invertible for almost all $\omega \in \Omega$;* (ii) $\mathfrak{R}(T^*(\omega)) = \mathfrak{X}^*$. *Furthermore, if the above conditions are satisfied, then $T^*(\omega)$ is invertible, and $(T^*(\omega))^{-1} = (T^{-1}(\omega))^*$. Moreover, if any of the operators $T(\omega)$, $T^{-1}(\omega)$, $T^*(\omega)$, $(T^{-1}(\omega))^*$ is measurable, then all four operators are measurable.*

F. Random operators on Hilbert spaces

Let H be a separable Hilbert space with inner product $(\cdot,\cdot)$, and let $\mathfrak{L}(H)$ denote the algebra of endomorphisms of H. Random operators on H to H can, of course, be defined using the notions introduced in Sect. 2.3B; however, a definition can also be given in terms of the inner product.

***Definition* 2.30.** A mapping $T(\omega)$: $\Omega \times H \to H$ is said to be a *random operator* on H if the function $(T(\omega)x, y)$ is a scalar-valued random variable for every $x, y \in H$. Obviously $T(\omega)$ is a random operator if and only if $T(\omega)x$ is an H-valued random variable for every $x \in H$.

We now prove the following results (cf. de Araya [13]).

THEOREM 2.17. *If $T(\omega)$ is a random endomorphism of H, then $\|T(\omega)\|$ is a nonnegative real-valued random variable.*

Proof. Let $x(\omega)$ be an H-valued random variable, and let $\{\varphi_1, \varphi_2, \ldots\}$ denote an orthonormal basis for H. Then $(x(\omega), \varphi_n)$ is a scalar-valued random variable for each n; and $\|x(\omega)\|^2 = \sum_{n-1}^{\infty} |(x(\omega), \varphi_n)|^2$ is a scalar-valued random variable. Now let S be a countable dense subset of the unit sphere in H. Then $\|T(\omega)\| = \sup_{x \in S} \|T(\omega)x\|$ is a scalar-valued random variable.

***Definition* 2.31.** The *Hilbert space adjoint* $T^*(\omega)$ of a random endomorphism $T(\omega)$ is defined, for almost all $\omega \in \Omega$, by the relation

$$(T(\omega)x, y) = (x, T^*(\omega)y).$$

Since $T(\omega)$ is a $\mathfrak{L}(H)$-valued random variable, it follows that $T^*(\omega)$ is also an $\mathfrak{L}(H)$-valued random variable.

The class of Hilbert–Schmidt operators (Def. 2.20) forms a Banach algebra (without identity) under the Hilbert–Schmidt norm. The involution $T \to T^*$ satisfies the identity $(TS, U) = (S, T^*U)$. Hence the class of Hilbert–Schmidt operators $\mathfrak{S}_2(H)$ is an H^*-algebra; and the results of Sect. 1.3F can be used to study random Hilbert–Schmidt operators as H^*-algebra-valued random variables. We refer to Kannan and Bharucha-Reid [29] for the definition of a stochastic integral which is a random variable with values in the class of Hilbert–Schmidt operators.

Also, in view of Theorem 2.12, random endomorphisms of H admit a random matrix representation; and these random matrix operators (especially their spectral properties) are of interest in many applied fields.

For other results on random operators on Hilbert spaces we refer to the paper of Haïnis [24].

G. Expectation and conditional expectation of random operators

Let $T(\omega)$ be an $\mathfrak{L}(\mathfrak{X}, \mathfrak{Y})$-valued random variable. Since $T(\omega)$ is a Banach space-valued random variable the results of Sect. 1.3E are applicable, and we can consider the expectation and conditional expectation of random operators.

In defining the expectation of a strong $\mathfrak{L}(\mathfrak{X},\mathfrak{Y})$-valued random variable, it is necessary to distinguish between the uniform and strong Bochner integrals used to define the expectation. If $T(\omega)$ is a uniform random operator and if $\mathscr{E}\{\|T(\omega)\|\} < \infty$, then $T(\omega) \in B_1(\Omega, \mathfrak{L}(\mathfrak{X},\mathfrak{Y}))$ and the results for the strong expectation given in Sect. 1.3E apply directly. In this case the *expectation* of $T(\omega)$ is given by

$$\mathscr{E}\{T(\omega)\} = U = (\mathrm{B})\int_{\Omega} T(\omega)\, d\mu, \tag{2.5}$$

and $U \in \mathfrak{L}(\mathfrak{X}, \mathfrak{Y})$. The Bochner integral in (2.5) is the limit in the uniform operator topology of the approximating integrals (B) $\int_{\Omega} T_n(\omega)\, d\mu$. If, however, $T(\omega)x \in B_1(\Omega, \mathfrak{Y})$ for each $x \in \mathfrak{X}$, then the earlier results simply assert that $\mathscr{E}\{T(\omega)x\} = Ux \in \mathfrak{Y}$.

We now state two results concerning the strong expectation of random operators. We refer to Hille and Phillips [28, pp. 85–86] for the proofs.

THEOREM 2.18. *If* $T(\omega) \in B_1(\Omega, \mathfrak{L}(\mathfrak{X},\mathfrak{Y}))$, *then* $T^*(\omega) \in B_1(\Omega, \mathfrak{L}(\mathfrak{Y}^*, \mathfrak{X}^*))$ *and* $(\mathscr{E}\{T(\omega)\})^* = \mathscr{E}\{T^*(\omega)\}$.

THEOREM 2.19. *If* $T(\omega)x \in B_1(\Omega, \mathfrak{Y})$ *for each* $x \in \mathfrak{X}$, *then*

$$Ux = \mathscr{E}\{T(\omega)\,x\}$$

defines a bounded linear operator on $\mathfrak{X}$ *to* $\mathfrak{Y}$.

We now prove a few results on the expectation of random endomorphisms on a Hilbert space (cf. Haïnis [24]).

THEOREM 2.20. *Let* $T(\omega)$ *be a random endomorphism of* H, *and let* $T^*(\omega)$ *be its Hilbert space adjoint. Then* $(\mathscr{E}\{T(\omega)\})^* = \mathscr{E}\{T^*(\omega)\}$.

Proof. We have $\mathscr{E}\{(T(\omega)x,y)\} = (\mathscr{E}\{T(\omega)x\},y)$ for all $x,y \in H$. Also

$$((\mathscr{E}\{T(\omega)\})^* x,y) = (x,\mathscr{E}\{T(\omega)\}y) = \mathscr{E}\{(x,T(\omega)y)\}$$
$$= \mathscr{E}\{(T^*(\omega)x,y)\} = (\mathscr{E}\{T^*(\omega)\}x,y).$$

Hence the result.

THEOREM 2.21. *Let $T(\omega)$ be a random endomorphism of H. If $\|T(\omega)\|$ is a scalar-valued random variable, then $\|\mathscr{E}\{T(\omega)\}\| \leqslant \mathscr{E}\{\|T(\omega)\|\}$.*

Proof. From the relation $\|\mathscr{E}\{T(\omega)x\}\| \leqslant \mathscr{E}\{\|T(\omega)x\|\}$, $x \in H$, it follows that

$$\sup_{\|x\|\leqslant 1} \|\mathscr{E}\{T(\omega)x\}\| \leqslant \sup_{\|x\|\leqslant 1} \mathscr{E}\{\|T(\omega)x\|\} \leqslant \mathscr{E}\{\sup_{\|x\|\leqslant 1} \|T(\omega)x\|\};$$

Hence $\|\mathscr{E}\{T(\omega)\}\| \leqslant \mathscr{E}\{\|T(\omega)\|\}$.

It is of interest to remark that it is also possible to define the expectation of a random closed linear operator. We know that a metric can be introduced in $\mathfrak{C}(\mathfrak{X},\mathfrak{Y})$ so as to make it into a metric space (cf. Sect. 2.2C). Also, a metric can be introduced in certain equivalence classes of closed linear operators which are infinitesimal generators of semigroups of operators (cf. Hille and Phillips [28, pp. 410–415]). In these cases the definition of expectation due to Doss [15] can be used to define the expectation of a random closed linear operator. If (M,d) is a metric space, then an M-valued random variable $x(\omega)$ is said to have *Doss expectation* $\xi \in M$ if

$$d(\xi,y) \leqslant \mathscr{E}\{d(x(\omega),y)\} \tag{2.6}$$

for every $y \in M$.

Concerning the conditional expectation of random operators, it is, of course, necessary to distinguish between the conditional expectations based on the uniform and strong Bochner integrals. In the first case, let $T(\omega) \in B_1(\Omega, \mathfrak{L}(\mathfrak{X},\mathfrak{Y}))$. Then, a strong random operator $\mathscr{E}\{T|\mathfrak{A}_0\}(\omega)$ is said to be the *conditional expectation of* $T(\omega) \in B_1(\Omega, \mathfrak{L}(\mathfrak{X},\mathfrak{Y}))$ *relative to* $\mathfrak{A}_0$ if and only if it satisfies the following conditions: (i) $\mathscr{E}\{T|\mathfrak{A}_0\}(\omega)$ is $\mathfrak{A}_0$-measurable and is an element of $B_1(\Omega, \mathfrak{L}(\mathfrak{X},\mathfrak{Y}))$, (ii) (B) $\int_A \mathscr{E}\{T|\mathfrak{A}_0\}(\omega)\,d\mu =$ (B) $\int_A T(\omega)\,d\mu$ for every $A \in \mathfrak{A}_0$.

If $T(\omega)x \in B_1(\Omega,\mathfrak{Y})$ for each $x \in \mathfrak{X}$, then $\mathscr{E}\{Tx|\mathfrak{A}_0\}(\omega)$ is a strong $\mathfrak{Y}$-valued random variable, and Def. 1.27 can be used to define the conditional expectation of $T(\omega)x$. We refer to Rao [44, 46] for a discussion of the conditional expectation of operator-valued random variables and its properties. The existence of the conditional expectation of operator-valued random variables enables us to define operator-valued martingales. These will be considered in Sect. 2.5.

H. Separable random operators

In this section we introduce the notion of a separable random operator, and consider certain properties of this class of random operators (cf. Mukherjea [38, 39], and Mukherjea and Bharucha-Reid [40]). Let $\mathfrak{X}$ be a separable Banach space; and let $\mathfrak{F}$ denote the countable class of all sets of the form $\{x:\|x-x_i\| \leqslant r\}$, $\{x:\|x-x_i\| \geqslant r\}$, and their finite intersections, where $\{x_i\}_{i=1}^{\infty}$ is dense in $\mathfrak{X}$ and r is a rational number.

***Definition* 2.32.** A random operator $T(\omega)$: $\Omega \times \mathfrak{X} \to \mathfrak{X}$ is said to be *separable* if there exists a countable set S in $\mathfrak{X}$ and a negligible set $N \in \mathfrak{A}$ such that

$$\{\omega:T(\omega)x \in K, \quad x \in F\} \triangle \{\omega:T(\omega)x \in K, \quad x \in F \cap S\} \subset N$$

for every compact set K and every $F \in \mathfrak{F}$.

It is clear that the above definition is simply that of a separable Banach space-valued random function (cf. Def. 1.36); that is, with reference to the mapping $T(\omega)x = T_x(\omega)$, x plays the role of a parameter, and the parameter set is a Banach space.

We now state a result which states that under certain conditions every random operator is equivalent to a separable random operator.

THEOREM 2.22. *Let $\mathfrak{X}$ be a separable Banach space, and let E be a compact subset of $\mathfrak{X}$. If $T(\omega)$: $\Omega \times \mathfrak{X} \to E$ is a random operator, then there exists a separable random operator $\tilde{T}(\omega)$: $\Omega \times \mathfrak{X} \to \mathfrak{X}$ such that*

$$\mu(\{\omega:T(\omega)x = \tilde{T}(\omega)x\}) = 1.$$

A sufficient condition for separability is given by the following theorem.

THEOREM 2.23. *Let $\mathfrak{X}$ be a separable Banach space, and let $T(\omega)$: $\Omega \times \mathfrak{X} \to \mathfrak{X}$ be a continuous random operator. Then $T(\omega)$ is separable.*

For a random endomorphism $T(\omega)$ on a separable Hilbert space we showed (Theorem 2.17) that $\|T(\omega)\|$ is a real-valued random variable. The next theorem establishes the connection between separability of a random operator and the measurability of its norm.

THEOREM 2.24. *Let $T(\omega)$: $\Omega \times \mathfrak{X} \to E$ be a separable random operator, where E is a compact subset of a separable Banach space $\mathfrak{X}$. Then $\|T(\omega)\|$ is a non-negative real-valued random variable.*

Proofs of the above theorems, as well as other results on separable random operators, are given by Mukherjea and Bharucha-Reid [40]. For some recent results on separable random operators we refer to Nashed and Salehi [41a].

2.4 Spectral Theory of Random Operators

A. Introduction

Let L be a linear operator on a Banach space $\mathfrak{X}$ to itself. Put $L(\lambda) = L - \lambda I$, where λ is a complex number. Then $L(\lambda)$ is a well-defined linear operator on $\mathfrak{D}(L) \subset \mathfrak{X}$.

***Definition* 2.33.** (i) The values of λ for which $L(\lambda)$ has a bounded inverse $(L - \lambda I)^{-1}$ with domain dense in $\mathfrak{X}$ form the *resolvent set* $\rho(L)$ of L. $R(\lambda; L) = (L - \lambda I)^{-1}$ is called the *resolvent operator* associated with L. (ii) The values of λ for which $L(\lambda)$ has an inverse whose domain is not dense in $\mathfrak{X}$ form the *residual spectrum* $R\sigma(L)$ of L. (iii) The values of λ for which $L(\lambda)$ has an unbounded inverse whose domain is dense in $\mathfrak{X}$ form the *continuous spectrum* $C\sigma(L)$ of L. (iv) The values of λ for which no inverse exists form the *point spectrum* $P\sigma(L)$ of L.

We state the following basic facts (cf. Hille and Phillips [28, Sect. 2.16], Taylor [49, Chap. 5]:

1. The four sets $\rho(L)$, $R\sigma(L)$, $C\sigma(L)$ and $P\sigma(L)$ are mutually exclusive; and $\rho(L) \cup R\sigma(L) \cup C\sigma(L) \cup P\sigma(L) = C$ (the complex plane).

$$R\sigma(L) \cup C\sigma(L) \cup P\sigma(L) = \sigma(L)$$

is the *spectrum* of the operator L.

2. $\rho(L)$ is an open set, and $\sigma(L)$ is a closed set. $\rho(L)$ is not empty when L is a bounded linear operator. If L is closed, then $\lambda \in \rho(L)$ if and only if $L^{-1}(\lambda)$ exists and $\mathfrak{R}(R(\lambda; L)) = \mathfrak{X}$. In this case $R(\lambda; L) \in \mathfrak{L}(\mathfrak{X})$.
3. $\lambda_0 \in P\sigma(L)$ is called an *eigenvalue* of L; and if $Lx_0 = \lambda_0 x_0$, $x_0 \neq \theta$, then x_0 is called an *eigenfunction* of L.
4. If L is an endomorphism of $\mathfrak{X}$, then the following limit exists: $\lim_{n\to\infty} \|L^n\|^{1/n} = r(L)$. $r(L)$ is called the *spectral radius* of L. The following inequality is always valid: $r(L) \leqslant \|L\|$.

The spectral theory of linear operators, which is of great importance in applications of operator theory, is concerned with the study of the relations among the operators L, $(L - \lambda I)^{-1}$, the sets $\rho(L)$ and $\sigma(L)$, and other operators and linear subspaces of $\mathfrak{X}$ which are related to the above. In this section we present some results on the spectral properties of random operators. The spectral theory of random operators is in its early stages of development, hence only a few general results are known. These results will be given in Sect. 2.4B. The spectral properties of random matrices have been studied extensively by statisticians and physicists. A survey of the spectral theory of random matrices is given in Sect. 2.4C. For other results on the eigenvalues of random operators we refer to Chap. 5.

B. Spectral analysis of random linear operators

Let $\mathfrak{X}$ be a separable Banach space, and let $\mathfrak{L}(\mathfrak{X})$ denote the algebra of endomorphisms of $\mathfrak{X}$. Let $T(\omega)$ be a random endomorphism of $\mathfrak{X}$; that is, $T(\omega)$ is a random variable with values in the measurable space $(\mathfrak{L}(\mathfrak{X}), \mathfrak{B})$. If $T(\omega)$ is a random endomorphism with $\mathfrak{D}(T(\omega)) = \mathfrak{X}$, then $T(\omega) - \lambda I$ is also a random endomorphism, where I denotes a random operator which is equivalent to the identity operator.

THEOREM 2.25. *Let $T(\omega)$ be a random endomorphism of $\mathfrak{X}$. Let*

$$\rho(T(\omega)) = \{(\lambda, \omega): (\lambda, \omega) \in R \times \Omega, \quad (T(\omega) - \lambda I)^{-1} \text{ exists and is bounded}\}$$

denote the resolvent set of $T(\omega)$. Then for every $\lambda \in R$, $\{\omega: (\lambda, \omega) \in \rho(T(\omega))\} \in \mathfrak{A}$.

Proof. Let L be an endomorphism of $\mathfrak{X}$. It is known that the set $B_\lambda = \{L: L \in \mathfrak{L}(\mathfrak{X}), \lambda \in \rho(L)\}$ is open for every $\lambda \in R$. Hence $B_\lambda \in \mathfrak{B}$. The assertion of the theorem follows from the equality

$$\{\omega: (\lambda, \omega) \in \rho(L(\omega))\} = \{\omega: T(\omega) \in B_\lambda\},$$

which holds for every $\lambda \in R$.

The above theorem, which is due to Hanš [27], establishes the fact that the λ-section of the resolvent set of a random endomorphism is $\mathfrak{A}$-measurable.

It is well known that if L is an endomorphism of $\mathfrak{X}$, then all λ such that $|\lambda| > \|L\|$ belong to $\rho(L)$, and for these λ the *resolvent operator* $R(\lambda; L) = (L - \lambda I)^{-1}$ admits the series representation

$$R(\lambda; L) = -\sum_{n=1}^{\infty} \lambda^{-n} L^{n-1}, \tag{2.7}$$

the series converging in the uniform topology of $L(\mathfrak{X})$ (cf. Taylor [49, pp. 260–261]). We now prove a result, due to Bharucha-Reid [6] (cf. also [5]) and Hanš [27], which establishes the existence and measurability of the resolvent operator of $T(\omega)$.

THEOREM 2.26. *Let $T(\omega)$ be a random endomorphism of $\mathfrak{X}$, and let $T(\omega) - \lambda I$ be invertible for each ω separately in the set $\Omega_0(\lambda) = \{\omega: |\lambda| > \|T(\omega)\|\}$. Then, the resolvent operator $R(\lambda; T(\omega))$ exists for all $\omega \in \Omega_0(\lambda)$ and is a random operator, measurable with respect to the σ-algebra $\Omega_0(\lambda) \cap \mathfrak{A}$.*

Proof. For an arbitrary, but fixed, ω_0 we can use the classical result which states that the resolvent operator $R(\lambda; T(\omega))$ exists for an arbitrary λ for which $|\lambda| > \|T(\omega_0)\|$, since

$$R(\lambda; T(\omega_0)) = -(1/\lambda)(I - T(\omega_0)\lambda^{-1})^{-1}$$
$$= -(1/\lambda) \sum_{n=0}^{\infty} \lambda^{-n} T^n(\omega_0),$$

this series converging for $\|T(\omega_0)\lambda^{-1}\| < 1$ (cf. Kolmogorov and Fomin [31, p. 236]). Now let $\Omega_0(\lambda) = \{\omega : |\lambda| > \|T(\omega)\|\} \subset \Omega$, where, by hypothesis, $T(\omega) - \lambda I$ is invertible separately for each $\omega \in \Omega_0(\lambda)$. Then $R(\lambda; T(\omega))$ exists for all $\omega \in \Omega_0(\lambda)$, and is measurable with respect to the reduced σ-algebra $(\Omega_0(\lambda) \cap \mathfrak{A})$; hence $R(\lambda; T(\omega))$ is a random endomorphism of $\mathfrak{X}$.

We remark that for all $\omega \in \Omega_0$, $\sigma(T(\omega))$ is contained in a circle of radius $\|T(\omega)\|$ with center at zero. Hence the radius of the circle enclosing the spectrum is a real-valued random variable.

In Sect. 3.3C we give another theorem on the invertibility of $T(\omega) - \lambda$, the proof of which is based on the contraction mapping theorem for random operators.

The next result, due to Bharucha-Reid [6], is a probabilistic analogue of the classical result for the series representation (2.7) of a resolvent operator.

THEOREM 2.27. *Let $T_0(\omega)$ denote the restriction of $T(\omega)$ to the set $\Omega_0(\lambda_0)$, λ_0 fixed. Then, for all λ such that $|\lambda| \geq |\lambda_0|$ belong to $\rho(T_0(\omega))$, and for these λ the resolvent operator admits the series representation*

$$R(\lambda; T_0(\omega)) = -\sum_{n=1}^{\infty} \lambda^{-n} T_0^{n-1}(\omega)$$

Proof. The existence of the resolvent operator for all λ such that $|\lambda| \geq |\lambda_0|$, and for all $\omega \in \Omega_0(\lambda_0)$, follows from Theorem 2.26, since for $|\lambda| \geq |\lambda_0|$ we have $\Omega_0(\lambda_0) \subset \Omega_0(\lambda)$. The other assertion of the theorem follows from the classical result.

Let $(C, \mathfrak{F})$ be a Borel space, where C is the complex plane and $\mathfrak{F}$ is the σ-algebra of all Borel sets of the complex plane. Consider the product measurable space $(C \times \Omega, \mathfrak{F} \times \mathfrak{A})$, where $\mathfrak{F} \times \mathfrak{A}$ is the minimal σ-algebra containing all rectangles $F \times A$, $F \in \mathfrak{F}$, $A \in \mathfrak{A}$. We now state and prove two theorems due to Ryll-Nardzewski [47].

THEOREM 2.28. *Let Z be a measurable subset of $C \times \Omega$, and let Z_ω and Z_λ denote the sets $\{\lambda : (\lambda, \omega) \in Z\}$ and $\{\omega : (\lambda, \omega) \in Z\}$, respectively. If the sets Z_ω are countable for almost all $\omega \in \Omega$, then $\mu(Z_\lambda) = 0$ for $\lambda \in C$ with the exception of a countable number of λ's, that is, the cardinality of the set $\{\lambda : \mu(Z_\lambda) > 0\} \leq \aleph_0$.*

Proof. Suppose the set $Q = \{\lambda : \mu(Z_\lambda) > 0\}$ is uncountable. $\mu(Z_\lambda)$, as a function of λ, is $\mathfrak{F}$-measurable for every rectangle $F \times A$, $F \in \mathfrak{F}$, $A \in \mathfrak{A}$. Consequently, $\mu(Z_\lambda)$ is measurable for every $Z \in \mathfrak{F} \times \mathfrak{A}$. Therefore $Q \in \mathfrak{F}$. According to a theorem of Alekandrov and Hausdorff (cf. Kuratowski [32, p. 355]) the uncountable Borel set Q contains a subset homeomorphic to the Cantor set. For the Cantor set one can construct a σ-finite measure ν vanishing for all one-point subsets and positive for the whole set (cf. Munroe [41, pp. 193–194]). Hence $\nu(Q) > 0$. An application of Fubini's theorem to the product measure $\nu \times \mu$ gives

$$(\nu \times \mu)(Z) = \int \mu(Z_\lambda)\, d\nu = \int_Q \mu(Z_\lambda)\, d\nu > 0.$$

On the other hand, since ν vanishes for all one-point subsets of C, we have

$$(\nu \times \mu)(Z) = \int \nu(Z_\omega)\, d\mu = 0.$$

This contradiction concludes the proof.

The above theorem, which is of independent interest, will now be used to prove a theorem which gives sufficient conditions that the point spectrum of a random endomorphism be countable.

THEOREM 2.29. *If* (i) *$T(\omega)$ is a weak random operator on a separable Banach space, and* (ii) *for almost all $\omega \in \Omega$ the set of λ's for which a bounded inverse of $(T(\omega) - \lambda I)$ does not exist is countable, then the set of λ's for which a bounded inverse of $T(\omega) - \lambda I$ on any Ω-set of positive measure does not exist is also countable.*

Proof. We first remark that the separability of $\mathfrak{X}$ implies that $T(\omega)$ is a strong random operator. Let

$$\begin{aligned} Z &= \{(\lambda, \omega) : (\lambda, \omega) \in C \times \Omega, \text{ a bounded inverse of } (T(\omega) - \lambda I) \text{ does not exist}\} \\ &= Z_1 \cup Z_2, \end{aligned}$$

where

$$Z_1 = \{(\lambda, \omega) : (\lambda, \omega) \in C \times \Omega,\ (T(\omega) - \lambda I)^{-1} \text{ exists but is not bounded}\}$$

$$Z_2 = \{(\lambda, \omega) : (\lambda, \omega) \in C \times \Omega,\ (T(\omega) - \lambda I)^{-1} \text{ does not exist}\}.$$

Let $\{x_n\}$ be a sequence dense in $\mathfrak{X}$. Then, we have

$$Z_1 = \bigcap_{n=1}^{\infty} \bigcap_{p=1}^{\infty} \bigcup_{m=1}^{\infty} \{(\lambda, \omega) : (\lambda, \omega) \in C \times \Omega,\quad \|(T(\omega) - \lambda I)\, x_m - x_n\| < 1/p\}$$

$$Z_2 = \bigcup_{p=1}^{\infty} \bigcap_{n=1}^{\infty} \{(\lambda, \omega) : (\lambda, \omega) \in C \times \Omega, \quad \|(T(\omega) - \lambda I) x_n\| \geqslant (1/p) \| x_n \| \}.$$

Thus, Z_1 and Z_2 are measurable with respect to the product σ-algebra $\mathfrak{F} \times \mathfrak{A}$, which, in turn, implies the measurability of Z. The assertion of the theorem now follows from condition (ii) and Theorem 2.28.

As pointed out earlier, if $T(\omega)$ is an $\mathfrak{L}(\mathfrak{X})$-valued random variable with $\mathfrak{D}(T(\omega)) = \mathfrak{X}$, then $T(\omega) - \lambda I$ is also. Let ν denote the probability measure on $(\mathfrak{L}(\mathfrak{X}), \mathfrak{B})$. Since $\mathfrak{L}(\mathfrak{X})$ is a Banach algebra on which we can define a probability measure, it is possible to utilize certain results from the theory of Banach algebras in order to formulate some measure-theoretic problems associated with resolvent operators of random operators (cf. Grenander [23, pp. 160–161]).

Let $\mathfrak{X}$ be a Banach algebra. An element $x \in \mathfrak{X}$ is said to be *regular* if x^{-1} exists, and is said to be *singular* if x^{-1} does not exist. Let $\mathfrak{R}$ denote the set of regular elements of $\mathfrak{X}$. The regular elements form an open set in $\mathfrak{X}$; and the inverse x^{-1} is a continuous function of x in $\mathfrak{R}$ (cf. Hille and Phillips [28, p. 118]).

Let $T(\omega)$ be an $\mathfrak{L}(\mathfrak{X})$-valued random variable, and consider the random operator $T(\omega) - \lambda I$. Since $\mathfrak{R}$ is an open set, $\mathfrak{R} \in \mathfrak{B}$, hence $\nu(\mathfrak{R})$ is the probability that $T(\omega) - \lambda I$ is regular, that is, invertible. If $\nu(\mathfrak{R}) > 0$, then we introduce the conditional probability of $(T(\omega) - \lambda I)^{-1}$ given that $T(\omega) - \lambda I \in \mathfrak{R}$:

$$\mathscr{P}\{(T(\omega) - \lambda I)^{-1} \in B | T(\omega) - \lambda I \in \mathfrak{R}\}$$

$$= \frac{\mathscr{P}\{(T(\omega) - \lambda I)^{-1} \in B, T(\omega) - \lambda I \in \mathfrak{R}\}}{\mathscr{P}\{T(\omega) - \lambda I \in \mathfrak{R}\}},$$

where $B \in \mathfrak{B}$. Since $(T(\omega) - \lambda I)^{-1}$ is a continuous function of $T(\omega)$, the set $\{(T(\omega) - \lambda I)^{-1} \in B, T(\omega) - \lambda I \in \mathfrak{R}\} \in \mathfrak{B}$. In view of the above, we can introduce the notion of the probability distribution of $(T(\omega) - \lambda I)^{-1}$.

Let $H(\lambda) = \mathscr{P}\{\lambda \in \sigma(T(\omega))\} = \mathscr{P}\{T(\omega) - \lambda I \notin \mathfrak{R}\}$. Consider the set $\{T(\omega) : (T(\omega) - \lambda I)^{-1}$ does not exist$\} \subset \mathfrak{L}(\mathfrak{X})$, and let $I(\lambda; T(\omega))$ denote the indicator or characteristic function of the above set. Then $H(\lambda) = \mathfrak{E}\{I(\lambda; T(\omega))\}$; and for any constant α the set $\{(\lambda, T(\omega)) : I(\lambda; T(\omega)) < \alpha\}$ is an open set in $C \times \mathfrak{L}(\mathfrak{X})$, which implies that $I(\lambda; T(\omega))$ is Borel measurable on $C \times \mathfrak{L}(\mathfrak{X})$.

In closing this subsection, we remark that the spectral radius of a random endomorphism, that is, $r(T(\omega)) = \lim_{n \to \infty} \|T^n(\omega)\|^{1/n}$, is a well-defined real-valued random variable; and the inequality $r(T(\omega)) \leqslant \|T(\omega)\|$ holds almost surely. Also, if $T(\omega)$ is an $\mathfrak{L}(H)$-valued random variable, where H is a separable Hilbert space, and $T(\omega) = T^*(\omega)$ a.s., then $r(T(\omega)) = \|T(\omega)\|$.

C. *Spectral theory of random matrices*

If M is a endomorphism of a finite-dimensional Banach space $\mathfrak{X}$ with $\mathfrak{D}(M)=\mathfrak{X}$, then M can be represented by an $n\times n$ matrix, say (m_{ij}). In this case $M-\lambda I$ is also represented by a matrix, and $\sigma(M)$ is composed of those scalars λ which are roots of the determinented equation

$$|M-\lambda I|=0. \tag{2.8}$$

Equation (2.8) is an algebraic polynomial of degree n in λ. Therefore, if the scalar field is complex, $\sigma(M)$ contains at least one point, and it may contain at most n points. If, however, the scalar field is real, $\sigma(M)$ may be empty.

Let M be an $n\times n$ random matrix; hence $M(\omega)=(m_{ij}(\omega))$ is an $\mathfrak{L}(R_n)$-valued random variable, and $M(\omega)-\lambda I$ is also. The spectrum of $M(\omega)$ is composed of those scalar-valued random variables which are the roots of the random algebraic equation

$$|M(\omega)-\lambda I|=0 \tag{2.9}$$

(cf. Bharucha-Reid [7]); hence $\sigma(M(\omega))$ will be a random subset of the complex plane.

The spectral theory of random matrices has been developed primarily by workers in mathematical statistics and mathematical physics. In multivariate statistical analysis random matrices of estimated variances, covariances and correlation coefficients are studied (cf. Anderson [1]). The matrix elements have a Wishart or normal distribution; and the random eigenvalues $\lambda_1(\omega),\lambda_2(\omega),\ldots,\lambda_n(\omega)$, which are used in the statistical analysis, form an n-dimensional random vector. In some cases the joint probability distribution, or density, can be determined explicitly.

In quantum mechanics the energy levels of a system are supposed to be described by the eigenvalues of a Hamiltonian operator H, which is a Hermitian (that is, linear symmetric) operator defined on an infinite-dimensional Hilbert space $\mathfrak{H}$. Because physicists are primarily interested in the discrete part of the energy level schemes of various quantum mechanical systems, the Hilbert space $\mathfrak{H}$ is approximated by a finite-dimensional Hilbert space $\mathfrak{H}_0$. The selection of a basis in $\mathfrak{H}_0$ permits H to be represented by a finite-dimensional matrix. Therefore, if the eigenvalue equation $(H-\lambda I)\psi=0$ can be solved, then the eigenvalues and eigenfunctions of the system can be determined.

Because of the complex nature of the Hamiltonian, probabilistic hypotheses on H are introduced. Hence H is represented by a random matrix. We refer to Mehta [36] and Porter [43, pp. 2–87] for detailed discussions of the physical aspects of the representation of the Hamiltonian operator by a random matrix.

We first consider the determination of the probability density of the eigenvalues of an $n\times n$ real symmetric random matrix. The method, which is based on Wishart distribution [1, 54] considers an $n \times n$ real symmetric matrix M whose elements are (i) statistically independent except for a symmetry condition so that m_{ij} and m_{rs} are independent if $(i,j) \neq (r,s)$ and $(i,j) \neq (s,r)$, $m_{ii}^2 = m_{ij}^2$, and (ii) the m_{ij} are normally distributed with means zero and variances $\sigma_{ii}^2 = 1$, $\sigma_{ij}^2 = \frac{1}{2}$ for $i \neq j$. In this case the joint probability density of the n eigenvalues is given by

$$f(\lambda_1, \lambda_2, \ldots, \lambda_n) = K \exp\left\{-\tfrac{1}{2} \sum_{i=1}^{n} \lambda_i^2\right\} \prod_{i<j} |\lambda_i - \lambda_j|, \tag{2.10}$$

(K a constant) in the region $\lambda_1 \geqslant \lambda_2 \geqslant \cdots \geqslant \lambda_n$, and 0 otherwise.

The determination of the asymptotic distribution of the eigenvalues of a random matrix is of great theoretical interest and of importance in many applied problems. Let us consider an $n \times n$ matrix $M_n = (m_{ij})_n$ whose elements are real-valued random variables. We assume that $m_{ij}(\omega) = m_{ji}(\omega)$ almost surely for all i, j, but that the m_{ij} are statistically independent for $i \geqslant j$. Also, we assume that the elements $m_{ij}(\omega)$ (for $i > j$) have the same distribution function F, while the diagonal elements m_{ii} have the same distribution function G. Finally, we assume that

$$\mathscr{E}\{m_{ij}^2(\omega)\} = \int x^2 \, dF = \sigma^2 < \infty \qquad (i \neq j).$$

Put

$$Q_n = (2\sigma\sqrt{n})^{-1} \; M_n, \tag{2.11}$$

and let $\lambda_{1,n}, \lambda_{2,n}, \ldots, \lambda_{n,n}$ denote the eigenvalues of Q_n. If we denote the empirical distribution function of the eigenvalues by W_n, then $W_n = n^{-1} N_n(x)$, where $N_n(x)$ is the number of eigenvalues less than x. The problem is to determine the asymptotic behavior of the sequence $\{W_n\}$ of random distribution functions.

The basic result is due to Wigner [52, 53], who proved the following result.

THEOREM 2.30. *If* (i) $\int |x|^k dF < \infty$ *and* $\int |x|^k dG < \infty$ *for all* $k = 1, 2, \ldots$, *and* (ii) F *and* G *are symmetric, then* $\lim_{n\to\infty} \mathscr{E}\{W_n(x)\} = W(x)$, *where* W *is an absolutely continuous distribution function with density*

$$w(x) = \begin{cases} (2/\pi)(1 - x^2)^{1/2} & \text{for} \quad |x| \leqslant 1 \\ 0 & \text{for} \quad |x| > 1. \end{cases} \tag{2.12}$$

Because of the form of $w(x)$, Theorem 2.30 is often referred to as Wigner's *semi-circle law*. Grenander [23, pp. 178–180, 209–210] observed that, under the hypotheses of Theorem 2.30, $\lim_{n\to\infty} W_n(x) = W(x)$ in probability. Arnold [2, 3] has proved the following result, which is both a weak (convergence in probability) and strong (convergence almost surely) generalization of Theorem 2.30. For related results, see Olson and Uppuluri [41b].

THEOREM 2.31. *Let* $\int x^2 dG < \infty$, $\int x^4 dF < 0$, *and* $\int x dF = 0$. *Then for the distribution function* W_n *of the eigenvalues of the random matrix* Q_n, $\lim_n W_n(x) = W(x)$ *in probability. If, moreover,* $\int x^4 dG < \infty$ *and* $\int x^6 dF < \infty$, *then* $\lim_n W_n(x) = W(x)$ *almost surely.*

It is of interest to note that Bohigas and Flores [8] have shown (numerically) that pronounced deviations from Wigner's semi-circle law are obtained if the two-body nature of the Hamiltonian is taken into account. In particular, they showed that the limiting density is approximately normal.

The literature dealing with random matrices and their properties is very extensive (cf. Mehta [36], Porter [43]); and readers interested in studying the spectral properties of random linear operators on Banach spaces should, as the first step, acquaint themselves with the techniques used to investigate the statistical properties of the eigenvalues of random matrices. Some interesting problems which we have not considered here concern (1) the study of the correlations between eigenvalues of a random matrix (Dyson [19], Mehta [37]), (2) the distribution of the trace of a random matrix, and (3) inequalities for the expectation of the roots of a random matrix (Cacoullos and Olkin [12]). In Sect. 5.2 we consider an eigenvalue problem for a random Fredholm operator with degenerate kernel (cf. also Umegaki and Bharucha-Reid [51]), and in Sect. 5.3 we consider some eigenvalue problems for random Fredholm operators with symmetric kernels (cf. also Boyce [9]).

2.5 Operator-Valued Random Functions

A. Introduction

Consider the measurable space $(\mathfrak{L}(\mathfrak{X}), \mathfrak{B})$ where $\mathfrak{L}(\mathfrak{X})$ is the algebra of endomorphisms of a separable Banach space $\mathfrak{X}$ and $\mathfrak{B}$ is the σ-algebra of Borel subsets of $\mathfrak{L}(\mathfrak{X})$; and let $(T, \mathfrak{T})$ be a measurable space where T is a subset of the extended real line $\bar{R}$ and $\mathfrak{T} = T \cap \bar{\mathfrak{R}}$, where $\bar{\mathfrak{R}}$ is the σ-algebra of Borel subsets of $\bar{R}$.

***Definition* 2.34.** An $\mathfrak{L}(\mathfrak{X})$-*valued random function* on T is a mapping $X(t, \omega): T \times \Omega \to \mathfrak{L}(\mathfrak{X})$ such that for every $t \in T$, X is an $\mathfrak{L}(\mathfrak{X})$-valued random variable.

In this section we will restrict our attention to random functions with values in the algebra of endomorphisms $\mathfrak{L}(\mathfrak{X})$; and since $\mathfrak{L}(\mathfrak{X})$ is a Banach space, all of the definitions of Banach space-valued random functions given in Sect. 1.4 can be used to define the basic classes of $\mathfrak{L}(\mathfrak{X})$-valued random functions.

Similarly, the notions of equivalence, separability, measurability, and convergence presented in Sect. 1.4 are applicable to $\mathfrak{L}(\mathfrak{X})$-valued random functions.

In the subsequent subsections we define and give some examples of some of the basic classes of random functions with values in $\mathfrak{L}(\mathfrak{X})$.

B. Stationary random functions

Operator-valued stationary random functions have been studied by several mathematicians; we refer, in particular, to Loynes [33], Mandrekar and Salehi [34], and Payen [42]. Let H_1 and H_2 be two separable Hilbert spaces, and let $(G,+)$ be a locally compact Abelian group.

***Definition* 2.35.** $X(t,\omega): G \times \Omega \to \mathfrak{S}_2(H_1, H_2)$ is said to be an $\mathfrak{S}_2(H_1,H_2)$-*valued weakly stationary random function* if $\mathscr{E}\{X^*(t,\omega)X(s,\omega)\}$ is a function of $s-t$

If $t \to X(t,\omega)$ is a continuous mapping, then $X(t,\omega)$ admits the integral representation (cf. [34, 42])

$$X(t,\omega) = \int_{\hat{G}} \langle t,\lambda\rangle \, dE(\lambda)\, X(0,\omega), \tag{2.13}$$

where $\hat{G}$ is the character group of G, and E is a spectral measure on the σ-algebra $\mathfrak{B}$ of Borel sets of $\hat{G}$ whose values are projection operators on H_2 into H_2. Put $\xi(A,\omega) = E(A)X(0,\omega)$. Then

$$\xi^*(A)\,\xi(B) = X^*(0,\omega)\,E(A \cap B)\,X(0,\omega) = M(A \cap B).$$

Since $X(0,\omega) \in \mathfrak{S}_2(H_1,H_2)$ and $E \in \mathfrak{L}(H_1,H_2)$, $\xi(A,\omega)$ is $\mathfrak{S}_2(H_1,H_2)$-valued. M is called the *spectral measure* of the random function.

We refer to Mandrekar and Salehi [34] and Payen [42] for detailed discussions of $\mathfrak{S}_2(H_1,H_2)$-valued weakly stationary random functions.

In Loynes [33] second-order stationary random functions with values in a so-called LVH-space.† Let $\mathfrak{H}$ be a LVH-space. An $\mathfrak{H}$-valued random variable is said to be a *second-order random variable* if $\mathfrak{E}\{[x(\omega),x(\omega)]\}$ is defined, where $[\cdot,\cdot]$ denotes a vector inner product. Let $\mathfrak{H}$ be the space of $r \times s$ (complex) matrices, and define $[A,B] = B^*A$, $A,B \in \mathfrak{H}$, where B^* denotes the Hermitian adjoint of B. Loynes has shown that $\mathfrak{H}$ is a LVH-space, and if $\{X_n(\omega)\}$ is a sequence of random $r \times s$ matrices and $\mathscr{E}\{X_m^*(\omega)X_{m+n}\} = R_n$ depends only on n, then

$$X_n(\omega) = \int_0^{2\pi} e^{in\lambda}\, dY(\lambda,\omega),$$

where $Y(\lambda,\omega)$, for each λ, is an $r \times s$ random matrix.

† We refer to Loynes [33] for the definition of a LVH-space.

C. Markov processes

Let $\mathfrak{L}(H)$ be the algebra of endomorphisms of a separable Hilbert space H, and let $\mathfrak{S}_2(H)$ be the class of all Hilbert–Schmidt operators in $\mathfrak{L}(H)$. Let $X(t,\omega)$, $t \in T$, be a $\mathfrak{S}_2(H)$-valued random function. Associated with $X(t,\omega)$ are the following subspaces: $M(X(t,\omega))$ denotes the closed subspace generated by $X(t,\omega)$ over $\mathfrak{L}(H)$; and $M(t)$ denotes the closed subspace generated by $X(\tau,\omega)$, $\tau \leqslant t$, over $\mathfrak{L}(\mathfrak{X})$. Mandrekar and Salehi [35] have introduced the following definition.

***Definition* 2.36.** An $\mathfrak{S}_2(H)$-valued random function is said to be a *wide-sense Markov process* if for $s \leqslant t$

$$(X(t,\omega)|M(s)) = (X(t,\omega)|X(s,\omega)), \tag{2.14}$$

where $(X(t,\omega)|M)$ denotes the projection of $X(t,\omega)$ onto M.

Let $A(s,t) = \tilde{\Gamma}^{-1}(s,s)\Gamma(t,s)$, where Γ is the covariance function defined by $\Gamma(s,t) = X^*(t,\omega)X(s,\omega)$, and $\tilde{\Gamma}^{-1}$ denotes the generalized inverse† of Γ. We denote by $\mathfrak{R}_{st}$ the range of $\Gamma(s,t)$. The following useful theorem gives a necessary and sufficient condition that an $\mathfrak{S}_2(H)$-valued random function be a wide-sense Markov process.

THEOREM 2.32. *Let $X(t,\omega)$ be an $\mathfrak{S}_2(H)$-valued random function such that $\mathfrak{R}_{ts} \subseteq \mathfrak{R}_{ss}$ with $A(s,t) = \tilde{\Gamma}(s,s)\Gamma(t,s)$. Then $X(t,\omega)$ is a wide-sense Markov process if and only if $A(s,u) = A(s,t)A(t,u)$, $t \in [s,u]$.*

We refer to Mandrekar and Salehi [35] for a proof of the above theorem, and for other results concerning $\mathfrak{S}_2(H)$-valued Markov processes and operator-valued random differential equations.

D. Wiener processes

The following generalization of the classical Wiener process is due to Cabaña [10]. Let $B_2(\Omega, \mathfrak{L}(H))$ denote the collection of second-order $\mathfrak{L}(H)$-valued random variables; that is, $B_2(\Omega, \mathfrak{L}(H))$ is the collection of all $\mathfrak{L}(H)$-valued random variables $X(\omega)$ such that $\mathscr{E}\{\|X(\omega)\|^2\} < \infty$.

***Definition* 2.37.** A random function $W(t,\omega)$: $[0,T] \to H_0$ (where T is finite or infinite) is said to be a *Wiener process* (or *operator*) if (i) the increments of $W(t,\omega)$ corresponding to disjoint intervals are independent; (ii) for every $s,t \in [0,T)$, $s < t$, $[\lambda(\Delta)]^{-1/2}(W(t,\omega) - W(s,\omega))$, where $\Delta = (s,t)$ and λ

† If $A \in \mathfrak{L}(\mathfrak{X})$, then A^{-1} is defined by $P_{\mathfrak{N}\perp(A)}A^{-1}P_{\mathfrak{R}(A)}$, where $\mathfrak{N}(A)$ and $\mathfrak{R}(A)$ are the null space and range of A, respectively.

is a finite measure on $[0,T)$; (iii) for every $x \in H$, $W(t,\omega)x$ is weakly continuous a.s. as a function of t.

The operator-valued Wiener process defined above has been used by Cabaña to define a stochastic integral of Itô type, this integral being required for the study of random differential and integral equations of Itô type in Hilbert spaces (cf. Sect. 7.3B). Cabaña [11] has given another definition of an operator-valued Wiener process.

E. Martingales

Let $\{X_n(\omega), \mathfrak{A}_n, n \geqslant 1\}$ be a sequence of strong $\mathfrak{L}(\mathfrak{X})$-valued random variables, where $\{\mathfrak{A}_n, n \geqslant 1\}$ is an increasing family of sub-σ-algebras of $\mathfrak{A}$. The following definition is due to Rao [45, 46].

***Definition* 2.38.** A sequence $\{X_n(\omega), \mathfrak{A}_n, n \geqslant 1\}$ of strong $\mathfrak{L}(\mathfrak{X})$-valued random variables is said to be a *strong operator-valued martingale* if $\mathscr{E}\{X_n(\omega)|\mathfrak{A}_m\} = X_m(\omega)$, for $m \leqslant n$, strongly a.s.

We remark that if $\mathfrak{A}_n$ is an increasing sequence and $X(\omega)$ is a strong and Bochner integrable $\mathfrak{L}(\mathfrak{X})$-valued random variable, and if $X_n(\omega) = \mathscr{E}\{X(\omega)|\mathfrak{A}_n\}$, then $\{X_n(\omega), \mathfrak{A}_n, n \geqslant 1\}$ is a strong operator-valued martingale; and if $\mathfrak{X}$ is separable, then $\{X_n(\omega)\}$ is a uniformly integrable martingale, that is, $\{\|X_n(\omega)\|, n \geqslant 1\}$ of real-valued random variables is uniformly integrable. If $\mathfrak{X}$ is not separable, then $\{\|X_n(\omega)x\|, n \geqslant 1\}$ is uniformly integrable for each $x \in \mathfrak{X}$.

A sufficient condition that a sequence of $\mathfrak{L}(\mathfrak{X})$-valued random variables be a martingale is given by the following theorem.

THEOREM 2.33. *Let $\{X_n(\omega), 1 \leqslant n \leqslant n_0\}$ (n_0 finite or infinite) be a sequence of strong $\mathfrak{L}(\mathfrak{X})$-valued random variables such that $\{x^*(X_n(\omega)x), \mathfrak{A}_n, 1 \leqslant n \leqslant n_0\}$, is a real-valued martingale for every $x^* \in \mathfrak{X}$ and $x \in \mathfrak{X}$. Then $\{X_n(\omega), \mathfrak{A}_n, 1 \leqslant n \leqslant n_0\}$ is a strong operator-valued martingale. If $\mathfrak{X}$ is separable, then $\{\|X_n(\omega)\|, \mathfrak{A}_n, 1 \leqslant n \leqslant n_0\}$ is a real-valued submartingale.*

Mandrekar and Salehi [35] have introduced the notion of a wide-sense $\mathfrak{S}_2(H)$-valued martingale, and have used this notion in the study of $\mathfrak{S}_2(H)$-valued Markov processes.

***Definition* 2.39.** $\{X(t,\omega), t \in T\}$, is said to be an *$\mathfrak{S}_2(H)$-valued wide-sense martingale* if $(X(t,\omega)|M(s)) = X(s,\omega)$.

2.6 Limit Theorems

A. Introduction

In Sect. 1.6 we considered some limit theorems for Banach space-valued random variables. Since in a Banach space addition is the only algebraic operation, the theorems considered in Sect. 1.6 might be termed *additive* limit theorems. For operator-valued random variables, we can study two general classes of limit theorems. $\mathfrak{L}(\mathfrak{X}, \mathfrak{Y})$ is a Banach space, hence for $\mathfrak{L}(\mathfrak{X},\mathfrak{Y})$ random variables the additive limit theorems of Sect. 1.6 are applicable. However, $\mathfrak{L}(\mathfrak{X})$ is also a Banach algebra; hence we can consider *multiplicative* limit theorems, as well as additive limit theorems, for $\mathfrak{L}(\mathfrak{X})$-valued random variables.

The study of multiplicative limit theorems was initiated by Bellman [4] who considered the asymptotic behavior of the product

$$T_n(\omega) = X_n(\omega)\, X_{n-1}(\omega) \ldots X_1(\omega), \tag{2.15}$$

where $\{X_n(\omega)\}$ is a stationary sequence of $k \times k$ random matrices. In particular, Bellman showed that if the $X_i(\omega)$ are independent and have strictly positive elements, then, under certain conditions a weak multiplicative law of large numbers holds; that is, if $t_{ij}^{(n)}(\omega)$ denotes an element of $T_n(\omega)$, then $\lim_{n\to\infty} (1/n)\,\mathfrak{E}\{t_{ij}^{(n)}(\omega)\}$ exists. The study of the asymptotic behavior of products of random matrices is of importance in the analysis of the limiting behavior of solutions of systems of differential and difference equations with random coefficients.

We now state a theorem due to Furstenberg and Kesten [22] which is a strong multiplicative law of large numbers for products of random matrices.

THEOREM 2.34. *If $\{X_i(\omega)\}$ is a stationary sequence of random matrices with values in $\mathfrak{L}(R_k)$, then* $\lim_{n\to\infty} (1/n)\,\mathscr{E}\{\log\|T_n(\omega)\|\} = \tau$ *exists (τ not necessarily finite). If, in addition the $X_i(\omega)$ are independent and identically distributed and* $\mathscr{E}\{\log^+\|X_1(\omega)\|\} < \infty$,† *then*

$$\limsup_{n\to\infty} (1/n) \log\|T_n(\omega)\| \leqslant \tau$$

almost surely.

The above theorem is for random variables with values in the algebra of matrices $\mathfrak{L}(R_k)$; however Grenander [23, Theorem 7.2.2] has shown that a theorem of the above type can be proved for a sequence of independent and identically distributed random variables with values in any separable Banach algebra. We refer to Furstenberg [21] for a detailed investigation of non-commuting products of independent and identically distributed random variables with values in an arbitrary group.

† $\log^+ x = \max(\log x, 0)$.

We now state and prove a multiplicative law of large numbers for random variables in a separable Banach algebra due to Grenander [23]. We state the theorem for $\mathfrak{L}(R_n)$-valued random variables, since only when $\mathfrak{X} = R_n$ is $\mathfrak{L}(\mathfrak{X})$ a separable Banach algebra.

THEOREM 2.35. *Let $\{X_i(\omega)\}$ be a sequence of independent and identically distributed $\mathfrak{L}(R_n)$-valued random variables such that $\mathfrak{E}\{\|X_1(\omega)\|\} < \infty$. Then*

$$T_n(\omega) = \left(I + \frac{1}{n} X_1(\omega)\right)\left(I + \frac{1}{n} X_2(\omega)\right) \cdots \left(I + \frac{1}{n} X_n(\omega)\right)$$

converges strongly in probability to $\tau = \exp\{\mathscr{E}\{X_1(\omega)\}\}$.

Proof. Put

$$S_1^{(n)}(\omega) = (1/n) \sum_{j=1}^{n} X_j(\omega),$$

$$S_2^{(n)}(\omega) = (1/n^2) \sum_{i \leq j < k \leq n} X_j(\omega)\, X_k(\omega),$$

etc. Then

$$T_n(\omega) = I + \sum_{i=1}^{n} S_i^{(n)}(\omega).$$

Now, by Theorem 1.32 (the additive strong law of large numbers for Banach space-valued random variables), $S_1^{(n)} \to \mathscr{E}\{X_1(\omega)\}$ strongly. $S_2^{(n)}$ can be written as

$$S_2^{(n)}(\omega) = \frac{1}{n} \sum_{k=1}^{n} \left(\frac{k-1}{n}\right) S_1^{(k)}(\omega)\, X_k(\omega),$$

where

$$S_1^{(k)}(\omega) = \frac{1}{k-1} \sum_{i=1}^{k-1} X_i(\omega).$$

However,

$$S_1^{(k)}(\omega) = \mathscr{E}\{X_1(\omega)\} + \epsilon_k(\omega) \qquad \text{a.s.},$$

where $\|\epsilon_k(\omega)\| \to 0$. Hence

$$S_2^{(n)}(\omega) = \frac{\mathscr{E}\{X_1(\omega)\}}{n} \sum_{k=1}^{n} \left(\frac{k-1}{n}\right) X_k(\omega) + \frac{1}{n} \sum_{k=1}^{n} \left(\frac{k-1}{n}\right) \epsilon_k(\omega)\, X_k(\omega).$$

Therefore

$$S_2^{(n)}(\omega) = \tfrac{1}{2}(\mathscr{E}\{X_1(\omega)\})^2 + \delta_n(\omega) \qquad \text{a.s.},$$

where $\|\delta_n(\omega)\| \to 0$. In general,

$$S_n^{(k)}(\omega) \to \frac{(\mathscr{E}\{X_1(\omega)\})^k}{k!} \qquad \text{a.s.}$$

Now

$$\|S_k^{(n)}(\omega)\| \leqslant \frac{1}{n^k} \sum_{1 \leqslant j_1 < j_2 < \cdots < j \leqslant n} \|X_{j_1}(\omega)\| \cdot \|X_{j_2}(\omega)\| \cdots \|X_{j_k}(\omega)\|,$$

so that

$$\mathscr{E}\{\|S(^{(n)}_k\omega)\|\} \leqslant (\mathscr{E}\{\|X_1(\omega)\|\})^k/k!.$$

It follows from the above results that for any $\alpha > 0$, $\mu(\{\omega : \|T_n(\omega) - \tau\| > \alpha\}) \to 0$, where

$$\tau = I + \mathscr{E}\{X_1(\omega)\} + \tfrac{1}{2}(\mathscr{E}\{X_1(\omega)\})^2 + \cdots = \exp\{\mathscr{E}\{X_1(\omega)\}\} \in \mathfrak{L}(R_n).$$

The next theorem which we state and prove is an additive strong law of large numbers for random variables with values in $\mathfrak{L}(H)$, where H is a separable Hilbert space. This result is due to Haïnis [24]. We first state the following result, also due to Haïnis.

THEOREM 2.36. *Let H be a separable Hilbert space, and let $\{T_i(\omega)\}_{i \in N}$ be a sequence of mutually independent $\mathfrak{L}(H)$-valued random variables with $U = \mathscr{E}\{T_i(\omega)\}$, $i \in N$. If there is a constant $M > 0$ such that for $i \in N$ and $x \in H$, $\mathscr{E}\{\|T_i(\omega)x\|\} \leqslant M$, then the sequence $\{Y_n(\omega)\}$, where*

$$Y_n(\omega) = (1/n) \sum_{i=1}^{n} X_i(\omega)$$

converges strongly a.s. to a unique limit operator U.

Let $\tilde{H} = \Sigma \otimes H$ be the space of all square summable sequences in H; that is, the sequence $Z = \{x_i\}_{i \in N} \in \tilde{H}$ if and only if $\sum \|x_i\|^2 < \infty$. For $T \in \mathfrak{L}(H)$, $\tilde{T}Z = T[x_i]$.

THEOREM 2.37. *If the hypotheses of Theorem 3.36 are satisfied, and if $\sum_{i=1}^{\infty} \mathscr{E}\{\|T_k(\omega)x_i\|^2\} < \infty$ for every $k \in N$, then the sequence of random operators $Y_n(\omega)$ converges strongly a.s. to a unique operator U.*

Proof. Let $\tilde{Y}_n(\omega) = (1/n) \sum_{i=1}^{n} \tilde{T}_i(\omega)$. If $Z = \{x_i\}_{i \in N} \in \tilde{H}$, then

$$\mathscr{E}\{\tilde{T}_k(\omega)Z\} = \{\mathscr{E}\{T_k(\omega)\}x_i\}_{i \in N} = \{U[x_i]\}_{i \in N} = \tilde{U}Z,$$

where $\tilde{U}$ is an endomorphism of $\tilde{H}$ defined by $U \in \mathfrak{L}(H)$.

From the relation

$$\mathscr{E}\{\|\tilde{T}_k(\omega)Z\|^2\} = \mathscr{E}\left\{\sum_i \|T_k(\omega)x_i\|^2\right\}$$

$$= \sum_i \mathscr{E}\{\|T_k(\omega)x_i\|^2\}$$

and Theorem 2.36, we conclude that there is a subset $\Omega_0 \subset \Omega$ such that $\mu(\Omega - \Omega_0) = 0$, and for every $\omega \in \Omega_0$ and $Z \in \tilde{H}$ the sequence $\{\|(\tilde{T}_n(\omega) - \tilde{U})Z\|\}_{n \in N} \to 0$.

Our last limit theorem is a martingale convergence theorem due to Rao [46]. Let $\mathfrak{X}$ and $\mathfrak{Y}$ be Banach spaces, where $\mathfrak{X}$ is arbitrary and $\mathfrak{Y}$ has the LRN-property.†

THEOREM 2.38. *Let $X_n(\omega)$: $\Omega \to \mathfrak{L}(\mathfrak{X},\mathfrak{Y})$, and let $\{X_n(\omega), \mathfrak{A}_n, n \geqslant 1\}$ be a strong operator-valued martingale. If* $\sup \int_\Omega \|X_n(\omega)x\|_{\mathfrak{Y}}\, d\mu = K_x < \infty$ *for each $x \in \mathfrak{X}$, then there exists a strong $\mathfrak{L}(\mathfrak{X},\mathfrak{Y})$-valued random variable $X_\infty(\omega)$ such that $X_n(\omega) \to X_\infty(\omega)$ strongly a.s. for each $x \in \mathfrak{X}$. If, moreover, $\mathfrak{X}$ is separable and* $\sup\{K_x : \|x\|_{\mathfrak{Y}} \leqslant 1\} < \infty$, *then $\|X_n(\omega) - X_\infty(\omega)\|_{\mathfrak{L}(\mathfrak{X},\mathfrak{Y})}$ is measurable and tends to zero a.s. as $n \to \infty$, even though $X_n(\omega)$ and $X_\infty(\omega)$ are necessarily uniform random operators.*

For other results on the convergence of operator-valued martingales, and their applications, we refer to Rao [46].

We close this section by remarking that Tutubalin [50] has proved some central limit theorems for products of random matrices.

References

1. Anderson, T. W., "Introduction to Multivariate Statistical Analysis." Wiley, New York, 1958.
2. Arnold, L., On the asymptotic distribution of the eigenvalues of random matrices. *J. Math. Anal. Appl.* **20** (1967), 262–268.
3. Arnold, L., "Zur Asymptotischen Verteilung der Eigenwerte Zufälliger Matrizen." Habilitationsschrift, Univ. Stuttgart, Stuttgart, 1969.
4. Bellman, R., Limit theorems for non-commutative operations. I. *Duke Math. J.* **21** (1954), 491–500.
5. Bharucha-Reid, A. T., On random operator equations in Banach space. *Bull. Acad. Polon. Sci. Sér. Sci. Math. Astronom. Phys.* **7** (1959), 561–564.
6. Bharucha-Reid, A. T., On random solutions of integral equations in Banach spaces. *Trans. 2nd Prague Conf. on Information Theory, Statist. Decision Functions, and Random Processes (1959)*, pp. 27–48, 1960.
7. Bharucha-Reid, A. T., Random algebraic equations. *In* "Probabilistic Methods in Applied Mathematics" (A. T. Bharucha-Reid, ed.), Vol. 2, pp. 1–52. Academic Press, New York, 1970.
8. Bohigas, O., and Flores, J., Two-body random Hamiltonian and level density. *Phys. Lett. B* **34** (1971), 261–263.

† A Banach space $\mathfrak{Y}$ is said to have the LRN (= Lebesgue–Radon–Nikodyn)-*property* relative to $(\Omega, \mathfrak{A}, \mu)$ if each countably additive measure $\nu: \mathfrak{A} \to \mathfrak{Y}$, which vanishes on sets of μ measure zero, has an integral representation relative to μ, that is, there exists a unique strong random variable $y(\omega)$: $\Omega \to \mathfrak{Y}$ such that

$$\nu(A) = (\mathrm{B}) \int_A y(\omega)\, d\mu, \qquad A \in \mathfrak{A}.$$

9. Boyce, W. E., Random eigenvalue problems. *In* "Probabilistic Methods in Applied Mathematics" (A. T. Bharucha-Reid, ed.), Vol. 1, pp. 1–73. Academic Press, New York, 1968.
10. Cabaña, E. M., Stochastic integration in separable Hilbert spaces. *Publ. Inst. Mat. Estadíst. Montevideo* **4** (1966), 49–80.
11. Cabaña, E. M., On stochastic differentials in Hilbert spaces. *Proc. Amer. Math. Soc.* **20** (1969), 259–265.
12. Cacoullos, T., and Olkin, I., On the bias of functions of characteristic roots of a random matrix. *Biometrika* **52** (1965), 87–94.
13. de Araya, J. A., A Radon–Nikodym theorem for vector and operator valued measures. *Pacific J. Math.* **29** (1969), 1–10.
14. Dinculeanu, N., "Vector Measures." Pergamon, Oxford, 1967.
15. Doss, S., Sur la moyenne d'un élément aléatoire dans un espace distancié. *Bull. Sci. Math.* **73** (1949), 48–72.
16. Dunford, N., On one parameter groups of linear transformations. *Ann. of Math.* **39** (1938), 569–573.
17. Dunford, N., and Schwartz, J. T., "Linear Operators. Part I: General Theory." Wiley (Interscience), New York, 1958.
18. Dunford, N., and Schwartz, J. T., "Linear Operators. Part II: Spectral Theory." Wiley (Interscience), New York, 1963.
19. Dyson, F. J., Correlations between eigenvalues of a random matrix. *Comm. Math. Phys.* **19** (1970), 235–250.
20. Frisch, U., Wave propagation in random media. *In* "Probabilistic Methods in Applied Mathematics" (A. T. Bharucha-Reid, ed.), Vol. 1, pp. 75–198. Academic Press, New York, 1968.
21. Furstenberg, H., Noncommuting random products. *Trans. Amer. Math. Soc.* **108** (1963), 377–428.
22. Furstenberg, H., and Kesten, H., Products of random matrices. *Ann. Math. Statist.* **31** (1960), 457–469.
23. Grenander, U., "Probabilities on Algebraic Structures." Wiley, New York, 1963.
24. Haïnis, J., Random variables with values in Banach algebras and random transformations in Hilbert spaces (Greek, French summary). *Bull. Soc. Math. Grèce* (N. S.) **7** (1966), 179–223.
25. Hanš, O., Generalized random variables. *Trans. 1st Prague Conf. on Information Theory, Statist. Decision Functions, and Random Processes (1956)*, pp. 61–103, 1957.
26. Hanš, O., Inverse and adjoint transforms of linear bounded random transforms. *Trans. 1st Prague Conf. on Information Theory, Statist. Decision Functions, and Random Processes (1956)*, pp. 127–133, 1957.
27. Hanš, O., Random operator equations. *Proc. 4th Berkeley Symp. on Math. Statist. and Probability (1960)*, Vol. II, pp. 185–202, 1961.
28. Hille, E., and Phillips, R. S., "Functional Analysis and Semi-Groups." Amer. Math. Soc., Providence, Rhode Island, 1957.
29. Kannan, D., and Bharucha-Reid, A. T., An operator-valued stochastic integral. *Proc. Japan Acad.* **47** (1971), 472–476.
30. Kato, T., "Perturbation Theory for Linear Operators." Springer-Verlag, Berlin and New York, 1966.
31. Kolmogorov, A. N., and Fomin, S. V., "Introductory Real Analysis," rev. Engl. ed., translated from the Russian. Prentice-Hall, Englewood Cliffs, New Jersey, 1970.
32. Kuratowski, K., "Topologie," Vol. I. Pánstwowe Wydawnictwo Naukowe, Warsaw, 1958.
33. Loynes, R. M., On a generalization of second-order stationarity. *Proc. London Math. Soc.*, **15** (1965), 385–398.

34. Mandrekar, V., and Salehi, H., The square-integrability of operator-valued functions with respect to a non-negative operator-valued measure and the Kolmogorov isomorphism theorem. *Indiana Univ. Math. J.* **20** (1970), 545–563.
35. Mandrekar, V., and Salehi, H., Operator-valued wide-sense Markov processes and solutions of infinite dimensional linear differential systems driven by white noise. *Math. Systems Theory* **4** (1970), 340–356.
36. Mehta, M. L., "Random Matrices and the Statistical Theory of Energy Levels." Academic Press, New York, 1967.
37. Mehta, M. L., A note on correlations between eigenvalues of a random matrix. *Comm. Math. Phys.* **20** (1971), 245–250.
38. Mukherjea, A., Random transformations on Banach spaces. Ph.D. Dissertation, Wayne State Univ., Detroit, Michigan, 1966.
39. Mukherjea, A., Transformations aléatoires séparables: Théorème du point fixe aléatoire. *C. R. Acad. Sci. Paris* **263** (1966), 393–395.
40. Mukherjea, A., and Bharucha-Reid, A. T., Separable random operators: I. *Rev. Roumaine Math. Pures Appl.* **14** (1969), 1553–1561.
41. Munroe, M. E., "Introduction to Measure and Integration." Addison-Wesley, Reading, Massachusetts, 1953.
41a. Nashed, M. Z., and Salehi, H., Measurability of generalized inverses of random linear operators. To be published.
41b. Olson, W. H., and Uppuluri, V. R. R., Asymptotic distribution of eigenvalues of random matrices. *Proc. 6th Berkeley Symp. on Math. Statist. and Probability (1970)*, to be published.
42. Payen, R., Fonctions aléatoires du second ordre à valeurs dans un espace de Hilbert. *Ann. Inst. H. Poincaré Sect. B* **3** (1967), 323–963.
43. Porter, C. E., "Statistical Theories of Spectra: Fluctuations." Academic Press, New York, 1965.
44. Rao, M. M., Abstract Lebesgue-Radon-Nikodym theorems. *Ann. Mat. Pura Appl.* **76** (1967), 107–132.
45. Rao, M. M., Prédictions non linéaires et martingales d'opérateurs. *C. R. Acad. Sci Paris Ser. A* **267** (1968), 122–124.
46. Rao, M. M., Abstract nonlinear prediction and operator martingales. *J. Multivariate Anal.* **1** (1971), 129–157.
47. Ryll-Nardzewski, C., An analogue of Fubini's theorem and its application to random linear equations. *Bull. Acad. Polon. Sci. Sér. Sci. Math. Astronom. Phys.* **8** (1960), 511–513.
48. Špaček, A., Sur l'inversion des transformations aléatoires presque sûrement linéaires. *Acta Math. Acad. Sci. Hungar.* **7** (1957), 355–358.
49. Taylor, A. E., "Introduction to Functional Analysis." Wiley, New York, 1958.
50. Tutubalin, V. N., Limit theorems for a product of random matrices (Russian). *Teor. Verojatnost. i Primenen.* **10** (1965), 19–32.
51. Umegaki, H., and Bharucha-Reid, A. T., Banach space-valued random variables and tensor products of Banach spaces. *J. Math. Anal. Appl.* **31** (1970), 49–67.
52. Wigner, E. P., Characteristic vectors of bordered matrices with infinite dimensions. *Ann. of Math.* **62** (1955), 548–564.
53. Wigner, E. P., On the distribution of the roots of certain symmetric matrices. *Ann. of Math.* **67** (1958), 325–326.
54. Wishart, J., The generalized product moment distribution in sampling from a normal multivariate population. *Biometrika* **20A** (1928), 32–52.
55. Zaanen, A. C., "Linear Analysis." North-Holland Publ., Amsterdam, 1953.
56. Zadeh, L. A., On a class of stochastic operators. *J. Math. and Phys.* **32** (1953), 48–53.

CHAPTER 3

Random Equations: Basic Concepts and Methods of Solution

3.1 Introduction

Let $\mathfrak{X}$ and $\mathfrak{Y}$ be two Banach spaces, and let T be an operator from $\mathfrak{X}$ to $\mathfrak{Y}$. The development of deterministic operator theory (indeed, many branches of functional analysis) was motivated to a large extent by questions which arose in applied mathematics in connection with the formulation of methods for solving *operator equations* of the form

$$Tx = y. \tag{3.1}$$

In Eq. (3.1), y is known, x is unknown, and T is a linear or nonlinear operator on $\mathfrak{X}$ to $\mathfrak{Y}$. An element $x_0 \in \mathfrak{X}$ is said to be a *solution* of Eq. (3.1) if $Tx_0 = y$; and the set $S = \{x_0 : Tx_0 = y, x_0 \in \mathfrak{X}\}$ is said to be the *solution set* of Eq. (3.1). If $S = \varnothing$ (the null set), then Eq. (3.1) does not possess a solution; and if $S \neq \varnothing$, Eq. (3.1) is said to be *solvable.*

One of the major problems in the study of operator equations is to determine conditions under which the equation is solvable. Problems of this type are referred to as *existence problems* for operator equations; and the theorems which establish such conditions are known as *existence theorems.* Hence an existence theorem for a given operator equation tells us when the solution set of the equation is not empty.

Of equal, if not greater importance in applications, is the problem of determining conditions such that an operator equation admits only one solution; that is, the solution set contains only one element. Problems of this type are called *uniqueness problems*; and the theorems that establish such conditions are known as *uniqueness theorems*.

An important special case of Eq. (3.1) is the *homogeneous equation*

$$Tx = \theta. \tag{3.2}$$

In this case S is a subspace of $\mathfrak{X}$, called the *solution space* of Eq. (3.2). If T is a linear operator, then the method for solving Eq. (3.1) can be described as follows: (1) Find all solutions of Eq. (3.2), (2) find one solution of Eq. (3.1), and (3) add the solutions.

A large number of problems in applied mathematics lead to operator equations of the form

$$(T - \lambda I)x = y \tag{3.3}$$

or

$$(T - \lambda I)x = \theta, \tag{3.4}$$

where λ is a scalar which can be real or complex. The problem of solving Eq. (3.3) is called the *eigenvalue problem* for the operator T. From Sect. 2.4A we know that the values of λ for which Eq. (3.3) has nontrivial solutions are the *eigenvalues* of T; and for each eigenvalue λ_0, the nonnull elements of $\mathfrak{X}$ which satisfy the equation $(T - \lambda_0 I)x = \theta$ are called the *eigenfunctions* of T belonging to λ_0.

There are a large number of methods available for the solution of linear and nonlinear operator equations (cf. Rall [52], Saaty [54]). When the formulation of a problem in applied mathematics leads to an operator equation of the form (3.1) or (3.3), the method used to solve the equation depends to a great extent on the Banach spaces $\mathfrak{X}$ and $\mathfrak{Y}$ (in particular, the space $\mathfrak{X}$ in which the solution is required) and the kinds of information which the solution is required to provide.

In this chapter we consider (1) the formulation of various types of random equations, and (2) methods of solving random equations. Section 3.2 is devoted to the formulation of random equations and the basic concepts associated with random equations. In Sect. 3.3 we consider methods for solving random equations. These methods are developed within the framework of probabilistic functional analysis. Finally, in Sect. 3.4 we consider some measure-theoretic and statistical problems associated with random equations and their solutions.

3.2 Random Equations: Basic Concepts and Examples

A. Introduction

We first consider the probabilistic analogues of Eqs. (3.1) and (3.3). It is clear that randomness can be introduced in the equations via the "known" function y (often called the input, forcing function, or nonhomogeneous term), the operator T, or both. Hence from Eq. (3.1) we can obtain the following random equations:

$$Tx = y(\omega) \tag{3.5}$$

$$T(\omega)x = y \tag{3.6}$$

$$T(\omega)x = y(\omega). \tag{3.7}$$

In order for Eqs. (3.5) and (3.6) to be consistent, we must assume that in Eq. (3.5) the deterministic operator T is an operator-valued random variable which assumes a given value with probability one. Similarly, in Eq. (3.6), we assume that y is a Banach space-valued random variable (or function) which assumes a given value with probability one.

As in the case of Eq. (3.1), we can obtain the following random equations from Eq. (3.3):

$$(T - \lambda I)x = y(\omega) \tag{3.8}$$

$$(T(\omega) - \lambda I)x = y \tag{3.9}$$

$$(T(\omega) - \lambda I)x = y(\omega). \tag{3.10}$$

We can refer to all of the above equations [that is, Eqs. (3.5)–(3.10)] as *random equations*; however we will use the term *random operator equation* when referring to equations of the form (3.6), (3.7), (3.9), and (3.10).

Since ω is an element of a measurable space $(\Omega, \mathfrak{A})$ on which there is defined a complete probability measure μ, it is clear that a random equation is actually a family of equations. This family of equations will have only one member in the case where μ is Dirac measure; that is $\mu(\{\omega_0\}) = 1$ and $\mu(\Omega - \{\omega_0\}) = 0$. In this case, of course, we are dealing with a deterministic equation. Hence we can state that the classical theory of deterministic equations is a special case of the theory of random equations.

Examples of random equations are numerous. We now discuss briefly a few examples of random equations that arise in various applied fields. Consider matrix equations of the form

$$Ax = y \qquad \text{and} \qquad (A - \lambda I)x = y, \tag{3.11}$$

where $A = (a_{ij})$ is an $n \times n$ matrix and x and y are n-vectors. If the matrix elements are subject to error, which is frequently the case in applied problems, then the a_{ij} can often be written in the form $a_{ij}(\omega) = \alpha_{ij} + \epsilon_{ij}(\omega)$, where the elements α_{ij} are assumed to be known and the perturbing elements $\epsilon_{ij}(\omega)$ are random variables. In this case Eqs. (3.11) lead to random equations of the form (3.6) and (3.9). In the analysis of systems of random linear equations it is of interest to consider the random eigenvalue equation

$$(A(\omega) - \lambda I)x = 0, \tag{3.12}$$

which, as we know, has a nontrivial solution if and only if λ satisfies the characteristic equation

$$|A(\omega) - \lambda I| = 0. \tag{3.13}$$

The characteristic polynomial obtained from (3.13) will be a random algebraic polynomial of degree n, say $F_n(\omega)$, and the solutions of the random algebraic equation $F_n(\omega) = 0$ are the eigenvalues of $A(\omega)$. Many problems in economics, linear programming, and physics lead to random matrix equations. We refer to Bharucha-Reid [11] for references to a number of studies. Systems of random linear equations are also encountered in the study of systems of random difference and differential equations.

Let τ denote the translation operator, that is, $\tau^n[x_k] = x_{k+n}$, $n = 0, 1, \ldots$. Consider the linear difference operator of order n

$$L = \sum_{k=0}^{n} a_k \tau^k,$$

where the coefficients a_k are constants. Linear difference equations of the form

$$Lx_k = y_k(\omega), \tag{3.14}$$

where the input $y_k(\omega)$ is a random sequence, have been studied rather extensively in mathematical economics (cf. Koopmans [40]) and in connection with stationary time series (cf. Grenander and Rosenblatt [28]).

Difference equations with random coefficients constitute an important class of random operator equations. While many random difference equations arise as mathematical models of concrete physical processes, most of those that have been studied are obtained as discrete-time analogues or approximate equations for differential equations with random coefficients.

Consider the system of random differential equations

$$dx(t)/dt = A_k(\omega)\, x(t), \qquad t \in [t_{k-1}, t_k], \quad k = 1, 2, \ldots, \tag{3.15}$$

where $x(t)$ is a real-valued n-vector and the $A_k(\omega)$ are $n \times n$ random matrices not depending on t. Equations of the form (3.15) arise in the study of systems of differential equations with coefficients that are random functions, but are piecewise constant with respect to t. If the $A_k(\omega)$ are bounded uniformly in k almost surely (that is, there exists a constant $M < \infty$ such that for all k, $\mu(\{\omega : \|A_k(\omega)\| < M\}) = 1$), then, for every fixed k, $A_k(\omega)$ is an $\mathfrak{L}(R_n)$-valued random variable.

Let $x(t_0, \omega) = x_0(\omega) = x_0$. Then the solution of Eq. (3.15) in the interval $[t_0, t_1)$ is

$$x(t, \omega) = \exp\{A_1(\omega)(t - t_0)\}\, x_0. \tag{3.16}$$

Put

$$\Phi_k(\omega) = \exp\{A_k(\omega)(t_k - t_{k-1})\}. \tag{3.17}$$

Then it can be shown that for $t \in [t_{k-1}, t_k]$, $k = 1, 2, \ldots$

$$x(t, \omega) = \exp\{A_k(\omega)(t - t_{k-1})\}\, \Phi_{k-1}(\omega) \cdots \Phi_1(\omega)\, x_0 . \tag{3.18}$$

If we restrict our attention to the solution at time $t = t_k$ only, then (3.18) can be rewritten as

$$x(t_k, \omega) = x_k(\omega) = \Phi_k(\omega)\, \Phi_{k-1}(\omega) \cdots \Phi_1\, x_0 . \tag{3.19}$$

Hence $x_k(\omega)$ satisfies the following first-order random difference equation

$$x_k(\omega) = \Phi_k(\omega)\, x_{k-1}(\omega), \qquad k = 1, 2, \ldots . \tag{3.20}$$

It is of interest to note that random difference equations of this form lead to an interesting class of limit theorems. From (3.19) it is clear that the study of $\lim_{k \to \infty} x_k(\omega)$ leads to the study of the limiting behavior of the product of random matrices $\Phi_k(\omega)\Phi_{k-1}(\omega) \cdots \Phi_1$. We refer to Sect. 2.6 for a discussion of multiplicative limit theorems for Banach algebra-valued random variables, the investigation of which was motivated by equations of the form (3.20).

Random difference equations of the general form

$$x_{n+1}(\omega) = \varphi(\omega, n, x_n(\omega), y_n(\omega)) \tag{3.21}$$

are used as models for *discrete-time probabilistic dynamical systems*. We refer to Åström [4] and Jazwinski [34] for treatments of equations of the above form.

For other results on random difference equations we refer to Grenander [27]; and we refer to Bharucha-Reid [12] for a survey of random difference equations and their applications in physics, engineering, economics, biology, psychology, and other fields.

The literature on random equations contains more studies on random differential equations and their applications than any other type of random equation. This is not at all surprising, for mathematical models formulated as deterministic differential equations have long played a fundamental role in virtually every branch of applied mathematics. And, as workers attempt to make their models more realistic and take into consideration the random nature of many of the processes and systems they study, it is only natural that models formulated as random differential equations play an increasingly important role.

In engineering and physics a number of problems lead to random ordinary differential equations of so-called Langevin type. The original *Langevin equation* is of the form

$$dv/dt = -\alpha v + F(t, \omega), \tag{3.22}$$

and is, to the best of our knowledge, the first random differential equation studied. We refer to Sect. 7.1 where random differential equations of Langevin type and their formulation as random integral equations is discussed.

Another type of random ordinary differential equation which has been studied by a large number of mathematicians, physicists, and engineers is the probabilistic analogue of a linear differential equation of order n:

$$L(\omega)[x(t)] = \sum_{k=0}^{n} a_k(t,\omega)\, d^k x/dt^k = y(t,\omega), \tag{3.23}$$

where the coefficients $a_k(t,\omega)$ and the input or forcing function $y(t,\omega)$ are random functions. It is generally assumed that $y(t,\omega)$ is independent of any of the coefficients, however the coefficients may be correlated with each other. We refer to Bharucha-Reid [10] for references to some papers devoted to methods of solving Eq. (3.23) and to investigations of the stability properties of the solution; and we refer to Åström [4, Chap. 3] for a discussion of random differential equations as models for physical processes and some brief historical comments about random differential equations.

The study of random partial differential equations was initiated by Kampé de Fériet (cf. [35]). In particular, he studied random solutions of the heat equation for an infinite rod when the initial temperature at the point x on the rod is a random function $f(x,\omega)$. Consider the random initial-value problem

$$\begin{aligned} \partial u/\partial t &= \partial^2 u/\partial x^2, \qquad x \in (-\infty,\infty) \\ u(0,x,\omega) &= f(x,\omega). \end{aligned} \tag{3.24}$$

The solution of (3.24) is

$$u(t,x,\omega) = \int_{-\infty}^{\infty} K(t, x-\xi) f(\xi,\omega)\, d\xi, \tag{3.25}$$

where $K(t,x) = (4\pi t)^{-1/2} \exp\{-x^2/4t\}$.

The problem considered by Kampé de Fériet is a special case of the Cauchy problem (with random initial data) for partial differential equations. We refer to Birkhoff *et al.* [13] for a systematic study of statistically well-posed Cauchy problems.

Gopalsamy and Bharucha-Reid [26] have studied the partial differential equation

$$\begin{aligned} (\partial/\partial t + L)u &= h(x)\,\xi(t,\omega), \qquad && x \in D, \quad t > 0, \quad h \in L_2(D) \\ u(0,x,\omega) &= g(x,\omega), && x \in D, \quad g \in L_2(D) \\ u(t,x,\omega) &= 0, && x \in \partial D, \quad t \geqslant 0. \end{aligned} \tag{3.26}$$

In the above D is a bounded open subset of R_n (n finite), ∂D denotes the boundary of D; and $L[u] = A[u] + a_0(x)u$, where

$$A[u] = -\sum_{i,j=1}^{n} \frac{\partial}{\partial x_i}\left(a_{ij}(x)\frac{\partial u}{\partial x_j}\right).$$

The solution of Eq. (3.26) is

$$u(t, x, \omega) = \int_0^t T(t-\tau) h(x) \xi(\tau, \omega) \, d\tau + T(t) g(x, \omega), \tag{3.27}$$

where $\{T(t), t \geqslant 0\}$ is a strongly continuous semigroup of contraction operators.† The solution was investigated in the case where the driving function $\xi(t, \omega)$ is a stochastic point process.

Random partial differential equations arise in quantum mechanics when applications lead to a Schrödinger equation with random potential function. We refer to Bharucha-Reid [10] for a brief discussion of some of these studies.

Since this book is devoted to random integral equations, we refer the reader to Chaps. 4–7 where many examples of random integral equations are given.

B. Solutions of random equations

In Sect. 3.1 we defined the solution of a deterministic operator equation. For random equations we can introduce two notions of solution, namely a wide-sense solution and a random solution. Consider, for example, Eq. (3.7).

***Definition* 3.1.** Any mapping $x(\omega)$: $\Omega \to \mathfrak{X}$ which satisfies the equality $T(\omega)x(\omega) = y(\omega)$ for every $\omega \in \Omega_0$, where $\mu(\Omega_0) = 1$ is said to be a *wide-sense solution* of Eq. (3.7).

If, in addition, a wide-sense solution is measurable, then we introduce the following definition.

***Definition* 3.2.** Any $\mathfrak{X}$-valued random variable $x(\omega)$ which satisfies the condition

$$\mu(\{\omega : T(\omega)\, x(\omega) = y(\omega)\}) = 1 \tag{3.28}$$

is said to be a *random solution* of Eq. (3.7). Random solutions have also been referred to as *strict-sense solutions*.

We now give an example of a wide-sense solution which is not a random solution. Let $\mathfrak{X} = R$, and let E be a nonmeasurable subset of Ω (that is, $E \notin \mathfrak{A}$). Let $T(\omega)$: $R \to R$ be a random operator defined for every $\omega \in \Omega$ and $x \in R$ as follows: $T(\omega)x = x^2 - 1$. In this case the real-valued random variable $\xi(\omega)$ such that $\xi(\omega) = 1$ for every $\omega \in \Omega$ is a random solution of the homogeneous

† We refer to Sect. 7.3D for the definition of a semigroup of operators.

equation $T(\omega)x = 0$. However, the real-valued random variable $\varphi(\omega)$ defined by

$$\varphi(\omega) = \begin{cases} 1, & \omega \in E \\ -1, & \omega \in \Omega - E \end{cases}$$

is only a wide-sense solution of $T(\omega)x = 0$.

It is of interest to consider the following questions: (1) If a random equation has at least one wide-sense solution, does this imply the existence of a random solution of the same equation? (2) Is a unique wide-sense solution also a random solution? (3) if $\mathfrak{X}$ and $\mathfrak{Y}$ are separable Banach spaces, then is the answer to (2) affirmative? The following example shows that the answer to all of the above questions is negative. Let $\Omega = R$ and let $\mathfrak{A}$ be the σ-algebra of all at most denumerable sets of real numbers. Let $\mathfrak{X} = \mathfrak{Y} = R$, and let $\mathfrak{B}_{\mathfrak{X}} = \mathfrak{B}_{\mathfrak{Y}} = \mathfrak{B}$ (that is, the σ-algebra of Borel subsets of R). Let

$$T(\omega)x = \begin{cases} 0 & \text{for every} \quad \omega = x \\ 1 & \text{for every} \quad \omega \neq x. \end{cases}$$

Then, the real-valued function $\xi(\omega)$ defined by $\xi(\omega) = \omega$ for every $\omega \in \Omega$ is the unique wide-sense solution of the equation $T(\omega)x = 0$. However, $\xi(\omega)$ is not a random solution since $\{\omega : \xi(\omega) \leqslant 0\} \notin \mathfrak{A}$.

In view of the above definitions we can introduce two types of solution sets, namely the set of wide-sense solutions and the set of random solutions. Throughout this book when we refer to a solution of a random equation we mean a random solution; hence the *solution set* S of the random equation $T(\omega)x(\omega) = y(\omega)$ is defined as follows:

$$S = \{x(\omega) : x(\omega) \in \mathfrak{X}, \qquad \mu(\{\omega : T(\omega)x(\omega) = y(\omega)\}) = 1\} \tag{3.29}$$

A random solution $x(\omega)$ will be called *unique* if it is the only random variable for which (3.28) holds. If $x_1(\omega)$ and $x_2(\omega)$ are two solutions of a random equation, then we say there exists a unique solution if $x_1(\omega)$ and $x_2(\omega)$ are equivalent.

3.3 The Solution of Random Equations

A. Introduction

The numerous methods (general or specific, analytic or numerical) that have been developed for solving deterministic operator equations have, in the main, been concerned with establishing the existence and uniqueness of solutions. Methods for solving random equations must not only establish existence and uniqueness, they must establish the measurability of the solutions. This is the essential difference between methods for solving deterministic equations and random equations. Classical (that is, deterministic) existence and uniqueness theorems serve as "models" for similar theorems for random operator equations; and the probabilistic version of a classical theorem is often obtained by using the classical result itself together with appropriate measure-theoretic hypotheses.

In this section we consider some methods for solving random equations, in particular random operator equations. In Sect. 3.3B we consider some random fixed point theorems. These theorems, which are probabilistic analogues of some classical fixed point theorems, play a main role in the developing theory of random operator equations. Section 3.3C is concerned with the use of inversion theorems to obtain solutions of random equations. In Sect. 3.3D we consider a perturbation method which is applicable to a large number of random equations. Finally, in Sect. 3.3E, we refer to several other methods which have been used to solve random equations.

B. Random fixed point theorems

Let $\mathfrak{X}$ be a Banach space, and let T be a linear or nonlinear operator mapping $\mathfrak{X}$ into itself.

***Definition* 3.3.** Any element $x \in \mathfrak{X}$ such that

$$Tx = x \tag{3.30}$$

is said to be a *fixed point* of T.

Let $\mathfrak{X} = R$, and consider the operator $T[x] = x^2$. Then it is clear that $x = 0$ and $x = 1$ are fixed points of T. Now, let $\mathfrak{X} = C[0, 1]$ and consider the operator

$$T[x] = x(0) + \int_0^t x(\xi)\, d\xi.$$

In this case any function $x(t)$ of the form $x(t) = ke^t$, $t \in [0, 1]$, k a real constant, is a fixed point of T.

Consider a concrete operator equation $Tx = y$ in a given Banach space $\mathfrak{X}$. Then it is clear that finding a fixed point of T is equivalent to obtaining a solution of the operator equation. Hence fixed point theorems constitute a general class of existence theorems for linear and nonlinear operator equations. In general, fixed point theorems fall into two classes: (1) *topological fixed point theorems*, and (2) *algebraic*, or *constructive*, *fixed point theorems*. Theorems of topological type are strictly existence theorems; that is, they establish conditions under which a fixed point exists but they do not provide a method for finding a fixed point (or solution) of the operator equation. On the other hand, theorems of algebraic type give a method for finding the fixed point which can be called an *iteration* or *sucessive approximation* procedure.

For detailed treatments of fixed point theorems and their applications we refer to the books of Anselone [3], Rall [52], and Saaty [54].

In this section we consider some probabilistic analogue of two well-known fixed point theorems. These fixed point theorems for random operators, called *random fixed point theorems*, will be used in Chaps. 4, 6, and 7 to establish the existence and measurability, and in some cases the uniqueness, of solutions of random integral equations.

The prototype of most algebraic fixed point theorems is the *contraction mapping theorem* or *principle* due to Banach [5]. This theorem can be considered as a result of a geometric interpretation and abstract formulation of the classical method of successive approximations due to Picard.

THEOREM 3.1. *If T is a contraction operator with $\|T\| \leqslant k < 1$ mapping a complete metric space $(\mathfrak{X}, d)$ into itself, then T has a unique fixed point (i.e., the equation $Tx = x$ has one and only one solution). The fixed point, say ξ, can be determined by the successive approximations*

$$x_{n+1} = Tx_n, \qquad n = 0, 1, \ldots, \tag{3.31}$$

(where x_0 is any arbitary element of $\mathfrak{X}$) which converge to ξ. The error estimate is given by

$$d(\xi, x_n) = \frac{k^n}{1-k}\, d(x_1, x_0). \tag{3.32}$$

When $\mathfrak{X}$ is a Banach space the contraction mapping principle can be stated as follows:

THEOREM 3.2. *If T is a contraction operator with $\|T\| \leqslant k < 1$ mapping a closed region E of a Banach space $\mathfrak{X}$ into itself, then, in E, T has a unique fixed point ξ, which is the limit of the successive approximations $x_{n+1} = Tx_n$. The error estimate is given by*

$$\|\xi - x_n\| \leqslant k^n(1-k)^{-1}\, \|x_1 - x_0\|, \qquad x_0 \in E.$$

Consider the mapping T of R_n into itself given by the system of linear equations

$$y_i = \sum_{i=1}^{n} a_{ij} x_j + b_i, \qquad i = 1, 2, \ldots, n. \tag{3.33}$$

Hence, for arbitrary $x = (x_1, x_2, \ldots, x_n)$, (3.33) is a matrix operator equation of the form $y = Tx$, where $Tx = Ax + b$. $A = (a_{ij})$ and $b = (b_1, b_2, \ldots, b_n)$. We wish to obtain a sufficient condition that T be a contraction operator on R_n. For any two points ξ_1, ξ_2 in R_n we have

$$\begin{aligned} d(T\xi_1, T\xi_2) &= \|T\xi_1 - T\xi_2\| = \|A\xi_1 - A\xi_2\| \\ &= \|A(\xi_1 - \xi_2)\| \leqslant \|A\|\, d(\xi_1, \xi_2). \end{aligned}$$

Hence, in order that T be a contraction we must have $\|A\| < 1$. Since in R_n we can define several metrics, the conditions under which A is a contraction depends on the choice of the metric. If we take

$$d(x,y) = \max_{1 \leqslant i \leqslant n} |x_i - y_i|,$$

then A is a contraction if $\sum_{j=1}^{n} |a_{ij}| < 1$, $i = 1,2,\ldots,n$. If

$$d(x,y) = \sum_{i=1}^{n} |x_i - y_i|,$$

then the contraction condition is $\sum_{i=1}^{n} |a_{ij}| < 1, j = 1,2,\ldots,n$. Finally, if

$$d(x,y) = \left(\sum_{i=1}^{n} (x_i - y_i)^2 \right)^{1/2}.$$

then A is a contraction if $\sum_{i=1}^{n} \sum_{j=1}^{n} a_{ij}^2 < 1$.

A useful generalization of the contraction mapping principle is the following result due to Kolmogorov and Fomin [39, pp. 70–71] (cf. also, Chu and Diaz [18]).

THEOREM 3.3. *If T is an operator mapping a complete metric space $\mathfrak{X}$ into itself, and if T^n is a contraction for some n (n is a positive integer), then T has a unique fixed point.*

The study of random fixed point theorems was initiated by Špaček [59] and Hanš [30]. Because of the wide applicability of Banach's contraction mapping theorem in the study of deterministic operator equations, Špaček and Hanš directed their attention to probabilistic versions of Banach's theorem, and applied their results to random linear Fredholm integral equations. The two theorems which we now state and prove are due to Hanš, who was the first to carry out a systematic investigation of random fixed point theorems [31] (cf. also Hanš [32]).

THEOREM 3.4. *Let $T(\omega)$ be a continuous random operator on $\Omega \times \mathfrak{X}$ to $\mathfrak{X}$, where $\mathfrak{X}$ is a separable Banach space. For every $\omega \in \Omega$ and $x \in \mathfrak{X}$, put*

$$T^1(\omega)x = T(\omega)x$$
$$T^{n+1}(\omega)x = T(\omega)[T^n(\omega)x], \qquad n = 1,2,\ldots.$$

If $T(\omega)$ satisfies the condition

$$\mu\left(\bigcup_{m=1}^{\infty} \bigcup_{n=1}^{\infty} \bigcap_{x_1 \in \mathfrak{X}} \bigcap_{x_2 \in \mathfrak{X}} \{\omega : \|T^n(\omega)x_1 - T^n(\omega)x_2\|\} \leqslant \left(1 - \frac{1}{m}\right) \|x_1 - x_2\| \right) = 1, \tag{3.34}$$

then there exists an $\mathfrak{X}$-valued random variable $\xi(\omega)$ which is the unique fixed point of $T(\omega)$; that is,

$$T(\omega)\,\xi(\omega) = \xi(\omega) \qquad \text{a.s.,} \tag{3.35}$$

and if $\varphi(\omega)$ is another $\mathfrak{X}$-valued random variable which satisfies (3.35) *then $\xi(\omega)$ and $\varphi(\omega)$ are equivalent.*

Proof. Let E denote the subset of Ω defined by the braces of (3.34) for which $T(\omega)$ is continuous. Clearly $E \in \mathfrak{A}$ and, by hypothesis, $\mu(E) = 1$. Let the mapping $\xi(\omega): \Omega \to \mathfrak{X}$ be defined as follows: For every $\omega \in E$, $\xi(\omega)$ equals the unique fixed point of $T(\omega)$, and for every $\omega \in \Omega - E$, put $\xi(\omega) = \theta$. Then (3.35) holds.

To establish the measurability of the fixed point $\xi(\omega)$ we proceed as follows. Let $x_0(\omega)$ be an arbitrary $\mathfrak{X}$-valued random variable. Put $x_1(\omega) = T(\omega)\,x_0(\omega)$. It follows from Theorem 2.14 that $x_1(\omega)$ is an $\mathfrak{X}$-valued random variable. Put $x_n(\omega) = T(\omega)\,x_{n-1}(\omega)$, $n = 1, 2, \ldots$. Repeated application of Theorem 2.14 establishes the fact that $\{x_n(\omega)\}$ is a sequence of $\mathfrak{X}$-valued random variables; hence it follows from Theorem 1.6 that $x_n(\omega)$ converges almost surely to $\xi(\omega)$, and $\xi(\omega)$ is an $\mathfrak{X}$-valued random variable.

The uniqueness of the fixed point follows from the uniqueness of $\xi(\omega)$ for every $\omega \in E$.

We now utilize Theorem 3.4 to establish the following *random contraction mapping theorem.*

THEOREM 3.5. *Let $T(\omega)$ be a continuous random operator on $\Omega \times \mathfrak{X}$ to $\mathfrak{X}$, where $\mathfrak{X}$ is a separable Banach space, and let $k(\omega)$ be a real-valued random variable such that $k(\omega) < 1$ almost surely and*

$$\|T(\omega)\,x_1 - T(\omega)\,x_2\| \leqslant k(\omega)\,\|x_1 - x_2\|$$

for every two elements $x_1, x_2 \in \mathfrak{X}$. Then there exists an $\mathfrak{X}$-valued random variable $\xi(\omega)$ which is the unique fixed point of $T(\omega)$.

Proof. Let

$$E = \{\omega : k(\omega) < 1\}$$
$$F = \{\omega : T(\omega)\,x \text{ is continuous in } x\}$$
$$G_{x_1, x_2} = \{\omega : \|T(\omega)\,x_1 - T(\omega)\,x_2\| \leqslant k(\omega)\,\|x_1 - x_2\|\}.$$

Since $\mathfrak{X}$ is separable, the intersections in the expression

$$\bigcap_{x_1 \in \mathfrak{X}} \bigcap_{x_2 \in \mathfrak{X}} \{G_{x_1, x_2} \cap E \cap F\}$$

can be replaced by intersections over a countable dense set of $\mathfrak{X}$. Therefore condition (3.34) of Theorem 3.4 is satisfied with $n = 1$.

Random contraction mapping theorems are of fundamental importance in probabilistic functional analysis in that they can be used to establish the existence, uniqueness, and measurability of solutions of random operator equations. In this book these theorems are applied to random integral equations. In addition to their applicability in the study of random operator equations, random contraction mapping theorems can be utilized in the study of stochastic approximation procedures (cf. Gardner [23], Hanš [31]); and have been used by Oza and Jury [49] to obtain an algorithm for the identification of a random linear discrete-time system described by a random difference equation. We refer to Grenander [27, p. 164] and Sehgal [58] for other random contraction mapping theorems. For some continuous analogues of random contraction mapping theorems and their applications, we refer to Driml and Hanš [20] and Hanš and Špaček [33].

Let $T(\omega)$ be a random contraction operator, and let $\xi(\omega)$ be its unique fixed point. $\xi(\omega)$ is an $\mathfrak{X}$-valued random variable; hence if its Bochner integral exists we can define the *expectation of the random fixed point*; that is,

$$\bar{\xi} = \mathscr{E}\{\xi(\omega)\} = (\mathrm{B})\int_\Omega \xi(\omega)\, d\mu. \tag{3.36}$$

Similarly, if $T(\omega)$ is a uniform random contraction operator, and if the Bochner integral of $T(\omega)x$ exists for all $x \in \mathfrak{X}$, then we can define the *expectation* of $T(\omega)$:

$$\bar{T}x = \mathscr{E}\{T(\omega)\,x\} = (\mathrm{B})\int_\Omega T(\omega)\,x\, d\mu. \tag{3.37}$$

Clearly $\bar{T}$ is a contraction operator; hence it has a unique fixed point, say ψ. ψ is called the *fixed point of the expectation of the random contraction operator* $T(\omega)$. A question of great interest is the following: Is $\bar{\xi} = \psi$? Consider the random operator $T(\omega)$ defined for every $\omega \in \Omega$ and $x \in \Omega$ by the formula

$$T(\omega)\,[x] = ax + z(\omega), \tag{3.38}$$

where $a < 1$ and $z(\omega)$ is an $\mathfrak{X}$-valued random variable whose expectation exists. In this case it is clear that $\bar{\xi} = \psi$; however, in general $\bar{\xi} \neq \psi$.

We now consider a topological fixed point theorem, due to Schauder [57], and a probabilistic version of this theorem.

THEOREM 3.6. *Let T be a continuous operator which transforms a compact, closed, convex subset E of a Banach space $\mathfrak{X}$ into E. Then there exists at least one element $\xi \in E$ such that $T\xi = \xi$.*

The following probabilistic version of Schauder's fixed point theorem is due to Mukherjea [48] (cf. also Mukherjea [46, 47]).

THEOREM 3.7. *Let $(\Omega, \mathfrak{A}, \mu)$ be an atomic probability measure space, and let E be a compact (or closed and bounded) convex subset of a separable Banach space $\mathfrak{X}$. Let $T(\omega)$ be a random compact operator mapping $\Omega \times E$ into E. Then, there exists an E-valued random variable $\xi(\omega)$ such that $T(\omega)\xi(\omega) = \xi(\omega)$.*

We need the following lemma, which we state without proof.

LEMMA. *If $x(\omega)$ is an $\mathfrak{X}$-valued random variable defined on an atomic probability measure space, then $x(\omega)$ is constant almost surely on every atom.*

Proof of Theorem. Let B_n be the atoms of $\mathfrak{A}$. If $\{x_i\}_{i=1}^{\infty}$ is dense in E, then it follows from the above lemma that $T(\omega)x_i$, for every i, is constant almost surely on every B_n; so that for every n we can find a $C_n \subset B_n$, with $\mu(B_n - C_n) = 0$ and such that

$$T(\omega)\,x_i = \sum_{n=1}^{\infty} \chi_{C_n}(\omega)\,T(\omega_n)\,x_i$$

for every i, where $\omega \in \bigcup_{n=1}^{\infty} C_n$ and $\omega_n \in C_n$. Let ω' and ω'' be any two points in C_n, and let x be any element in E. We claim that $T(\omega_1)x = T(\omega_2)x$. If this relation did not obtain, then $\|T(\omega_1)x - T(\omega_2)x\| > k > 0$. Since $T(\omega)$ is continuous, we can find x_i such that $\|T(\omega_j)x_i - T(\omega_j)x_i\| < k/2$, $j = 1,2$, which together with the fact that $T(\omega_1)x_i = T(\omega_2)x_i$ yields a contradiction. Hence

$$T(\omega)\,x = \sum_{n=1}^{\infty} \chi_{C_n}(\omega)\,T(\omega_n)\,x \qquad \text{a.s.}$$

for every $x \in E$. We can now apply Schauder's theorem to each of the operators $T(\omega_n)$; hence there exists at least one point φ_n for each n such that $T(\omega_n)\varphi_n = \varphi_n$. Put

$$\xi_n(\omega) = \begin{cases} \varphi_n & \text{for} \quad \omega \in C_n \\ 0 & \text{for} \quad \omega \notin C_n. \end{cases}$$

Then $T(\omega)\xi(\omega) = \xi(\omega)$ almost surely.

Theorem 3.6 is used in Sect. 6.5 to establish the existence and measurability of solutions of random nonlinear integral equations of Uryson type.

It is of interest to state a topological fixed point theorem due to Krasnosel'skiĭ [41] which yields the theorems of Banach and Schauder as special cases.

THEOREM 3.8. *Let E be a closed, bounded, convex subset of a Banach space $\mathfrak{X}$. Let S and T be operators on $\mathfrak{X}$ to itself such that* (i) *$Sx + Ty \in E$ for $x, y \in E$,* (ii) *S is a contraction operator with contraction constant $k \in (0,1)$,* (iii) *T is compact. Then there exists at least one element $\xi \in E$ such that $S\xi + T\xi = \xi$.*

A random fixed point theorem of Krasnosel'skiĭ type would be very useful in the theory of random operator equations.†

C. Inversion theorems

We first consider random equations of the form

$$Tx = y(t, \omega) \tag{3.39}$$

$$(T - \lambda I)\, x = y(t, \omega); \tag{3.40}$$

that is, equations of the form (3.5) and (3.8) when the input is a random function. Random equations of the form (3.39) and (3.40) have been studied by a large number of workers; and clearly they are among the simplest types of random equations since the operators T and $T - \lambda I$ are deterministic. Hence, inversion theorems for deterministic operators can be used to obtain conditions for the existence of a random solution.

We first consider Eq. (3.39). If T is an endomorphism of a separable Banach space $\mathfrak{X}$, then the following result is an immediate consequence of the fundamental theorem giving necessary and sufficient conditions for the existence of the inverse of an endomorphism of $\mathfrak{X}$ (cf. Rall [52, Theorem 9.1]).

THEOREM 3.9. *Consider the equation $Tx = y(t, \omega)$, where $T \in \mathfrak{L}(\mathfrak{X})$ and $y(t, \omega)$ is an $\mathfrak{X}$-valued random function. Then the random solution $x(t, \omega)$ exists if and only if there is an operator $S \in \mathfrak{L}(\mathfrak{X})$ such that* (i) *S^{-1} exists and* (ii) $\|I - ST\| < 1$. *We have*

$$x(t, \omega) = T^{-1} y(t, \omega)$$

$$= \sum_{n=0}^{\infty} (I - ST)^n S y(t, \omega), \tag{3.41}$$

where the sum converges in $\mathfrak{L}(\mathfrak{X})$. The solution $x(t, \omega)$ is an $\mathfrak{X}$-valued random function.

Similarly, the existence of a random solution of Eq. (3.40) follows from a basic result on the existence of $(T - \lambda I)^{-1}$ (cf. Sect. 2.4B, and Taylor [60, pp. 260–261].

† Prakasa Rao [51a] has recently obtained a probabilistic analogue of Krasnosel'skiĭ's theorem.

THEOREM 3.10. *Consider the equation $(T-\lambda I)x = y(t,\omega)$, where $T \in \mathfrak{L}(\mathfrak{X})$ and $y(t,\omega)$ is an $\mathfrak{X}$-valued random function. Then the random solution $x(t,\omega)$ exists if $|\lambda| > \|T\|$. In this case $\lambda \in \rho(T)$, and*

$$\begin{aligned} x(t,\omega) &= (T-\lambda I)^{-1}\, y(t,\omega) \\ &= R(\lambda; T)\, y(t,\omega) \\ &= -\sum_{n=1}^{\infty} \lambda^{-n} T^{n-1}\, y(t,\omega), \end{aligned} \tag{3.42}$$

where the series converges in $\mathfrak{L}(\mathfrak{X})$. The solution $x(t,\omega)$ is an $\mathfrak{X}$-valued random function.

Equations of the form

$$T(\omega)\, x = y(t), \tag{3.43}$$

$$(T(\omega) - \lambda I)\, x = y(t), \tag{3.44}$$

$$T(\omega)\, x = y(t,\omega), \tag{3.45}$$

and

$$(T(\omega) - \lambda I)\, x = y(t,\omega) \tag{3.46}$$

involve random operators; hence the inversion theorems for deterministic operators are not applicable, and we require probabilistic versions of the classical theorems in order to solve random operator equations of the above types. At the present time there are no known results which give necessary and sufficient conditions for the existence of the inverse of a random operator $T(\omega) \in \mathfrak{L}(\mathfrak{X})$; hence we will not consider equations of the form (3.43) and (3.45). However, results on the inversion of random operators of the form $T(\omega) - \lambda I$ are known (cf. Sect. 2.4B), and these results can be used to establish the existence of random solutions of Eqs. (3.44) and (3.46).

We first give an inversion theorem due to Hanš [32], the proof of which utilizes the random contraction mapping principle.

THEOREM 3.11. *Let $T(\omega)$ be a random contraction operator on a separable Banach space $\mathfrak{X}$, and let $k(\omega)$ be a real-valued random variable such that $k(\omega) < 1$ almost surely. Then for every real number $\lambda \neq 0$ such that $k(\omega) < |\lambda|$ almost surely there exists a random operator $S(\omega)$ which is the inverse of $(T(\omega) - \lambda I)$.*

Proof. Since $\lambda \neq 0$, $T(\omega) - \lambda I$ is invertible whenever the random operator $(1/\lambda)T(\omega) - I$ is invertible, and vice versa. However, for every $\xi \in \mathfrak{X}$, the random operator $T_\xi(\omega)$ defined, for every $\omega \in \Omega$ and $x \in \mathfrak{X}$, by

$$T_\xi(\omega)\, x = (1/\lambda)\, T(\omega)\, x - \xi$$

is a random contraction operator. Therefore, by Theorem 3.5, there exists a unique random fixed point $x_\xi(\omega)$ satisfying the relation

$$x_\xi(\omega) = (1/\lambda)\, T(\omega)\, x_\xi(\omega) - \xi$$

almost surely. However the above statement is equivalent to the invertibility of the random operator $(1/\lambda)T(\omega) - I$, and therefore the invertibility of $(T(\omega) - \lambda I)$.

The applicability of the above theorem to equations of the form (3.44) and (3.46) is clear.

We now state another result on the solution of random equations of the form (3.44) and (3.46). The main content of this theorem is due to Hanš [32], the proof of which follows easily from the classical result.

THEOREM 3.12. *Consider the equation* $(T(\omega) - \lambda I)x = y(t, \omega)$ *where* $T(\omega) \in \mathfrak{L}(\mathfrak{X})$ *and* $y(t, \omega)$ *is an* $\mathfrak{X}$*-valued random function. Then for every real number* $\lambda \neq 0$ *such that*

$$\mu\left(\bigcup_{n=1}^{\infty} \{\omega : \|T^n(\omega)\| < |\lambda|^n\}\right) = 1$$

there exists a random endomorphism $S(\omega)$ *which is the inverse of* $(T(\omega) - \lambda I)$; *and*

$$\mu\left(\bigcap_{x \in \mathfrak{X}} \{\omega : S(\omega)x = -(1/\lambda) \sum_{n=0}^{\infty} \lambda^{-n} T^n(\omega)\, x\}\right) = 1.$$

Therefore, the random solution $x(t, \omega)$ *is of the form* $x(t, \omega) = S(\omega)y(t, \omega)$, *and* $x(t, \omega)$ *is an* $\mathfrak{X}$*-valued random function.*

Finally, we state a result which is an immediate consequence of Theorem 2.26.

THEOREM 3.13. *Consider the equation* $(T(\omega) - \lambda I)x = y(t, \omega)$ *where* $T(\omega) \in \mathfrak{L}(\mathfrak{X})$ *and* $y(t, \omega)$ *is an* $\mathfrak{X}$*-valued random function. Then, for every* $\lambda \neq 0$, *the resolvent operator* $R(\lambda; T(\omega))$ *exists for every* $\omega \in \Omega(\lambda) = \{\omega : |\lambda| > \|T(\omega)\|\}$ *and admits the representation*

$$R(\lambda; T(\omega)) = -\sum_{n=1}^{\infty} \lambda^{-n} T^{n-1}(\omega). \tag{3.47}$$

Finally, the solution $x(t, \omega)$ *exists for every* $\omega \in \Omega_0(\lambda)$ *and is of the form*

$$x(t, \omega) = R(\lambda; T(\omega))\, y(t, \omega). \tag{3.48}$$

The random solution $x(t, \omega)$ *is an* $\mathfrak{X}$*-valued random function, and for every fixed* t *we have* $x(t, \omega) \in \Omega_0(\lambda) \cap \mathfrak{A}$.

D. A perturbation method and semigroups of operators†

A large number of random operator equations in applied mathematics are of the form

$$dx/dt = L(\omega)x$$
$$= (A + B(\omega))x, \tag{3.49}$$

where $x(t,\omega)$ is for every fixed t a random variable with values in a separable Banach space $\mathfrak{X}$. The operator A is assumed to be a (deterministic) closed linear operator on $\mathfrak{X}$ which is the infinitesimal generator of strongly continuous semigroup of contraction operators‡ of class (C_0), say

$$S(t) = e^{At}, \qquad t \geqslant 0, \tag{3.50}$$

$B(\omega)$ is a random endomorphism of $\mathfrak{X}$ with $\|B(\omega)\| \leqslant M$, and

$$\mathfrak{D}(A + B(\omega)) = \mathfrak{D}(A) \qquad \text{a.s.}$$

Phillips [51] has proved the following result: *Let A be the infinitesimal generator of a strongly continuous semigroup of operators $\{S(t), t \geqslant 0\}$ on $\mathfrak{X}$. Let $B \in \mathfrak{L}(\mathfrak{X})$. Then there is a unique one-parameter semigroup of bounded linear operators $\{T(t), t \geqslant 0\}$ on $\mathfrak{X}$, strongly continuous on $[0,\infty]$ such that $T(0) = I$ and for $x \in \mathfrak{D}(A)$, $T(t)x$ is strongly continuous differentiable and*

$$dT(t)x/dt = (A + B)T(t)x. \tag{3.51}$$

The solution $T(t)$ admits the representation

$$T(t) = \sum_{n=0}^{\infty} T_n(t), \tag{3.52}$$

where

$$T_0(t) = S(t) \quad \text{and} \quad T_n(t) = \int_0^t S(t-\tau)BS_{n-1}(\tau)\,d\tau.$$

It follows from the above result that for almost all $\omega \in \Omega$ the random closed linear operator $L(\omega) = A + B(\omega)$ is the infinitesimal generator of a strongly continuous *semigroup of random operators* of class (C_0), say $T(t,\omega)$, on $\mathfrak{X}$; and $T(t,\omega)$ is given by the following perturbation formula:

$$\begin{aligned} T(t,\omega) &= \exp\{(A + B(\omega))t\} \\ &= \exp\{At\} + \int_0^t \exp\{A(t-\tau)\}\,B(\omega)\exp\{A\tau_1\}\,d\tau \\ &\quad + \int_0^t \int_0^{\tau_1} \exp\{A(t-\tau_1)\}\,B(\omega)\exp\{A(t-\tau_2)\} \\ &\quad \times B(\omega)\exp\{A\tau_2\}\,d\tau_1\,d\tau_2 + \cdots. \end{aligned} \tag{3.53}$$

† See Frisch [22].
‡ We refer to Sect. 7.3D for the semigroup terminology used in this section.

Since $S(t) = e^{At}$ is, for every fixed t, a contraction operator, we have $\|S(t)\| = \|e^{At}\| \leqslant 1$; hence from (3.53) we obtain the following estimate:

$$\|T(t,\omega)\| = \|\exp\{(A + B(\omega))\,t\}\| \leqslant 1 + Mt + \frac{M^2 t^2}{2!} + \cdots$$
$$= e^{Mt}. \tag{3.54}$$

Hence the perturbation expansion (3.53) converges uniformly in any bounded interval $[0, a]$; and it can be shown that the series (3.53) converges to a random operator $T(t,\omega)$ which is the *solution operator* of Eq. (3.49). Hence

$$x(t,\omega) = T(t,\omega)\,[x(0,\omega)]. \tag{3.55}$$

E. Some other methods for solving random equations†

In this subsection we discuss briefly three methods which have been proposed for solving random equations, and have been applied to certain types of random equations.

1. ***Direct integral method.*** Lions [42] (cf. also Frisch [22, p. 186]) has proposed a method for solving random equations in Hilbert spaces that is based on the notion of the direct integral of Hilbert spaces. Consider, for example, a random equation of the form $T(\omega)x = y$. The general idea is to associate with every $\omega \in \Omega$ a separable Hilbert space $H(\omega)$. We denote by $\mathfrak{H}$ the *direct* or *Hilbert integral* of the $H(\omega)$ (cf. Dixmier [19]); that is,

$$\mathfrak{H} = \int^{\oplus} H(\omega)\, d\mu(\omega). \tag{3.56}$$

When all of the Hilbert spaces $H(\omega)$ are the same, it is known that $\mathfrak{H} = B_2(\Omega, H(\omega))$. Suppose $H(\omega)$ is, for every ω, the Hilbert space $L_2[a,b]$ of functions $x(t)$. In this case $\mathfrak{H}$ can be thought of as the space of all second-order random functions $x(t,\omega)$. Lions has shown that random equations which are formulated in $H(\omega)$, for every $\omega \in \Omega$, can be reformulated as deterministic equations in $\mathfrak{H}$. The advantage of this reformulation is that it simplifies the problem of establishing the measurability of solutions.

2. ***The method of random Mikusiński operators.*** Ullrich [61] has developed a method for solving random equations which is a probabilistic analogue of the Mikusiński operational calculus (cf. Mikusiński [45]). This method has been applied to random ordinary and partial differential equations [26]; however, it should have wider applicability in the study of random equations.

3. ***Approximation methods.*** In recent years an increasing number of papers have appeared which are devoted to the development of methods for

† We refer to Nashed and Salehi [48a] who have studied *best approximate solutions* of random linear operator equations.

obtaining approximate solutions of linear and nonlinear operator equations (cf. Kantorovich and Krylov [36], Rall [52], Saaty [54]). In particular, the literature on approximate solutions of differential and integral equations is very extensive. In the development of the theory of random operator equations it is clear that approximation methods will play an important role. Approximation methods are required because of the difficulties encountered when one attempts to solve a random operator equation; and they are required for computational solutions of random operator equations. At the present time only a few rather general approximation methods have been developed for random operator equations, and most of these are concerned with finding statistical properties of the solution. These methods will be discussed in Sect. 3.4C.

In this subsection we consider two well-known approximation methods, namely Galerkin's method and Newton's method, which may prove useful in obtaining approximate solutions of random operator equations.

Galerkin's method can be described as follows: Let $\mathfrak{X}$ be a Banach space with a Schauder basis $\{e_n\}$, and let $\mathfrak{X}_1 \subset \mathfrak{X}_2 \subset \cdots$ be a sequence of finite-dimensional subspaces of increasing dimension such that $\bigcup_{n=1}^{\infty} \mathfrak{X}_n$ is dense in $\mathfrak{X}$. Let $P_1, P_2, \ldots$ be a sequence of linear projection operators with uniformly bounded norms. Let $P_n \mathfrak{X} = \mathfrak{X}_n$. Consider the operator equation

$$Tx = x, \tag{3.57}$$

where T is a compact operator and assume that Eq. (3.57) has an isolated solution ξ. Since $\mathfrak{X}$ has a basis, every element $x \in \mathfrak{X}$ can be represented as a series $x = \sum_{i=1}^{\infty} \eta_i(x) e_i$, where the $\eta_i(x)$ are (continuous) coefficient functionals. It can be shown that the projection operator P_n defined by

$$P_n x = \sum_{i=1}^{n} \eta_i(x) e_i \tag{3.58}$$

have uniformly bounded norms. Hence the subspaces $\mathfrak{X}_n$ are generated by the first n basis elements $(e_1, e_2, \ldots, e_n)$. The solutions x_n of the equations

$$P_n Tx = x \tag{3.59}$$

are called the *Galerkin approximate solutions* of Eq. (3.57); and we have $x_n \to \xi$.

Galerkin's method has been used extensively to obtain approximate solutions of differential and integral equations; however in applications to random operator equations a suitable choice of subspaces is difficult. A modification of Galerkin's method, called the *method of moments* (cf. Vorobyev [62]), is a formal method by which the equation to be solved determines the sequences of subspaces to be used. The iteration process given by the method of moments differs from the Galerkin process in that the first step is not selecting a basis in $\mathfrak{X}$, but selecting a single element, say $\varphi_0 \in \mathfrak{X}$. An infinite sequence of elements $\{\varphi_n\}$ is generated from φ_0 by the recurrence relation $\varphi_n = T\varphi_{n-1}$, $n = 1, 2, \ldots$. If the φ_n are linearly independent, then the φ_n, like the e_n in Galerkin's method, can be used to define an increasing sequence of subspaces of $\mathfrak{X}$.

McCoy [44] has used the method of moments to obtain approximate solutions of some random differential equations. In particular, he showed that, under certain conditions, the iteration process obtained involves deterministic systems of linear algebraic equations.

Newton's method is concerned with finding a solution $x = \xi$, say, of the equation

$$Tx = \theta, \tag{3.60}$$

where T is an operator on a Banach space. Let x_0 denote an appropriately selected initial point. If $(T'x_0)^{-1}$ exists, where T' is the derivative† of T, put

$$x_1 = x_0 - (T'x_0)^{-1}Tx_0.$$

If $(T'x_n)^{-1}$ exists for $n = 0, 1, \ldots$, the *Newton sequence* $\{x_n\}$ is defined by

$$x_{n+1} = x_n - (T'x_n)^{-1}Tx_n. \tag{3.61}$$

The iterative process for construction of the sequence defined by (3.61) is called *Newton's method* for solving Eq. (3.60).

Bharucha-Reid [9] has considered a probabilistic analogue of the Newton sequence defined by

$$x_{n+1}(\omega) = x_n(\omega)(T'(\omega)x_n)^{-1}T(\omega)x_n(\omega), \qquad n = 0, 1, \ldots, \tag{3.62}$$

where the derivative of $T(\omega)$ is defined in the mean-square sense, and $x_0(\omega)$ is a suitable chosen $\mathfrak{X}$-valued random variable. Some results were obtained on the measurability and convergence of the iterative process $\{x_n(\omega)\}$; however much more research needs to be done on approximation methods of Newton type for solving random operator equations.

3.4 Some Measure-Theoretic and Statistical Problems Associated with Random Equations

A. *Introduction*

The main objectives of the theory of random equations might be stated as follows: (1) to establish the existence, uniqueness and measurability of solutions of various types and classes of random equations; (2) to determine

† If an endomorphism L of $\mathfrak{X}$ exists such that

$$\lim_{\|\Delta x\| \to 0} \|T(x_0 + \Delta x) - Tx_0 - L\Delta x\| = 0,$$

then T is said to be differentiable at x_0, and $L = T'x_0$ is called the first derivative of T at x_0.

(a) the probability distributions (or laws) of random solutions, and (b) the properties of functionals of random solutions; (3) the determination of various statistical properties of a random solution, such as its expectation, variance, and higher moments.

In Sect. 3.3B we presented some results on the existence, uniqueness and measurability of solutions of certain general types of random equations; and the results on probability measures on Banach spaces surveyed in Sect. 1.5 are of fundamental importance in the study of probability distributions of random solutions. In the second subsection of this section we consider some measure-theoretic problems associated with solutions of random equations; in particular, we consider some questions concerning the transformation of probability measures on function spaces.

Although results on the existence, uniqueness, and measurability of solutions are of theoretical interest, and are required for the systematic development of the theory of random equations, we are obliged to recognize that in most applied problems it is not the solution that is of great importance but the statistical properties of the solution. In particular, it is of interest to establish the relationship between the statistical properties of the solution and the statistical properties of the random quantities (for example, initial or boundary data, coefficients, kernels, forcing functions) introduced into the equation. In the third subsection of this section we consider some methods for studying the statistical properties of the solutions of random equations.

B. Measure-theoretic problems

Consider a random equation of the form

$$Lx(t) = y(t, \omega); \tag{3.63}$$

that is, an operator equation with input a known random function $y(t, \omega)$. If L is invertible, then the solution $x(t, \omega)$ is the random function obtained by the transformation of $y(t, \omega)$ by L^{-1}:

$$x(t, \omega) = L^{-1}\, y(t, \omega). \tag{3.64}$$

The random function $x(t, \omega)$ is frequently referred to as the *solution* or *output process*.

Two measure-theoretic problems associated with Eq. (3.63), and its solution (3.64), can be stated as follows: Let ν_1 denote the probability measure associated with the known $\mathfrak{X}$-valued random function $y(t, \omega)$, and let L be a known deterministic operator on $\mathfrak{X}$ to itself. (1) Determine the probability measure ν_2 associated with the solution process $x(t, \omega)$. (2) Determine conditions such that ν_2 is absolutely continuous with respect to ν_1. The first problem is, of course, a fundamental one; namely, given $y(t, \omega)$ and L, to determine the probability measure ν_2 generated by the solution process $x(t,\omega)$ on the minimal σ-algebra containing all cylinder sets of all functions defined on the interval $[a,b]$, say, with values in the Banach space $\mathfrak{X}$. The absolute continuity of probability measures was considered in Sect. 1.5E. As pointed out in that section, if $\nu_2 \ll \nu_1$, then the solution process $x(t,\omega)$ will satisfy almost surely all of those properties which the input process $y(t,\omega)$ satisfies almost surely. Hence the second problem is not only of theoretical interest, but is important in many applications of random equations to concrete problems.

We now survey some of the contributions to the solution of the first problem for various types of random equations. Most of the results that have been obtained are for concrete random equations; hence they are not general measure-theoretic results, but give the distribution function (or density function) of the random solution, or determine the class of random functions to which the solution process belongs.

The simplest case to consider is that of an operator equation with random initial conditions. A standard method for problems of this type can be described as follows: Let f denote the density function of the random initial condition $x(0,\omega)$. Then, the density function g of the random solution $x(t,\omega)$ is given by $g = fJ$, where J is the Jacobian of the transformation which maps $x(0,\omega)$ to $x(t,\omega)$.

Consider the first-order differential equation

$$dx/dt = c, \qquad t \in [0, \infty), \tag{3.65}$$

where c is a nonnegative constant. The solution of Eq. (3.65) is

$$x(t) = x(0) + ct. \tag{3.66}$$

Let us now assume that the initial value is a random variable $x(0,\omega)$ which is normally distributed with mean 0 and variance σ^2. Hence the density function of $x(0,\omega)$ is

$$f(\xi) = (2\pi)^{-1/2}\,\sigma^{-1}\exp\{-\xi^2/2\sigma^2\}, \qquad \xi \in (-\infty, \infty). \tag{3.67}$$

The distribution function of the random solution $x(t,\omega)$ can be determined as follows: Using (3.66)and (3.67), we have

$$\begin{aligned} G(\gamma) &= \mathscr{P}\{x(t,\omega) < \gamma\} = \mathscr{P}\{x(0,\omega) + ct < \gamma\} \\ &= \mathscr{P}\{x(0,\omega) < \gamma - ct\} = \int_{-\infty}^{\gamma - ct} f(\xi)\,d\xi \\ &= \int_{-\infty}^{\gamma} f(\lambda - ct)\,d\lambda = \int_{-\infty}^{\gamma} g(\lambda, t)\,d\lambda. \end{aligned}$$

Hence the density function of $x(t,\omega)$ is given by

$$g(\lambda, t) = (2\pi)^{-1/2}\,\sigma^{-1}\exp\{-(\lambda - ct)^2/2\sigma^2\}; \tag{3.68}$$

and we see that $x(t,\omega)$ is a real-valued normal (or Gaussian) random function with mean ct and variance σ^2.

As another example, consider the system of differential equations

$$dx/dt = Ax, \tag{3.69}$$

where $x = (x_1, x_2, \ldots, x_n) \in R_n$ and $A = (a_{ij})$ is the $n \times n$ coefficient matrix. Let us assume that the initial condition is a random vector $x(0,\omega)$ with known density function f. The solution of Eq. (3.69) is

$$x(t,\omega) = x(0,\omega)\, e^{At}. \tag{3.70}$$

The inverse relation

$$x(0,\omega) = e^{-At}\, x(t,\omega)$$

enables us to compute the Jacobian

$$J = \exp\{-\mathrm{Tr}(A)\, t\}.$$

Hence given f, we have $g = Jf$.

A number of random initial-value problems for partial differential equations lead to solutions of the form

$$u(t,x,\omega) = T(t) f(x,\omega), \tag{3.71}$$

where $\{T(t),\, t \geqslant 0\}$ is a semigroup of operators on some concrete Banach space $\mathfrak{X}$, and $f(x,\omega) = u(0,x,\omega)$. Kampé de Fériet [35], who was the first to study random initial-value problems for partial differential equations, showed that when the initial temperature $f(x,\omega)$ is a weakly stationary random process, then the solution $u(t,x,\omega)$, as a function of x, is also a weakly stationary random process.

Frequently the solution of a random initial-value problem is of the form

$$x(t,\omega) = T(t_0, t)\, x(t_0, \omega), \qquad x(t_0,\omega) \in \mathfrak{X}, \tag{3.72}$$

where the linear operators $T(s,t)$ have the properties (almost surely)

$$\begin{aligned} T(t_0, \tau)\,[T(\tau, t)\, x(t_0, \omega)] &= T(t_0, t)\, x(t_0, \omega) \\ T(t, t)\, x(t_0, \omega) &= x(t_0, \omega). \end{aligned} \tag{3.73}$$

In the study of stochastic dynamical systems it is often of interest to know if the solution process $x(t,\omega)$ is a random function of Markov type. $x(t,\omega)$ will be a *Markov process* if, for any bounded, Borel measurable function φ,

$$\mathscr{E}\{\varphi(x(t,\omega))\} = \mathscr{E}\{\mathscr{E}\{(\varphi(T(\tau,t)\,\xi) |\, T(t_0,\tau)\, x(t_0,\omega) = \xi\}\} \tag{3.74}$$

where $\tau \in (t_0, t)$.

We now consider random equations of the form

$$Tx = y(t,\omega). \tag{3.75}$$

This type of random equation has received a considerable amount of attention. Since T is deterministic, classical operator theory and methods of solving deterministic operators can be employed to solve equations of the form (3.75). Since, if T is invertible, the solution of Eq. (3.75) is of the form

$$x(t,\omega) = T^{-1}\,y(t,\omega), \tag{3.76}$$

the solution process is obtained by a transformation of the input process $y(t,\omega)$. As is well known, an important problem in communication theory and control theory is the characterization of the response of systems to random inputs.

Difference equations with random inputs (that is, T is the difference operator defined by

$$T[x_k] = \sum_{i=0}^{n} a_i\,\tau^i[x_k],$$

where $\tau^i[x_k] = x_{k+i}$) arise in mathematical economics, time series analysis, and in the study of discrete-time dynamical systems. If we assume that the input $y_k(\omega)$ is a weakly stationary sequence of uncorrelated random variables with means 0 and variances σ^2, then

$$y_k(\omega) = \int_{-\pi}^{\pi} e^{ik\lambda}\,dz(\lambda,\omega),$$

where $z(\lambda,\omega)$ is a process with orthogonal increments. It is known (cf. Gihman and Skorohod [25, pp. 240–241]) that the solution sequence $x_k(\omega)$ is weakly stationary, and admits the representation

$$x_k(\omega) = \int_{-\pi}^{\pi} e^{ik\lambda}[\varphi(e^{i\lambda})]^{-1}\,dz(\lambda,\omega), \tag{3.77}$$

where $\varphi(s) = \sum_{j=0}^{n} a_j s^j$.

Random ordinary and partial differential equations with random inputs arise in biology, physics, and in many branches of engineering. As in the case considered above, since the differential operators are assumed to be deterministic, classical methods can be used to obtain the solution process. For example, consider the solution of Eq. (3.75), where T is a linear differential operator of order n with constant coefficients, when certain boundary conditions are imposed. Then the formal solution of Eq. (3.75) is of the form

$$x(t,\omega) = \int_a^b G(t,\tau)\,y(\tau,\omega)\,d\tau, \tag{3.78}$$

where $G(t,\tau)$ is the *Green's function* of the differential boundary-value problem (cf. Sect. 4.1). We refer to Papoulis [50, Chap. 9] and Saaty [54, Chap. 8], where many concrete examples of ordinary differential equations with random inputs are considered. In Sect. 7.2 we consider the random differential equations of Langevin type which arise in physics and engineering;

in particular, we study the Itô random integral equation formulation of Langevin type equations and the solution process generated by these equations.

As an example of a partial differential equation driven by a random process, we refer to Eq. (3.26). In [26] it was shown that the solution process (3.27) is approximately Gaussian when the random input is a Poisson process.

The solution processes generated by linear integral equations with random inputs are studied in Sect. 4.2; hence they will not be discussed here.

A problem of general interest in many applications is to characterize those linear operators that preserve stationarity. Let $y(t,\omega)$ be a weakly stationary random function, and consider the linear integral operator S defined by

$$S[y(t,\omega)] = \int_{-\infty}^{\infty} K(t,\tau)\, y(\tau,\omega)\, d\tau, \tag{3.79}$$

where the kernel $K(t,\tau)$ is real valued, Lebesgue measurable as a function of t, and such that

$$\int_{-\infty}^{\infty} |K(t,\tau)|\, d\tau < M < \infty$$

for all t. Franaszek [21] has proved the following result: *A necessary and sufficient condition that the operator S defined by* (3.79) *transform weakly stationary random functions into weakly stationary random functions is that the kernel $K(t,\tau)$ be of the form*

$$K(t,\tau) = (2\pi)^{-1} \int_{-\infty}^{\infty} H(\lambda) \exp\{i[f(\lambda)\, t + g(\lambda)]\}\, e^{-i\lambda\tau}\, d\lambda,$$

where $H(\lambda)$, $f(\lambda)$, and $g(\lambda)$ are real-valued functions, with $f(\lambda)$ and $g(\lambda)$ odd, and $H(\lambda)$ even.

We refer to Blanc-Lapierre *et al.* [14] for results related to the one stated above, and to Lugannani and Thomas [43] for a discussion of a class of random functions which are closed under linear transformations.

We now consider the case of random operator equations of the form

$$T(\omega)\, x(t) = y(t,\omega). \tag{3.80}$$

Unfortunately, very little is known about the solution processes of equations of the above form. Some results, however, are known for random ordinary differential equations (cf. Samuels and Eringen [55]). Consider the equation

$$L(\omega)\, x(t) = \sum_{k=0}^{n} a_k(t,\omega)\, d^k x/dt^k = y(t,\omega), \tag{3.81}$$

where the coefficients $a_k(t,\omega)$ and the input $y(t,\omega)$ are random functions. It is assumed that $y(t,\omega)$ is independent of the coefficients $a_k(t,\omega)$; however the coefficients may be correlated with each other.

The solution of Eq. (3.81) can be obtained using perturbation methods (cf. Bellman [6, pp. 2–4]). Assume that the coefficients are of the form

$$a_k(t,\omega) = \alpha_k(t) + \epsilon\beta_k(t,\omega), \tag{3.82}$$

where the $\beta_k(t,\omega)$ are random functions with $\mathscr{E}\{\beta_k(t,\omega)\} = 0$. Hence $\mathscr{E}\{a_k(t,\omega)\} = \alpha_k(t)$. With the coefficients as defined by (3.82), a solution of Eq. (3.81) is sought of the form

$$x(t,\omega) = \sum_{k=0}^{\infty} \epsilon^k x_k(t,\omega). \tag{3.83}$$

If we now substitute (3.83) in (3.81) and equate the coefficients of ϵ^k, we obtain the following pair of operator equations:

$$\begin{aligned} Mx_0(t,\omega) &= y(t,\omega) \\ Mx_k(t,\omega) &= -N(\omega)\,x_{k-1}(t,\omega), \qquad i = 1, 2, \ldots, \end{aligned} \tag{3.84}$$

where the linear differential operators M and $N(\omega)$ denote the operator $L(\omega)$ with the coefficients $a_k(t,\omega)$ replaced by $\alpha_k(t)$ and $\beta_k(t,\omega)$, respectively. Hence M is a deterministic operator and $N(\omega)$ is a random operator. The formal solution of (3.84) is

$$\begin{aligned} x_0(t,\omega) &= \sum_{i=1}^{n} c_i\,\xi_i(t) + \int_0^t G(t,\tau)\,y(\tau,\omega)\,d\tau \\ x_k(t,\omega) &= -\int_0^t G(t,\tau)\,N(\omega)\,x_{k-1}(\tau,\omega)\,d\tau, \qquad k = 1, 2, \ldots. \end{aligned} \tag{3.85}$$

In (3.85), $\{\xi_i(t)\}$ is a set of fundamental solutions of the homogeneous deterministic equation $Mx = 0$, the c_i are constants to be determined by the initial conditions, and $G(t,\tau)$ is the one-sided Green's function associated with M.

We consider the case where the solution is terminated after two terms, and (i) the $y(t,\omega) = y(t)$ (deterministic) and (ii) the $\beta_k(t,\omega)$ are independent Gaussian random functions. In this case it is easy to show that the truncated solution $x(t,\omega) = x_0(t,\omega) + \epsilon x_1(t,\omega)$ is a Gaussian process. If the input $y(t,\omega)$ is assumed to be a Gaussian process, then it is clear that $x_0(t,\omega)$ is Gaussian; however further assumptions are needed to characterize $x_1(t,\omega)$.

The above methods can be applied to difference equations with random coefficients and to some integral equations with random kernels; however the perturbation approach is only useful when ϵ is so small that powers greater than some small number can be neglected. Hence other methods are required for the solution of equations of the form (3.80), and for the investigation of the solution process.

We refer to Saaty [54, Chap. 8] for a survey of studies on differential equations with random coefficients.

In the study of random equations in Hilbert spaces we frequently encounter transformations of the form $x(\omega) = Ly(\omega)$, where $y(\omega)$ (the input) is an H-valued random variable and $L \in \mathfrak{L}(H)$. Let us assume that $y(\omega)$ is a Gaussian random variable, and let m_1 and S_1 denote the mean and covariance operator, respectively, of the probability measure ν_1 induced by $y(\omega)$. It follows from Theorem 1.26 that $x(\omega)$ is an H-valued random variable; and if ν_2 denotes the probability measure induced by $x(\omega)$, then $m_2 = Lm_1$ and $S_2 = LS_1L^*$ are the mean and covariance operator of ν_2, respectively. We can, without loss of generality, assume $m_1 = 0$. Then, it follows from Theorem 1.29 that a necessary condition for the absolute continuity of ν_2 with respect to ν_1 is that there exists a $\lambda > 0$ such that $\lambda^{-1} \leqslant (LS_1L^*\xi, \xi)[(S_1\xi, \xi)]^{-1} \leqslant \lambda$, $\xi \in H$.

As pointed out in Sect. 1.5E, the paper of Gihman and Skorohod [24] gives a systematic account of studies on the absolute continuity of probability measures on function spaces. For a result on the absolute continuity of probability measures associated with the solutions of some random differential equations we refer to Šatašvili [56]. In Sect. 6.6 we consider the absolute continuity of the solution measure of a random nonlinear integrodifferential equation with respect to the input measure.

Virtually nothing is known about the probability measures associated with solutions of random operator equations of the form $T(\omega)x = y(t, \omega)$; hence this is an open area for research, and should lead to interesting theoretical results which have important applications.

C. Statistical problems

1. *Introduction.* Let $x(t, \omega)$ denote the solution of a random equation in a separable Banach space $\mathfrak{X}$. Since $x(t, \omega)$ is an $\mathfrak{X}$-valued random function, for every fixed t the solution is an $\mathfrak{X}$-valued random variable. Hence, the expectation of $x(t, \omega)$, or the *expected solution*, is given by

$$m(t) = \mathscr{E}\{x(t, \omega)\} = (\mathrm{B})\int_\Omega x(t, \omega)\, d\omega, \tag{3.86}$$

provided the Bochner integral of $x(t, \omega)$ exists for every fixed t. It follows from Theorem 1.8 that the expected solution exists if and only if $\mathscr{E}\{\|x(t, \omega)\|\} < \infty$ for every fixed t. Higher moments of the random solution can be defined as in Sect. 1.3E; in particular, the *variance of the random solution* is given by

$$\mathrm{Var}\{x(t, \omega)\} = \mathscr{E}\{\|x(t, \omega) - m(t)\|^2\}. \tag{3.87}$$

We now consider the formal expressions for the expected solutions of various types of random equations. Consider the equation

$$Tx(t) = y(t, \omega). \tag{3.88}$$

Let us assume that T is linear, independent of t, and invertible. Put $T^{-1} = S$. Then (cf. Theorem 2.4), $T \in \mathfrak{L}(\mathfrak{X})$ implies $S \in \mathfrak{L}(\mathfrak{X})$; and $T \in \mathfrak{C}(\mathfrak{X})$ implies $S \in \mathfrak{C}(\mathfrak{X})$. Therefore

$$x(t, \omega) = Sy(t, \omega). \tag{3.89}$$

Now, if $\mathscr{E}\{y(t,\omega)\}$ exists and $\mathscr{E}\{Sy(t,\omega)\}$ exists, then (cf. Theorem 1.9)

$$m(t) = \mathscr{E}\{x(t, \omega)\} = \mathscr{E}\{Sy(t, \omega)\} = S\mathscr{E}\{y(t, \omega)\}. \tag{3.90}$$

Hence the expected solution of Eq. (3.88) is obtained by a linear transformation of the expectation of the random input. We refer to Papoulis [50, Chaps. 9 and 10] for some applications of the above result to some random differential equations.

We now consider the equation

$$T(\omega)\,x(t) = y(t), \tag{3.91}$$

where $T(\omega)$ is an $\mathfrak{L}(\mathfrak{X})$-valued random variable. In this case, if $T(\omega)$ is invertible, then the solution of Eq. (3.91) is of the form

$$x(t, \omega) = S(\omega)\,y(t), \tag{3.92}$$

where $S(\omega) = T^{-1}(\omega)$. Now, if $S(\omega)y(t) \in B_1(\Omega, \mathfrak{X})$ for every $y \in \mathfrak{X}$, then it follows from Theorem 2.19 that

$$m(t) = \mathscr{E}\{x(t, \omega)\} = (\mathrm{B})\int_{\tilde{\Omega}} S(\omega)\,y(t)\,d\mu = Uy(t), \tag{3.93}$$

where $U \in \mathfrak{L}(\mathfrak{X})$. In (3.93) $\tilde{\Omega}$ denotes the subset of Ω for which $S(\omega)$ exists.

Finally, we consider an equation of the form

$$T(\omega)\,x(t) = y(t, \omega). \tag{3.94}$$

The formal solution of Eq. (3.94) is

$$x(t, \omega) = S(\omega)\,y(t, \omega). \tag{3.95}$$

It follows from Theorem 2.14 (the composition theorem) that $x(t,\omega)$ is an $\mathfrak{X}$-valued random function. Now

$$m(t) = \mathscr{E}\{x(t, \omega)\} = \int_{\tilde{\Omega}} S(\omega)\,y(t, \omega)\,d\mu. \tag{3.96}$$

To the best of our knowledge, nothing is known about Bochner integrals of functions of the form $S(\omega)y(t,\omega)$. However, if $S(\omega)$ and $y(t,\omega)$ are stochastically independent, and if their expectations exist, then we have

$$m(t) = \mathscr{E}\{x(t, \omega)\} = \mathscr{E}\{S(\omega)\}\,\mathscr{E}\{y(t, \omega)\} \tag{3.97}$$

It is clear that the above results are applicable to random equations of the form

$$(T - \lambda I)\,x(t) = y(t, \omega), \qquad (T(\omega) - \lambda I)\,x(t) = y(t),$$

and

$$(T(\omega) - \lambda I)\,x(t) = y(t, \omega).$$

A problem of fundamental importance in the theory of random equations concerns the relationship between the expected solution of a random equation and the solution of the deterministic equation obtained by replacing the random quantities in the equation by their expected values. Consider the random operator equation

$$T(\omega)\,x(t) = y(t, \omega). \tag{3.98}$$

Let $\tilde{T} = \mathscr{E}\{T(\omega)\}$ and $\tilde{y}(t) = \mathscr{E}\{y(t, \omega)\}$; and consider the deterministic operator equation

$$\tilde{T}\tilde{x}(t) = \tilde{y}(t). \tag{3.99}$$

The question can now be stated as follows: Is

$$\mathscr{E}\{x(t, \omega)\} = \tilde{x}(t)? \tag{3.100}$$

It is clear that in general (3.100) will not obtain. When (3.100) does hold, we can conclude that the random solution simply takes into account the random fluctuations about the deterministic solution, which in view of (3.100) is identical with the expected solution. As an example of a random solution for which (3.100) holds, consider the differential equation $dx/dt = c$ with a random initial condition $x(0) = x(0, \omega)$. Using (3.66), we have

$$x(t, \omega) = x(0, \omega) + ct.$$

Let $\mathscr{E}\{x(0, \omega)\} = x_0$. Then

$$\begin{aligned}\mathscr{E}\{x(t, \omega)\} &= \mathscr{E}\{x(0, \omega) + ct\}\\ &= x_0 + ct.\end{aligned}$$

The solution of the differential equation with initial condition $x(0) = x_0$, yields

$$\tilde{x}(t) = x_0 + ct = \mathscr{E}\{x(t, \omega)\}.$$

We close this subsection by remarking that it would be of interest to study the conditional expectation of random solutions, since some information is generally available about the random quantities that occur in random equations.

In the remaining two subsections we consider two techniques for solving random equations and investigating their statistical properties. Keller [38] has classified various approximation methods for solving random operator equations into two types: (1) "honest," and (2) "dishonest." The first step in an honest method is to obtain a solution $x(t, \omega)$ for each $\omega \in \Omega$. Hence, in this procedure for obtaining a random solution, randomness plays no role; and the advantages of using a random equation as a model are not available. The second step in an honest method is to compute the expected value of $x(t,\omega)$, as well as higher moments, from the set of explicit solutions obtained in the first step.

In a dishonest method the probabilistic nature of the equation is taken into consideration before the solution $x(t,\omega)$ is obtained. As an example, if one is primarily interested in the expected solution $\mathscr{E}\{x(t,\omega)\}$, then the random equation may be "averaged" to yield an equation for the expectation of $x(t,\omega)$. In most cases the "averaged equation" will involve terms of the form $\mathscr{E}\{x^2(t,\omega)\}$, $\mathscr{E}\{\alpha(\omega)x(t,\omega)\}$, etc. The next step involves the "dishonest" procedure of making certain assumptions about the above expectations; for example,

$$\mathscr{E}\{x^2(t,\omega)\} = (\mathscr{E}\{x(t,\omega)\})^2,$$
$$\mathscr{E}\{\alpha(\omega)\,x(t,\omega)\} = \mathscr{E}\{\alpha(\omega)\}\,\mathscr{E}\{x(t,\omega)\},$$

etc. Hence some unproved assumptions are made concerning the statistical properties of the random solution. Dishonest methods have one advantage over honest methods, namely they generally lead to a simpler problem which is solvable. Unfortunately, the solutions obtained using a dishonest method are at least suspect, and in many cases they are incorrect. Keller was primarily concerned with the solutions of random equations which arise in the study of wave propagation in random media; however, the two types of methods have been employed to study properties of other random equations, for example Boyce [15] has employed honest and dishonest methods in the study of random eigenvalue problems.

In the second subsection we consider a perturbation method for determining the expected solutions of random operator equations. We refer to Adomian [1, 2] for a discussion of other methods for investigating the statistical properties of random solutions. A method of particular interest involves establishing a hierarchy of equations from which the statistical properties of the solution of a random equation can be obtained. Hierarchy methods have been used in the study of random differential equations (cf. Richardson [53]), and in the study of random eigenvalue problems for random differential equations (cf. Boyce [15] and Haines [29]).

2. *Perturbation methods.* Consider the random linear operator equation

$$T(\omega)\,x = y \tag{3.101}$$

in a separable Banach space $\mathfrak{X}$. We assume, at present, that $y = y(t)$ is a deterministic function. We now consider a perturbation method which leads to an equation for the expected solution of a random operator equation. In general, perturbation methods can be utilized to study Eq. (3.101) if we assume that the random variations in the operator $T(\omega)$ are sufficiently small so corrections to the deterministic solution are of low order. In case $T(\omega)$ is a differential operator this would require small random variations in the coefficients, and if $T(\omega)$ is an integral operator, random variations in the kernel would have to be small.

We now assume that $T(\omega)$ depends on a small parameter ϵ, and that when $\epsilon = 0$ the random operator $T(\omega, \epsilon)$ reduces to a deterministic operator T. If we now expand $T(\omega, \epsilon)$ in powers of ϵ, we have

$$T(\omega, \alpha) = T + \epsilon T_1(\omega) + \epsilon^2 T_2(\omega) + O(\epsilon^3) \tag{3.102}$$

Hence $T(\omega, \alpha)$ is the sum of a deterministic operator T and two random operators, $T_1(\omega)$ and $T_2(\omega)$, which represent random perturbations of T. Consider the case when $\epsilon = 0$. If the deterministic operator equation $Tx = y$ is solvable (that is, T^{-1} exists), then $x_0(t) = T^{-1} y(t)$. Using (3.102), Eq. (3.101) is of the form

$$[T + \epsilon T_1(\omega) + \epsilon^2 T_2(\omega) + O(\epsilon^3)]\, x = y(t). \tag{3.103}$$

Hence

$$x(t, \omega) = x_0(t) - T^{-1}[\epsilon T_1(\omega) + \epsilon^2 T_2(\omega)]\, x(t, \omega) + O(\epsilon^3). \tag{3.104}$$

In order to solve Eq. (3.104) we use the method of iterations or sucessive substitutions†:

$$x(t, \omega) = x_0(t) + \epsilon x_1(t, \omega) + \epsilon^2 x_2(t, \omega) + \cdots. \tag{3.105}$$

Substitution of (3.105) in (3.104) yields

$$x(t, \omega) = x_0(t) - \epsilon T^{-1} T_1(\omega)\, x_0(t) + \epsilon^2 T^{-1}[T_1(\omega) T^{-1} T_1(\omega) - T_2(\omega)]\, x_0(t) + O(\epsilon^3). \tag{3.106}$$

If we now apply the expectation operator to (3.106), we have

$$\mathscr{E}\{x(t, \omega)\} = x_0(t) - \epsilon T^{-1} \mathscr{E}\{T_1(\omega)\}\, x_0(t) + \epsilon^2 T^{-1}[\mathscr{E}\{T_1(\omega) T^{-1} T_1(\omega)\} - \mathscr{E}\{T_2(\omega)\}]\, x_0(t) + O(\epsilon^3). \tag{3.107}$$

To obtain an equation for the expected solution, we eliminate the deterministic solution $x_0(t)$ from (3.107). It follows from (3.107) that

$$\begin{aligned} x_0(t) &= \mathscr{E}\{x(t, \omega)\} + \epsilon T^{-1} \mathscr{E}\{T_1(\omega)\}\, x_0(t) + O(\epsilon^2) \\ &= \mathscr{E}\{x(t, \omega)\} + \epsilon T^{-1} \mathscr{E}\{T_1(\omega)\}\, \mathscr{E}\{x(t, \omega)\} + O(\epsilon^2). \end{aligned} \tag{3.108}$$

Substituting (3.108) in (3.107) we obtain

$$\begin{aligned} \mathscr{E}\{x(t, \omega)\} = x_0(t) &- \epsilon T^{-1} \mathscr{E}\{T_1(\omega)\}\, \mathscr{E}\{x(t, \omega)\} \\ &+ \epsilon^2 T^{-1}[\mathscr{E}\{T_1(\omega) T^{-1} T_1(\omega)\} - \mathscr{E}\{T_1(\omega)\} T^{-1} \mathscr{E}\{T_1(\omega)\} \\ &- \mathscr{E}\{T_2(\omega)\}]\, \mathscr{E}\{x(t, \omega)\} + O(\epsilon^3). \end{aligned} \tag{3.109}$$

† See Bellman [6, p. 47].

If we now multiply (3.109) by T, we obtain

$$\{T + \epsilon\mathscr{E}\{T_1(\omega)\} + \epsilon^2[\mathscr{E}\{T_1(\omega)\}\,T^{-1}\,\mathscr{E}\{T_1(\omega)\} \\ - \mathscr{E}\{T_1(\omega)\,T^{-1}\,T_1(\omega)\} + \mathscr{E}\{T_2(\omega)\}]\}\,\mathscr{E}\{x(t,\omega)\} = y + O(\epsilon^3). \tag{3.110}$$

If $\mathscr{E}\{T_1(\omega)\} = \mathscr{E}\{T_2(\omega)\} = \Theta$ (the null operator), then (3.110) becomes

$$\{T - \epsilon^2\,\mathscr{E}\{T_1(\omega)\,T^{-1}\,T_1(\omega)\}\}\,\mathscr{E}\{x(t,\omega)\} = y + O(\epsilon^3). \tag{3.111}$$

If we drop the term $O(\epsilon^3)$ from Eqs. (3.109), (3.110) or (3.111), we obtain an explicit equation for $\mathscr{E}\{x(t,\omega)\}$.

It is of interest to consider a dishonest approach which yields Eq. (3.111). Let us put $T_2(\omega) = \Theta$ in Eq. (3.103), and then apply the expectation operator. We have

$$T\mathscr{E}\{x(t,\omega)\} + \epsilon\mathscr{E}\{T_1(\omega)\,x(t,\omega)\} + O(\epsilon^3) = y(t). \tag{3.112}$$

We now assume $\mathscr{E}\{T_1(\omega)\} = \Theta$. In order to determine $\mathscr{E}\{T_1(\omega)\,x(t,\omega)\}$ we first multiply (3.104) by $T_1(\omega)$, and then apply the expectation operator. We have

$$\mathscr{E}\{T_1(\omega)\,x(t,\omega)\} = -\epsilon\mathscr{E}\{T_1(\omega)\,T^{-1}\,T_1(\omega)\,x(t,\omega)\} + O(\epsilon^3). \tag{3.113}$$

We now assume that

$$\mathscr{E}\{T_1(\omega)\,T^{-1}\,T_1(\omega)\,x(t,\omega)\} = \mathscr{E}\{T_1(\omega)\,T^{-1}\,T_1(\omega)\}\,\mathscr{E}\{x(t,\omega)\}. \tag{3.114}$$

Hence (3.113) becomes

$$\mathscr{E}\{T_1(\omega)\,x(t,\omega)\} = -\epsilon\mathscr{E}\{T_1(\omega)\,T^{-1}\,T_1(\omega)\}\,\mathscr{E}\{x(t,\omega)\} + O(\epsilon^3).$$

If the above expression is substituted in (3.112) we obtain (3.111). Relation (3.114) is called the assumption of *local independence*.

The case when the input is a random function $y(t,\omega)$ does not cause any difficulties. It is, however, necessary to assume that $T(\omega)$ and $y(t,\omega)$ are stochastically independent. For a discussion of this case we refer to Adomian [2].

Perturbation methods have been applied to a number of random equations in applied mathematics, especially in the study of wave propagation in random media (cf. for example, Chen [17], Chen and Soong [16], Karal and Keller [37], and Keller [38]).

References

1. Adomian, G., Theory of random systems. *Trans. 4th Prague Conf. Information Theory, Statist. Decision Functions, and Random Processes* (*1965*), pp. 205–222, 1967.
2. Adomian, G., Random operator equations in mathematical physics. I. *J. Math. Phys.* **11** (1970), 1069–1084.
3. Anselone, P. M., ed., "Nonlinear Integral Equations." The Univ. of Wisconsin Press, Madison, Wisconsin, 1964.

4. Åström, K. J., "Introduction to Stochastic Control Theory." Academic Press, New York, 1970.
5. Banach, S., Sur les opérations dans les ensembles abstraits et leur applications aux équations intégrales. *Fund. Math.* **3** (1922), 133–181.
6. Bellman, R., "Perturbation Techniques in Mathematics, Physics, and Engineering." Holt, New York, 1964.
7. Bharucha-Reid, A. T., On random operator equations in Banach space. *Bull. Acad. Polon. Sci. Sér. Sci. Math. Astronom. Phys.* **7** (1959), 561–564.
8. Bharucha-Reid, A. T., On random solutions of integral equations in Banach spaces. *Trans. 2nd Prague Conf. Information Theory, Statist. Decision Functions, and Random Processes (1959)*, pp. 27–48, 1960.
9. Bharucha-Reid, A. T., Approximate solutions of random operator equations. *Notices Amer. Math. Soc.* **7** (1960), 361.
10. Bharucha-Reid, A. T., On the theory of random equations. *Proc. Symp. Appl. Math. 16th, 1963*, pp. 40–69. Amer. Math. Soc., Providence, Rhode Island, 1964.
11. Bharucha-Reid, A. T., Random algebraic equations. *In* "Probabilistic Methods in Applied Mathematics" (A. T. Bharucha-Reid, ed.), Vol. 2, pp. 1–52. Academic Press, New York, 1970.
12. Bharucha-Reid, A. T., Random difference equations. *In* "Probabilistic Methods in Applied Mathematics" (A. T. Bharucha-Reid, ed.), Vol. 4. To be published.
13. Birkhoff, G., Bona, J., and Kampé de Fériet, J., Statistically well-set Cauchy problems. *In* "Probabilistic Methods in Applied Mathematics" (A. T. Bharucha-Reid, ed.), Vol. 3. Academic Press, New York, 1973.
14. Blanc-Lapierre, A., Krinvine, B., and Sultan, R., Transformations linéaires conservant la stationnarité de second ordre. *C. R. Acad. Sci. Paris Ser. A***268** (1969), 1115–1117.
15. Boyce, W. E., Random eigenvalue problems. *In* "Probabilistic Methods in Applied Mathematics" (A. T. Bharucha-Reid, ed.), Vol. 1, pp. 1–73. Academic Press, New York, 1968.
16. Chen, K. K., and Soong, T. T., Covariance properties of wave propagating in a random medium. *J. Acoust. Soc. Amer.* **49** (1971), 1639–1642.
17. Chen, Y. M., On scattering of waves by objects imbedded in random media: Stochastic linear partial differential equations and scattering of waves by a conducting sphere imbedded in random media. *J. Math. Phys.* **5** (1964), 1541–1546.
18. Chu, S. C., and Diaz, J. B., Remarks on a generalization of Banach's principle of contraction mappings. *J. Math. Anal. Appl.* **11** (1965), 440–446.
19. Dixmier, J., "Les algèbres d'opérateurs dans l'espace Hilbertien (Algèbres de von Neumann)," 2nd ed. Gauthier-Villars, Paris, 1969.
20. Driml, M., and Hanš, O., Continuous stochastic approximations. *Trans. 2nd Prague Conf. on Information Theory, Statist. Decision Functions, and Random Processes (1959)*, pp. 113–122, 1960.
21. Franaszek, P. A. On linear systems which preserve wide sense stationary. *SIAM J. Appl. Math.* **15** (1967), 1481–1484.
22. Frisch, U., Wave propagation in random media. *In* "Probabilistic Methods in Applied Mathematics" (A. T. Bharucha-Reid, ed.), Vol. 1, pp. 75–198. Academic Press, New York, 1968.
23. Gardner, L. A., Stochastic approximation and its application to prediction and control synthesis. *Internat. Symp. Nonlinear Differential Equations and Nonlinear Mechanics* (J. P. La Salle and S. Lefschetz, eds.), pp. 241–258. Academic Press, New York, 1963.
24. Gihman, I. I., and Skorohod, A. V., On the densities of probability measures in function spaces (Russian). *Uspehi Mat. Nauk* **21** (1966), 83–152.

25. Gihman, I. I., and Skorohod, A. V., "Introduction to the Theory of Random Processes," translated from the Russian. Saunders, Philadelphia, Pennsylvania, 1969.
26. Gopalsamy, K., and Bharucha-Reid, A. T., On a class of partial differential equations driven by stochastic point processes. To be published.
27. Grenander, U., "Probabilities on Algebraic Structures." Wiley, New York, 1963.
28. Grenander, U., and Rosenblatt, M., "Statistical Analysis of Stationary Time Series." Wiley, New York, 1957.
29. Haines, C. W., Hierarchy methods for random vibrations of elastic strings and beams. *J. Engrg. Math.* **1** (1967), 293–305.
30. Hanš, O., Reduzierende zufällige Transformationen. *Czechoslovak Math. J.* **7** (1957), 154–158.
31. Hanš, O., Random fixed point theorems. *Trans. 1st Prague Conf. Information Theory, Statist. Decision Functions, and Random Processes (1956)*, pp. 105–125, 1957.
32. Hanš, O., Random operator equations, *Proc. 4th Berkeley Symp. Math. Statist. Probability (1960)*, Vol. II, pp. 185–202, 1961.
33. Hanš, O., and Špaček, A., Random fixed point approximation by differentiable trajectories. *Trans. 2nd Prague Conf. Information Theory, Statist. Decision Functions, and Random Processes (1959)*, pp. 203–213, 1960.
34. Jazwinski, A. H., "Stochastic Processes and Filtering Theory." Academic Press, New York, 1970.
35. Kampé de Fériet, J., Random solutions of partial differential equations, *Proc. 3rd Berkeley Symp. on Math. Statist. Probability (1955)*, Vol. III, pp. 199–208, 1956.
36. Kantorovich, L. V., and Krylov, V. I., "Approximate Methods of Higher Analysis," translated from the Russian. Wiley (Interscience), New York, 1958.
37. Karal, F. C., and Keller, J. B., Elastic, electromagnetic, and other waves in a random medium. *J. Math. Phys.* **5** (1964), 537–547.
38. Keller, J. B., Stochastic equations and wave propagation in random media. *Proc. Symp. Appl. Math., 16th, 1963*, pp. 145–170. Amer. Math. Soc., Providence, Rhode Island, 1964.
39. Kolmogorov, A. N., and Fomin, S. V., "Introductory Real Analysis," rev. Engl. ed., translated from the Russian. Prentice-Hall, Englewood Cliffs, New Jersey, 1970.
40. Koopmans, T., "Statistical Inference in Dynamic Economic Models." Wiley, New York, 1950.
41. Krasnosel'skiĭ, M. A., Two remarks on the method of successive approximations (Russian). *Uspehi Mat. Nauk* **11** (1955), No. 1, 123–127.
42. Lions, J. L., Unpublished notes. Personal communication, 1966.
43. Lugannani, R., and Thomas, J. B., On a class of stochastic processes which are closed under linear transformations. *Information and Control* **10** (1967), 1–21.
44. McCoy, J. J., An application of the method of moments to stochastic equations. *Quart. Appl. Math.* **26** (1969), 521–536.
45. Mikusiński, J., "Operational Calculus," translated from the Polish. Pergamon, Oxford, 1959.
46. Mukherjea, A., Random transformations on Banach spaces. Ph.D. Dissertation, Wayne State Univ., Detroit, Michigan, 1966.
47. Mukherjea, A., Transformations aléatoires séparables: Théorème du point fixe aléatoire. *C. R. Acad. Sci. Paris Ser.* **263** (1966), 393–395.
48. Mukherjea, A., On a random integral equation of Uryson type. I. To be published.
48a. Nashed, M. Z., and Salehi, H., Measurability of generalized inverses of random linear operators. To be published.
49. Oza, K. G., and Jury, E. I., System identification and the principle of random contraction mapping. *SIAM J. Control* **6** (1968), 244–257.

50. Papoulis, A., "Probability, Random Variables, and Stochastic Processes." McGraw-Hill, New York, 1965.
51. Phillips, R. S., Perturbation theory for semi-groups of linear operators. *Trans. Amer. Math. Soc.* **74** (1953), 199–221.
51a. Prakasa Rao, B. L. S., Stochastic integral equations of mixed type. II. To be published.
52. Rall, L. B., "Computational Solution of Nonlinear Operator Equations." Wiley, New York, 1969.
53. Richardson, J. M., The application of truncated hierarchy techniques in the solution of a stochastic linear differential equation. *Proc. Symp. Appl. Math., 16th*, pp. 290–302. Amer. Math. Soc., Providence, Rhode Island, 1964.
54. Saaty, T. L., "Modern Nonlinear Equations." McGraw-Hill, New York, 1967.
55. Samuels, J. C., and Eringen, A. C., On stochastic linear systems. *J. Math. and Phys.* **38** (1959), 83–103.
56. Šatašvili, A. D., On measure densities corresponding to the solutions of certain differential equations with random functions (Russian). *Dokl. Akad. Nauk SSSR* **194** (1970), 275–277.
57. Schauder, J., Der Fixpunktsatz in Funktionenräumen. *Studia Math.* **2** (1930), 171–182.
58. Sehgal, V. M., Some fixed point theorems in functional analysis and probability. Ph.D. Dissertation, Wayne State Univ., Detroit, Michigan, 1966.
59. Špaček, A., Zufällige Gleichungen. *Czechoslovak Math. J.* **5** (1955), 462–466.
60. Taylor, A. E., "Introduction to Functional Analysis." Wiley, New York, 1958.
61. Ullrich, M., Random Mikusiński operators. *Trans. 2nd Prague Conf. Information Theory, Statist. Decision Functions, and Random Processes (1959)*, pp. 639–659, 1960.
62. Vorobyev, Ju. V., "Method of Moments in Applied Mathematics," translated from the Russian. Gordon & Breach, New York, 1965.

CHAPTER 4

Random Linear Integral Equations

4.1 Introduction

Let $K(x,y)$ be a complex-valued function of two real variables, where $x \in [a,b]$, $y \in [a,b]$; and, for example, let $f(x) \in L_2[a,b]$. The transformation L defined by

$$L[f(x)] = \int_a^b K(x,y)f(y)\,dy \tag{4.1}$$

is a *linear integral operator* on L_2, since $L[f_1+f_2]=L[f_1]+L[f_2]$ and $L[\alpha f]=\alpha L[f]$; and we say that the function $K(x,y)$, called the *kernel* of L, generates the integral operator L. Linear integral equations, that is, linear operator equations in which the operator is of the form (4.1), are divided into two basic types: (1) *Fredholm equations*, which have a fixed domain of integration; and (2) *Volterra equations*, in which the upper limit of integration is variable. Further, if the unknown function appears under the integral sign only, the equation is called an *integral equation of the first kind*; and if the unknown function also appears outside the integral, we refer to the equation as an *integral equation of the second kind.* In view of the above classification the linear integral equations

$$\int_a^b K(x,y)f(y)\,dy = g(x) \tag{4.2}$$

$$\int_a^b K(x,y)f(y)\,dy - \lambda f(x) = g(x) \tag{4.3}$$

are nonhomogeneous *Fredholm equations of the first and second kind*, respectively; and the linear integral equations

$$\int_a^x K(x,y)f(y)\,dy = g(x) \tag{4.4}$$

$$\int_a^x K(x,y)f(y)\,dy - \lambda f(x) = g(x) \tag{4.5}$$

are *nonhomogeneous Volterra equations of the first and second kind*, respectively. From the above classification of linear equations, Eqs. (4.2) and (4.4), and Eqs. (4.3) and (4.5), it is clear that a Volterra equation is a special case of a Fredholm equation with kernel

$$\tilde{K}(x,y) = \begin{cases} K(x,y), & x > y \\ 0, & x < y. \end{cases} \tag{4.6}$$

Linear integral equations, which constitute a very important class of linear operator equations, are of fundamental importance in applied mathematics. In many applied problems they arise as mathematical formulations of physical processes; in particular, linear integral equations frequently arise in physical problems as a result of the possibility of superimposing the effects due to several causes. Certain integral equations can be deduced from differential equations, and in these cases the integral equations are alternative formulations of problems whose initial mathematical formulation leads to differential equations.

To illustrate the above we consider three examples which indicate the relationship between differential and integral equations.

1. Consider the *initial-value problem* consisting of the second order differential equation

$$\frac{d^2x}{dt^2} + a\frac{dx}{dt} + bx = f(t), \tag{4.7}$$

together with the prescribed initial conditions

$$x(0) = x_0, \qquad x'(0) = v_0. \tag{4.8}$$

In (4.7) a and b may be functions of t. If we first rewrite Eq. (4.7) as

$$\frac{d^2x}{dt^2} = -a\frac{dx}{dt} - bx + f(t),$$

and integrate over the interval $(0,t)$, we obtain, using (4.8),

$$\begin{aligned}\frac{dx}{dt} &= -\int_0^t a\frac{dx}{dt}\,d\tau - \int_0^t bx\,d\tau + \int_0^t f\,d\tau \\ &= -ax - \int_0^t (b-a')\,x\,d\tau + \int_0^t f\,d\tau + a(0)\,x_0 + v_0.\end{aligned}$$

Integrating the above we obtain the relation

$$x(t) = x_0 - \int_0^t a(\tau)\, x(\tau)\, d\tau - \int_0^t \int_0^t [b(\tau) - a'(\tau)]\, x(\tau)\, d\tau\, d\tau$$

$$+ \int_0^t \int_0^t f(\tau)\, d\tau\, d\tau + [a(0)\, x_0 + v_0]\, t,$$

which can be written in the form

$$x(t) = -\int_0^t \{a(\tau) + (t - \tau)\, [b(\tau) - a'(\tau)]\}\, x(\tau)\, d\tau$$

$$+ \int_0^t (t - \tau) f(\tau)\, d\tau + [a(0)\, x_0 + v_0]\, t + x_0 .$$

The above can now be rewritten as

$$x(t) - \int_0^t K(t, \tau)\, x(\tau)\, d\tau = g(t), \tag{4.9}$$

where

$$K(t, \tau) = (\tau - t)\, [b(\tau) - a'(\tau)] - a(\tau)$$

$$g(t) = \int_0^t (t - \tau) f(\tau)\, d\tau + [a(0)\, x_0 + v_0]\, t + x_0 .$$

Hence we have shown that the initial value problem (4.7) and (4.8) admits an integral equation formulation (4.9) as a Volterra equation of the second kind.

2. Consider the *differential boundary-value problem*

$$d^2 x/dt^2 + \lambda x = 0, \qquad x(0) = 0, \quad x(a) = 0 \tag{4.10}$$

Proceeding as in the first example, integration over the interval $(0, t)$ gives the relation

$$dx/dt = -\lambda \int_0^t x(\tau)\, d\tau + x'(0),$$

where $x'(0)$ is unknown. Integrating again over $(0, t)$ and using the condition $x(0) = 0$, we obtain

$$x(t) = -\lambda \int_0^t (t - \tau)\, x(\tau)\, d\tau + x'(0)\, t. \tag{4.11}$$

Imposing the second condition $x(a) = 0$, we have

$$x'(0) = (\lambda/a) \int_0^a (a - \tau)\, x(\tau)\, d\tau.$$

Hence (4.11) can be rewritten as

$$x(t) = -\lambda \int_0^t (t - \tau)\, x(\tau)\, d\tau + t(\lambda/a) \int_0^a (a - \tau)\, x(\tau)\, d\tau$$

$$= (\lambda/a) \int_0^t \tau(a - t)\, x(\tau)\, d\tau + (\lambda/a) \int_t^a t(a - \tau)\, x(\tau)\, d\tau. \tag{4.12}$$

If we now put

$$K(t,\tau)=\begin{cases}(\tau/a)(a-t) & \text{for} \quad \tau<t\\ (t/a)(a-\tau) & \text{for} \quad \tau>t,\end{cases}$$

Eq. (4.12) can be written as

$$x(t)=\lambda\int_0^a K(t,\tau)\,x(\tau)\,d\tau. \tag{4.13}$$

Hence, the integral equation formulation of the differential boundary-value problem (4.10) leads to a Fredholm equation of the second kind.

3. Consider the ordinary linear differential operator of second order

$$L[x]=\frac{d}{dt}\left[p(t)\frac{dx}{dt}\right]+q(t)\,x, \tag{4.14}$$

where $p(t)\geqslant 0$. We will consider functions $x(t)$ which, at the ends of a given interval (a,b), satisfy the homogeneous boundary conditions

$$\alpha x(a)+\beta x'(a)=0, \qquad \gamma x(b)+\delta x'(b)=0; \tag{4.15}$$

and inside (a,b) are continuous and have a continuous first derivative. In order that $L[x]$ have a meaning, we impose the condition of uniqueness on the second derivative of $x(t)$. We also assume that the unique solution $x(t)$ of the equation $Lx=0$, which satisfies the boundary conditions (4.15), and such that $x(t)$ and $x'(t)$ are continuous, is the trivial solution $x(t)\equiv 0$.

The *Green's function*, or *influence function*, associated with the differential operator L and the boundary conditions is the function $G(t,\tau)$ with the following properties:

(i) $G(t,\tau)$ is continuous for $t,\tau\in[a,b]$;

(ii) In each of the intervals $[a,\tau]$ and $(\tau,b]$, the derivatives $\partial G/\partial t$ and $\partial G/\partial\tau$ are continuous;

(iii) $G(t,\tau)$ is continuous at $t=\tau$;

(iv) The derivative of G has a discontinuity of magnitude $-1/p(\tau)$ at $t=\tau$; that is,

$$\left.\frac{\partial G}{\partial\tau}\right]_{t=\tau^+}-\left.\frac{\partial G}{\partial\tau}\right]_{t=\tau^-}=\frac{1}{p(\tau)};$$

(v) For fixed τ, $G(t,\tau)$ satisfies the equation $L[G]=0$ in each of the intervals $[a,\tau)$ and $(\tau,b]$;

(vi) As a function of t, $G(t,\tau)$ satisfies the boundary conditions (4.15).

In order to determine the Green's function we construct integrals $u(t)$ and $v(t)$ of $L[x]=0$ satisfying the Cauchy conditions

$$u(a)=\beta, \qquad u'(a)=-\alpha$$
$$v(b)=\delta, \qquad v'(b)=-\gamma.$$

The integrals $u(t)$ and $v(t)$ are linearly independent; and from the theory of linear differential equations we have the identity

$$p(t)\,[u(t)\,v'(t) - u'(t)\,v(t)] = c \neq 0.$$

We can now define a function $G(t,\tau)$ by

$$G(t,\tau) = \begin{cases} u(t)\,v(\tau)/c, & \text{for} \quad t \in [a,\tau] \\ u(\tau)\,v(t)/c, & \text{for} \quad t \in [\tau,b]; \end{cases} \tag{4.16}$$

and it is not difficult to verify that the function defined by (4.16) satisfies the properties (i)–(vi) of a Green's function. It follows from (4.16) that $G(t,\tau)$ enjoys the following property:

(vii) $G(t,\tau) = G(\tau,t)$; that is $G(t,\tau)$ is a symmetric function.

We now state two results which are of fundamental importance in the reduction of differential equations of the form $Lx = f(t)$ to Fredholm equations.

THEOREM 4.1. *Let $f(t)$ be a continuous function defined on the interval $[a,b]$. If $x(t)$ is a solution of the differential equation*

$$Lx + f(t) = 0 \tag{4.17}$$

satisfying the boundary conditions (4.15), *then $x(t)$ can be written in the form*

$$x(t) = \int_a^b G(t,\tau)\,f(\tau)\,d\tau. \tag{4.18}$$

THEOREM 4.2. *The function $x(t)$ given by* (4.18) *satisfies Eq.* (4.17) *together with the boundary conditions* (4.15).

Consider the differential equation

$$Lx + \lambda h(t)\,x = 0, \tag{4.19}$$

together with the boundary conditions (4.15). It follows from the above results that the differential boundary-value problem consisting of Eq. (4.10) and boundary conditions (4.15) can be reduced, using the Green's function, to the integral equation

$$x(t) = \lambda \int_a^b h(\tau)\,G(t,\tau)\,x(\tau)\,d\tau. \tag{4.20}$$

Equation (4.20) is a homogeneous integral equation; nonsymmetric except in the case when $h(\tau)$ is a constant. If we put

$$\sqrt{h(t)}\,x(t) = y(t) \qquad \text{and} \qquad \sqrt{h(t)\,h(\tau)}\,G(t,\tau) = K(t,\tau),$$

then Eq. (4.20) can be rewritten as

$$y(t) = \lambda \int_a^b K(t, \tau) y(\tau) \, d\tau, \tag{4.21}$$

which is a homogeneous Fredholm equation of the second kind with symmetric kernel.

The three examples given above stressed the integral equation formulation of certain types of differential equations which arise in many physical problems. From the point of view of analysis, there are several good reasons for utilizing the integral equation formulation: (1) the integral operator is usually bounded (and in many cases compact), whereas the differential operator is unbounded; (2) the integral equation formulation provides a model which is suitable for numerical analysis; and (3), as illustrated by Examples 2 and 3, the integral equation incorporates the boundary conditions imposed on a differential equation; hence every solution of the integral equation automatically satisfies these boundary conditions.

There is a voluminous literature on linear integral equations and their applications. For expositions of the theory of linear integral equations and their applications, and for references to the literature, we refer to the books by Corduneanu [11a], Courant and Hilbert [12, Chap. II], Green [17], Hildebrand [21], Kanwal [25a], Mihlin [32], Pogorzelski [34], Riesz and Sz.-Nagy [35, Chap. IV], Saaty [36, Chap. 6], Stakgold [42, Chap. 3], Tricomi [44], Yosida [45], Zaanen [46, Part III].

B. Random linear integral equations. General remarks

Consider, for example, linear operator equations of the form

$$Lf = g \tag{4.22a}$$

$$(L - \lambda I) f = g, \tag{4.22b}$$

where L is the *Fredholm operator*

$$L[f(x)] = \int_a^b K(x, y) f(y) \, dy \tag{4.23}$$

or the *Volterra operator*

$$L[f(x)] = \int_a^x K(x, y) f(y) \, dy. \tag{4.24}$$

Probabilistic analogues of Eqs. (4.22a) and (4.22b) can be obtained as follows:

1. *The function g (right-hand side, free, forcing, or nonhomogeneous term) is a random function.* In this case the integral equations are of the form $Lf = g(x, \omega)$ and $(L - \lambda I) f = g(x, \omega)$; and the stochastic properties of the solutions $f(x, \omega)$ depend on the stochastic properties of $g(x, \omega)$ since the operators are deterministic.

2. *The integral operator L is a random operator.* In this case the integral equations are of the form $L(\omega)f = g$ and $(L(\omega) - \lambda I)f = g$. We shall show that Fredholm (or Volterra) operators defined over a *random domain* $I_\omega \subset [a,b]$ (or $I_\omega = [a,\omega]$), or equivalently, the kernel K is a *random kernel* $K(x,y,\omega)$, are random operators; and the stochastic properties of the solutions $f(x,\omega)$ depend on the stochastic properties of the random operator $L(\omega)$.

3. *The integral operator L is a random operator and g is a random function.* In this case the integral equations are of the general form $L(\omega)f = g(x,\omega)$ and $(L(\omega) - \lambda I)f = g(x,\omega)$. Clearly the two previous cases are special cases of these general random integral equations. The probabilistic analysis of the solutions $f(x,\omega)$ in this case is much more involved than the previous cases, since the stochastic properties of the solutions will depend on the stochastic properties of the random operator $L(\omega)$ and the random forcing function.

In addition to random integral equations that arise as direct probabilistic analogues of Fredholm and Volterra equations, random linear integral equations also arise as alternative formulations of random differential equations; in particular, as integral equation formulations of random initial-value and random boundary-value problems for differential equations.

In Sect. 4.2 we study Fredholm and Volterra equations with random forcing functions; and in Sect. 4.3 we consider Fredholm and Volterra equations with random kernels. We do not consider Case 3 mentioned above since in this case very little is known about equations of this general type, except, of course, some general results concerning existence, uniqueness, and measurability, which follow from the theorems of Chap. 3. Finally, in Sect. 4.4 we consider some random Fredholm and Volterra equations arising in applied fields.

In Chap. 5 we study some eigenvalue problems for Fredholm equations; in particular, we consider (1) the integral equation formulation of a random differential boundary-value problem and (2) the eigenvalue problem for Fredholm equations with random degenerate kernels.

4.2 Fredholm and Volterra Integral Equations with Random Forcing Functions

A. Introduction

Consider the random linear integral equation

$$f(x,\omega) - Lf(y,\omega) = g(x,\omega), \tag{4.25}$$

where L is a Fredholm (or Volterra) operator over $[a,b]$ (a and b finite), and $g(x,\omega)$, $x \in [a,b]$, is a second-order random function which is assumed to be continuous in mean square; that is, $g(x,\omega)$ is such that

$$\text{(i)} \quad \mathscr{E}\{|g(x,\omega)|^2\} < \infty \qquad \text{for every} \quad x \in [a,b],$$

and

$$\text{(ii)} \quad \lim_{h \to 0} \mathscr{E}\{|g(x+h,\omega) - g(x,\omega)|^2\} = 0 \qquad \text{for every} \quad x \in [a,b].$$

Equation (4.25) is a deterministic linear operator equation with random forcing function; hence the solution of the equation defines a new random function $f(x,\omega)$, the stochastic properties of which will depend on the stochastic properties of $g(x,\omega)$. Since the Fredholm operator is deterministic, the solution of Eq. (4.25) can be obtained by well-known classical methods; in particular, under the assumption that the kernel is "small," the equation can be solved by iteration (cf. Yosida [45, pp. 115–118]). In this section, the results of which are due to Anderson [2], we obtain the solution $f(x,\omega)$ of Eq. (4.25), calculate the covariance function of $f(x,\omega)$, and investigate the continuity of $f(x,\omega)$.

B. Solution of the integral equation

We now state and prove the following theorem.

THEOREM 4.3. *If* (i) $K(x,y)$, $x,y \in [a,b]$, *is a Fredholm kernel such that* $|b-a| \max |K(x,y)| < 1$ *and* (ii) $g(x,\omega)$, $x \in [a,b]$, $\omega \in \Omega$, *is a second-order random function which is continuous in mean square, then the random function* $f(x,\omega)$ *defined by*

$$f(x,\omega) = g(x,\omega) - \int_a^b \Gamma(x,y)\, g(y,\omega)\, dy, \tag{4.26}$$

$x \in [a,b]$, $\omega \in \Omega$, *satisfies the Fredholm equation* (4.25) *on* $[a,b] \times \Omega$.

Proof. The reciprocal form of Eq. (4.25) is

$$g(x,\omega) = f(x,\omega) - \int_a^b \Gamma(x,y)\, g(y,\omega)\, dy, \tag{4.27}$$

where $\Gamma(x,y)$, the *resolvent kernel* of $K(x,y)$, is given by the *Neumann series*

$$\Gamma(x,y) = -\sum_{n=1}^{\infty} K^{(n)}(x,y). \tag{4.28}$$

The iterated kernels $K^{(1)}(x,y)$, $K^{(2)}(x,y), \ldots$ are defined as follows:

$$K^{(1)}(x,y) = K(x,y)$$

$$K^{(2)}(x,y) = \int_a^b K(x,z)\, K(z,y)\, dz,$$

and in general

$$K^{(n)}(x,y) = \int_a^b K^{(n-1)}(x,z)\, K(z,y)\, dz,$$

$n = 3, 4, \ldots$. Under the assumption that the kernel is small [that is, $(b-a)\max|K(x,y)| < 1$] the Neumann series (4.28) converges uniformly, by comparison with a geometric series.

Since $g(x,\omega)$ is a second-order random function,

$$\int_a^b |g(x,\omega)|^2\,dx < \infty \tag{4.29}$$

almost surely. Also, since the resolvent kernel $\Gamma(x,y)$ is an L_2-kernel† on $[a,b]$,

$$\int_a^b |\Gamma(x,y)|^2\,dy < \infty$$

for every $x \in [a,b]$. Hence the integral

$$\int_a^b \Gamma(x,y)\,g(y,\omega)\,dy$$

exists on $[a,b] \times \Omega$. From the above we can conclude that

$$f(x,\omega) = g(x,\omega) - \int_a^b \Gamma(x,y)\,g(y,\omega)\,dy$$

is well defined on $[a,b] \times \Omega$.

We now show that

$$\int_a^b |\Gamma(x,y)\,g(y,\omega)|\,dy \in L_2[a,b]$$

for almost every $\omega \in \Omega$. An application of Hölder's inequality yields

$$\left(\int_a^b |\Gamma(x,y)\,g(y,\omega)|\,dy\right)^2 \leqslant \int_a^b |\Gamma(x,y)|^2\,dy \int_a^b |g(y,\omega)|^2\,dy.$$

Hence

$$\begin{aligned}\int_a^b \left\{\int_a^b |\Gamma(x,y)\,g(y,\omega)\,dy\right\}^2 dx &\leqslant \int_a^b \left\{\int_a^b |\Gamma(x,y)|^2\,dy \int_a^b |g(y,\omega)|^2\,dy\right\} dx \\ &= \int_a^b\int_a^b |\Gamma(x,y)|^2\,dx\,dy \int_a^b |g(y,\omega)|^2\,dy < \infty\end{aligned}$$

almost surely.

Finally, we show that the random function $f(x,\omega)$ defined by (4.26) satisfies the random Fredholm equation (4.25) on $[a,b] \times \Omega$ almost surely. Consider the identity (cf. Yosida [45, p. 117])

$$K(x,y) + \Gamma(x,y) - \int_a^b K(x,z)\,\Gamma(z,y)\,dz = 0. \tag{4.30}$$

† A function $\Phi(x,y)$ on $[a,b] \times [a,b]$ is said to be an *L_2-kernel* if (i) $\int_a^b \int_a^b |\Phi(x,y)|^2\,dx\,dy < \infty$, (ii) for each $x \in [a,b]$, $\Phi(x,y)$ is a measurable function of y such that $\int_a^b |\Phi(x,y)|^2\,dy < \infty$, and (iii) for each $y \in [a,b]$, $\Phi(x,y)$ is a measurable function of x such that $\int_a^b |\Phi(x,y)|^2 dx < \infty$. We refer to Tricomi [44] for a discussion of L_2-kernels.

If we multiply (4.30) by $g(y,\omega)$, and then integrate over the interval $[a,b]$, we obtain

$$\int_a^b g(y,\omega)\left\{K(x,y)+\Gamma(x,y)-\int_a^b K(x,z)\,\Gamma(z,y)\,dz\right\}dy=0;$$

which, upon rearrangement, yields

$$\int_a^b g(y,\omega)\,K(x,y)\,dy-\int_a^b\left\{\int_a^b K(x,z)\,\Gamma(z,y)\,g(y,\omega)\,dz\right\}dy=-\int_a^b \Gamma(x,y)\,g(y,\omega)\,dy. \tag{4.31}$$

Since

$$\int_a^b |\Gamma(z,y)\,g(y,\omega)|^2\,dy\in L_2[a,b]$$

almost surely, and since $|K(x,z)|\in L_2[a,b]$ for every $x\in[a,b]$, we have

$$\int_a^b |K(x,z)|\,dz\int_a^b |\Gamma(z,y)\,g(y,\omega)|\,dy<\infty$$

almost surely. An application of Tonelli's theorem† to the second term on the left-hand side of (4.31) yields

$$\int_a^b\left\{\int_a^b K(x,z)\,\Gamma(z,y)\,g(y,\omega)\,dz\right\}dy=\int_a^b\left\{K(x,z)\int_a^b \Gamma(z,y)\,g(y,\omega)\,dy\right\}dz. \tag{4.32}$$

If we now change the variable of integration in the first term on the left-hand side of (4.31) from y to z and then use (4.32), we can rewrite (4.31) as

$$\int_a^b K(x,z)\left\{g(z,\omega)-\int_a^b \Gamma(z,y)\,g(y,\omega)\,dy\right\}dz=-\int_a^b \Gamma(x,y)\,g(y,\omega)\,dy.$$

Using the definition of $f(x,\omega)$, as given by (4.26), the above expression reduces to

$$\int_a^b K(x,z)\,f(z,\omega)\,dz=-\int_a^b \Gamma(x,y)\,g(y,\omega)\,dy. \tag{4.33}$$

Rewriting (4.33) as

$$\int_a^b K(x,z)\,f(z,\omega)\,dz=g(x,\omega)-g(x,\omega)-\int_a^b \Gamma(x,y)\,g(y,\omega)\,dy,$$

and using the definition of $f(x,\omega)$, we see that (4.33) is equivalent to the random Fredholm equation (4.25).

The above result can be easily specialized to the case in which $K(x,y)$ is a Volterra kernel. In this case Eq. (4.25) is of the form

$$f(x,\omega)-\int_a^x K(x,y)\,f(y,\omega)\,dy=g(x,\omega); \tag{4.34}$$

† We refer to Dunford and Schwartz [14, p. 194] for a statement of Tonelli's theorem.

and the solution of the random Volterra equation is

$$f(x, \omega) = g(x, \omega) - \int_a^x \Gamma(x, y) g(y, \omega)\, dy. \tag{4.35}$$

We now state, without proof, the following result.

COROLLARY 4.1. *If $K(x,y)$ is a Volterra kernel on $[0,\tau] \times [0,\tau]$, $\tau > 0$, and if $g(x,\omega)$ is a second-order random function on $[0,\infty)$ which is continuous in mean square, then the random function $f(x,\omega)$ defined by (4.35) on $[0,\infty) \times \Omega$ satisfies the random Volterra equation (4.34) on $[0,\infty) \times \Omega$.*

C. Covariance function of the solution

Let

$$R_f(x_1, x_2) = \mathscr{E}\{f(x_1, \omega) f(x_2, \omega)\}, \qquad x_1, x_2 \in [a, b] \tag{4.36}$$

denote the *covariance function* of the solution $f(x,\omega)$ of the random integral equation (4.25). To establish the existence of $R_f(x_1, x_2)$ we have only to show that $\mathscr{E}\{|f(x,\omega)|^2\} < \infty$, for every $x \in [a,b]$; that is, $f(x,\omega)$ is a second-order random function. It follows from Hölder's inequality that

$$\left|\int_a^b \Gamma(x, y) g(y, \omega)\, dy\right|^2 \leqslant \int_a^b |\Gamma(x, y)|^2\, dy \int_a^b |g(y, \omega)|^2\, dy.$$

and

$$\mathscr{E}\left\{\left|\int_a^b \Gamma(x, y) g(y, \omega)\, dy\right|^2\right\} \leqslant \int_a^b |\Gamma(x, y)|^2\, dy\, \mathscr{E}\left\{\int_a^b |g(y, \omega)|^2\, dy\right\} < \infty$$

for every $x \in [a,b]$, since $g(x,\omega)$ is continuous in mean square. Hence, it follows from (4.26) that $\mathscr{E}\{|f(x,\omega)|^2\} < \infty$ for every $x \in [a,b]$; which establishes the existence of the covariance function $R_f(x_1, x_2)$.

The calculation of $R_f(x_1, x_2)$ is straightforward. From (4.36) and (4.26) we have

$$\begin{aligned} R_f(x_1, x_2) &= \mathscr{E}\left\{\left(g(x_1, \omega) - \int_a^b \Gamma(x_1, y) g(y, \omega)\, dy\right)\left(\overline{g(x_2, \omega)}\right.\right. \\ &\qquad \left.\left. - \int_a^b \overline{\Gamma(x_2, y) g(y, \omega)}\, dy\right)\right\} \\ &= \mathscr{E}\{g(x_1, \omega)\overline{g(x_2, \omega)}\} - \mathscr{E}\left\{g(x_1, \omega) \int_a^b \overline{\Gamma(x_2, y) g(y, \omega)}\, dy\right\} \\ &\qquad - \mathscr{E}\left\{\overline{g(x_2, \omega)} \int_a^b \Gamma(x_1, y) g(y, \omega)\, dy\right\} \\ &\qquad + \mathscr{E}\left\{\int_a^b \Gamma(x_1, y) g(y, \omega)\, dy \int_a^b \overline{\Gamma(x_2, y) g(y, \omega)}\, dy\right\} \end{aligned}$$

$$R_f(x_1, x_2) = R_g(x_1, x_2) - \int_a^b (\overline{\Gamma(x_2, y)}\, \mathscr{E}\{g(x_1, \omega)\overline{g(y, \omega)}\})\, dy$$
$$- \int_a^b (\Gamma(x_1, y)\, \mathscr{E}\{g(y, \omega)\, \overline{g(x_2, \omega)}\})\, dy$$
$$- \int_a^b \int_a^b (\Gamma(x_1, y_1)\, \overline{\Gamma(x_2, y_2)}\, \mathscr{E}\{g(x_1, \omega)\, \overline{g(x_2, \omega)}\})\, dy_1\, dy_2$$
$$= R_g(x_1, x_2) - \int_a^b \Gamma(x_2, y)\, R_g(x_1, y)\, dy - \int_a^b \Gamma(x_1, y)\, R_g(y, x_2)\, dy$$
$$- \int_a^b \int_a^b \Gamma(x_1, y_1)\, \overline{\Gamma(x_2, y_2)}\, R_g(y_1, y_2)\, dy_1\, dy_2 .$$

Put

$$H(x_1, x_2) = R_g(x_1, x_2) - \int_a^b \overline{\Gamma(x_2, y)}\, R_g(x_1, y)\, dy. \tag{4.37}$$

A simple calculation yields the following representation of $R_f(x_1, x_2)$ in terms of the covariance function $R_g(x_1, x_2)$ of the input random function $g(x, \omega)$:

$$R_f(x_1, x_2) = H(x_1, x_2) - \int_a^b \Gamma(x_1, y)\, H(y, x_2)\, dy. \tag{4.38}$$

Since $g(x, \omega)$ is continuous in mean square, its covariance function $R_g(x_1, x_2)$ is a symmetric nonnegative function continuous on $[a, b] \times [a, b]$. Hence, by Mercer's theorem (see Loève [31]),

$$R_g(x_1, x_2) = \sum_{n=1}^{\infty} \lambda_n\, \varphi_n(x_1)\, \overline{\varphi_n(x_2)}, \tag{4.39}$$

where the series converges absolutely and uniformly on $[a, b] \times [a, b]$. In (4.39) $\{\varphi_n(x)\}$ is the sequence of normalized eigenfunctions of $R_g(x_1, x_2)$, and $\{\lambda_n\}$ is the sequence of associated nonnegative eigenvalues; that is, for all integers m and n,

$$\lambda_n\, \varphi_n(x) = \int_a^b R_g(\tau, x)\, \varphi_n(\tau)\, d\tau, \qquad x \in [a, b]$$

$$\int_a^b \varphi_m(x)\, \varphi_n(x)\, dx = \delta_{m,n}$$

where $\delta_{m,n}$ is the Kronecker delta.

Put

$$\xi_n(\omega) = \int_a^b g(x, \omega)\, \varphi_n(x)\, dx, \qquad n = 1, 2, \ldots . \tag{4.40}$$

The random variables $\xi_n(\omega)$ are well defined since $\int_a^b |g(x, \omega)|^2\, dx < \infty$ almost surely, and the eigenfunctions are continuous on $[a, b]$. The sequence $\{\xi_n(\omega)\}_{n=1}^{\infty}$ is orthonormal on Ω; and for every $x \in [a, b]$ the series

$$\sum_{n=1}^{\infty} \lambda_n^{1/2}\, \xi_n(\omega)\, \varphi_n(x) \tag{4.41}$$

is a representation for $g(x,\omega)$ in the sense that

$$\lim_{n\to\infty} \mathscr{E}\left\{\left|g(x,\omega)-\sum_{n=1}^{N}\lambda_n^{1/2}\xi_n(\omega)\varphi_n(x)\right|^2\right\}=0.$$

Now, let $\psi_n(x)$, $n=1,2,\ldots$, be the solutions of the (deterministic) integral equations

$$\psi_n(x)-\int_a^b K(x,y)\psi_n(y)\,dt=\varphi_n(x); \tag{4.42}$$

that is,

$$\psi_n(x)=\varphi_n(x)-\int_a^b \Gamma(x,y)\psi_n(y)\,dy, \tag{4.43}$$

where, as before, $\Gamma(x,y)$ is the resolvent kernel of the Fredholm (or Volterra) kernel $K(x,y)$. It can be shown that $f(x,\omega)$, the solution of Eq. (4.25), admits the orthogonal representation

$$f(x,\omega)=\sum_{n=1}^{\infty}\lambda_n^{1/2}\xi_n(\omega)\psi_n(x), \qquad x\in[a,b]; \tag{4.44}$$

from which it follows that the covariance function $R_f(x_1,x_2)$ admits the orthogonal representation

$$R_f(x_1,x_2)=\sum_{n=1}^{\infty}\lambda_n\psi_n(x_1)\overline{\psi_n(x_2)}. \tag{4.45}$$

D. *Mean-square continuity of the solution*

We now show that the random function $f(x,\omega)$ is continuous in mean square if the kernel $K(x,y)$ of the integral operator is continuous.

THEOREM 4.4 *Let $K(x,y)$ be a Fredholm kernel on $[a,b]\times[a,b]$, and let $\Gamma(x,y)$ denote the associated resolvent kernel. If $K(x,y)$ is continuous on $[a,b]\times[a,b]$, then the solution $f(x,\omega)$ of the integral equation* (4.25) *is continuous in mean square on $[a,b]$.*

Proof. Let $x_0\in[a,b]$. It follows from (4.26) and an application of Minkowski's inequality that

$$\begin{aligned}(\mathscr{E}\{|f(x,\omega)-f(x_0,\omega)|^2\})^{1/2}&\\ =\left(\mathscr{E}\left\{\left|g(x,\omega)-g(x_0,\omega)+\int_a^b g(y,\omega)[\Gamma(x,y)-\Gamma(x_0,y)]^2\,dy\right|^2\right\}\right)^{1/2}&\\ \leqslant(\mathscr{E}\{|g(x,\omega)-g(x_0,\omega)|^2\})^{1/2}+\left(\mathscr{E}\left\{\left|\int_a^b g(y,\omega)[\Gamma(x,y)-\Gamma(x_0,y)]\,dy\right|^2\right\}\right)^{1/2}.&\end{aligned}$$

Since $g(x,\omega)$ is continuous in mean square,

$$\lim_{x\to x_0} \mathscr{E}\{|g(x,\omega)-g(x_0,\omega)|^2\}=0;$$

hence it is sufficient to show that

$$\lim_{x\to x_0}\left\{\left|\int_a^b g(y,\omega)\,[\Gamma(x_0,y)-\Gamma(x,y)]\,dy\right|^2\right\}=0. \tag{4.46}$$

An application of Hölder's inequality yields

$$\left|\int_a^b g(y,\omega)\,[\Gamma(x_0,y)-\Gamma(x,y)]\,dy\right|^2$$
$$\leqslant \int_a^b |g(y,\omega)|^2\,dy \int_a^b |\Gamma(x_0,y)-\Gamma(x,y)|^2\,dy.$$

Since $g(y,\omega)$ is continuous in mean square,

$$\mathscr{E}\left\{\int_a^b |g(y,\omega)|^2\,dy\right\}=M<\infty;$$

hence

$$\mathscr{E}\left\{\left|\int_a^b g(y,\omega)\,[\Gamma(x_0,y)-\Gamma(x,y)]\,dy\right|^2\right\}\leqslant M\int_a^b |\Gamma(x_0,y)-\Gamma(x,y)|^2\,dy.$$

Therefore, it remains to show that

$$\lim_{x\to x_0}\int_a^b |\Gamma(x_0,y)-\Gamma(x,y)|^2\,dy=0.$$

Let $\epsilon>0$. By hypothesis $K(x,y)$ is continuous on $[a,b]\times[a,b]$; hence its resolvent kernel $\Gamma(x,y)$ is also continuous on $[a,b]\times[a,b]$. Since $\Gamma(x,y)$ is uniformly continuous on $[a,b]\times[a,b]$, we can pick a $\delta>0$ such that

$$\int_a^b |\Gamma(x,y)-\Gamma(x_0,y)|^2\,dy<\epsilon$$

for $|x-x_0|<\delta$. This establishes (4.46).

E. A concrete example: A Volterra integral equation with Wiener input

In this subsection we consider as a concrete example of the type of integral equation studied in this section, a Volterra integral equation with input a Wiener process. Consider the integral equation (4.25) on $[0,1]$ with kernel

$$K(x,y)=\begin{cases}-1 & \text{for} \quad x\geqslant y\\ 0 & \text{for} \quad x<y.\end{cases} \tag{4.47}$$

In this case Eq. (4.25) is of the form

$$f(x,\omega)+\int_0^x f(y,\omega)\,dy=g(x,\omega). \tag{4.48}$$

The resolvent kernel $\Gamma(x,y)$, which is easily calculated, is given by the Neumann series

$$-\sum_{n=1}^{\infty} K^{(n)}(x,y) = \Gamma(x,y) = \begin{cases} e^{-(x-y)} & \text{for } x \geqslant y \\ 0 & \text{for } x < y. \end{cases}$$

It now follows from Theorem 4.3 (and Corollary 4.1) that the random function $f(x,\omega)$ on $[0,1]$ which satisfies Eq. (4.48) is given by

$$f(x,\omega) = g(x,\omega) - \int_0^x e^{-(x-y)} g(y,\omega)\, dy. \tag{4.49}$$

In order to calculate the covariance function $R_f(x_1,x_2)$ we use (4.37) and the fact that the covariance function $R_g(x_1,x_2)$ of the Wiener process $g(x,\omega)$ is $R_g(x_1,x_2) = \min(x_1,x_2)$, $x_1,x_2 \in [0,1]$. From (4.37) and (4.49) we have

$$H(x_1,x_2) = \begin{cases} x_1 - \int_0^{x_1} \exp[-(x_2-\xi)]\,\xi\, d\xi - x_1 \int_{x_1}^{x_2} \exp[-(x_2-\xi)]\, d\xi & \text{for } x_1 < x_2 \\ x_2 - \int_0^{x_2} \exp[-(x_2-\xi)]\,\xi\, d\xi & \text{for } x_1 \geqslant x_2. \end{cases}$$

Substitution of the above expression for $H(x_1,x_2)$ in (4.38) yields

$$R_f(x_1,x_2) = \tfrac{1}{2}\exp\{-|x_1-x_2|\} - \tfrac{1}{2}\exp\{-(x_1+x_2)\}. \tag{4.50}$$

4.3 Fredholm and Volterra Integral Equations with Random Kernels

A. Introduction

Consider the random integral equation

$$L(\omega)f - \lambda f = g, \tag{4.51}$$

where f and g are elements of a Banach space $\mathfrak{X}$, and $L(\omega)$ is a random Fredholm or Volterra operator on $\mathfrak{X}$; that is, for every $f \in \mathfrak{X}$, $L(\omega)f$ is a generalized random variable with values in $\mathfrak{X}$. Random integral operators arise when we consider integral operators with *random kernels*, or when the integrals are defined over *random domains*. We shall show, however, that in certain instances these two cases are equivalent; that is, we can consider an integral equation over a random domain as an integral equation with a random kernel, and vice versa.

In this section we study integral equations with random kernels. Since in this case the stochastic properties of a solution depend on the stochastic properties of the random integral operator $L(\omega)$, the problems of existence, uniqueness, and measurability of the solutions are a little more involved than in the case considered in the last section, that is, integral equations with random forcing functions.

In Sect. 4.3B we consider Fredholm integral equations with random degenerate kernels. As in the theory of deterministic integral equations, this class of random integral equations is one of the easiest to handle; however, they constitute an important class of random integral equations which are useful in many applied problems. Sections 4.3C and D are devoted to random Fredholm (and Volterra) equations in the space of continuous functions and in Orlicz spaces, respectively. In Sect. 4.3E we consider a class of random Volterra equations which arise in the study of ordinary differential equations with random coefficients.

B. Fredholm equations with random degenerate kernels

In this section we consider the Fredholm integral equation of second kind

$$\int_0^1 K(x,y)f(y)\,dy - \lambda f(x) = g(x) \tag{4.52}$$

in the case when the kernel $K(x,y)$ is a random degenerate kernel. A Fredholm kernel $K(x,y)$ is said to be *degenerate*† if it is of form

$$K(x,y) = \sum_{i=1}^{n} \alpha_i(x)\,\beta_i(y), \tag{4.53}$$

where $\{\alpha_i(x)\}_{i=1}^n$ and $\{\beta_i(y)\}_{i=1}^n$ are two independent sets of linearly independent $L_2(0,1)$-functions.‡ In this case the Fredholm integral equation (4.52) is equivalent to a system of n linear algebraic equations in n unknowns. If we put

$$\xi_j = \int_0^1 \beta_j(x) f(x)\,dx, \quad j = 1, 2, \ldots, n, \tag{4.54}$$

Eq. (4.52) with kernel (4.53) becomes

$$\sum_{j=1}^{n} \xi_j\,\alpha_j(x) - \lambda f(x) = g(x). \tag{4.55}$$

The ξ_j are unknown constants, since the function $f(x)$ is unknown. From Eq. (4.55) we obtain

$$f(x) = (1/\lambda)\left\{\sum_{j=1}^{n} \xi_j\,\alpha_j(x) - g(x)\right\}; \tag{4.56}$$

hence the problem reduces to the determination of the constants ξ_j.

If we now multiply Eq. (4.55) by $\beta_i(x)$, $i = 1,2,\ldots,n$, and then integrate, we obtain

$$\sum_{j=1}^{n} \xi_j \int_0^1 \alpha_j(x)\,\beta_i(x)\,dx - \lambda\xi_i = \int_0^1 \beta_i(x)\,g(x)\,dx;$$

† Degenerate kernels are frequently referred to as *separable*, or *Pincherle–Goursat* kernels, and kernels of *finite rank*.

‡ It is of interest to note that a degenerate kernel generates a bounded linear transformation L on $L_2(0,1)$, and since its range is finite dimensional, L is compact.

that is,

$$\sum_{j=1}^{n} a_{ij}\xi_j - \lambda\xi_i = b_i, \qquad i = 1, 2, \ldots, n, \tag{4.57}$$

where

$$a_i = \int_0^1 \alpha_j(x)\,\beta_i(x)\,dx \tag{4.58}$$

$$b_i = \int_0^1 \beta_i(x)\,g(x)\,dx. \tag{4.59}$$

Rewriting Eq. (4.57) in matrix form, we have

$$(A - \lambda I)\,\xi = b, \tag{4.60}$$

where $A = (a_{ij})$ is an $n \times n$ matrix, and ξ and b are n-vectors. The system (4.57) is equivalent to Eq. (4.52). Hence, if ξ_i is a solution of Eq. (4.57), the corresponding solution of Eq. (4.52) is given by Eq. (4.56).

Fredholm equations with degenerate kernels are of independent interest; however, they are often utilized to obtain approximate solutions of Fredholm equations in which the kernel can be approximated by a polynomial in x and y. Also, the general case of a continuous Fredholm kernel can be reduced to the case of a degenerate kernel by utilizing the Weierstrass approximation theorem, which, we recall, states that a continuous function of two variables defined on a closed rectangle can be uniformly approximated by polynomials (cf. Petrovskiĭ [33, p. 28]).

We refer to Courant and Hilbert [12, Chap. III], Kantorovich and Krylov [25, Chap. II], Mihlin [32, Chap. I], Pogorzelski [34, Chap. II], Stakgold [42, Chap. 3] and Tricomi [44, Chap. II] for detailed discussions of Fredholm equations with degenerate kernels and their applications.

Consider the *random degenerate kernel*

$$K(x, y, \omega) = \sum_{i=1}^{n} \alpha_i(x, \omega)\,\beta_i(y). \tag{4.61}$$

In (4.61) $\{\alpha_i(x,\omega)\}_{i=1}^n$ is a family of almost surely independent $L_2(0,1)$-random functions, and $\{\beta_i(y)\}_{i=1}^n$ is a set of independent $L_2(0,1)$-determinate functions. Clearly, for every fixed $x, y \in (0,1)$ the kernel $K(x,y,\omega)$ is a measurable function of ω. Put

$$\xi_i = \int_0^1 \beta_i(x)\,f(x)\,dx, \qquad i = 1, 2, \ldots, n. \tag{4.62}$$

Then, proceeding as in the deterministic case, the Fredholm equation

$$\int_0^1 K(x, y, \omega)\,f(y)\,dy - \lambda f(x) = g(x) \tag{4.63}$$

with random degenerate kernel $K(x,y,\omega)$ as given by (4.61) can be reduced to the following system of random linear algebraic equations:

$$\sum_{j=1}^{n} a_{ij}(\omega)\,\xi_j - \lambda\xi_i = b_i, \qquad i = 1,2,\ldots,n. \tag{4.64}$$

In (4.64)

$$a_{ij}(\omega) = \int_0^1 \alpha_j(x,\omega)\,\beta_i(x)\,dx, \qquad i,j = 1,2,\ldots,n \tag{4.65}$$

and

$$b_i = \int_0^1 \beta_i(x)\,g(x)\,dx, \qquad i = 1,2,\ldots,n. \tag{4.66}$$

The integrals in (4.65) and (4.66) are well defined; and the Riemann integrability of $\beta_i(x_1)\beta_i(x_2)\,R_j(x_1,x_2)$, where $R_j(x_1,x_2)$ is the covariance function of the $\alpha_j(x,\omega)$-process, is sufficient to ensure that the integral in (4.65) exists in mean square, and defines, for every pair i,j, a real-valued random variable $a_{ij}(\omega)$.

Equation (4.64) can be rewritten as the random operator equation

$$(A(\omega) - \lambda I)\,\xi = b, \tag{4.67}$$

where $A(\omega)$ is an $n \times n$ random matrix with elements $a_{ij}(\omega)$ defined by (4.65), and ξ and b are n-vectors. We remark that Eq. (4.67) can be interpreted as a random operator equation in the Euclidean space R_n or the Hilbert space $l_2(n)$. In [6] the existence, uniqueness and measurability of the solution $\xi(\omega)$ of Eq. (4.67) was established using the Špaček–Hanš probabilistic analogue of the Banach contraction mapping theorem (Theorem 3.5). However, an application of the result of Bharucha-Reid and Hanš (Theorem 3.13) on the invertibility of linear random operators of the form $L(\omega) - \lambda I$ enables us to state the following result for Eq. (4.67).

THEOREM 4.5. *Let $\lambda \neq 0$ be a real number such that*

$$\mu(\Omega(\lambda)) = \mu\left(\left\{\omega : \left[\sum_{i,j=1}^{n} a_{ij}^2(\omega)\right]^{1/2}\right\} < |\lambda|\right) = 1.$$

Then the random matrix $A(\omega) - \lambda I$ is invertible; and the solution

$$\xi(\omega) = (A(\omega) - \lambda I)^{-1}\,b$$

of Eq. (4.67), equivalently the solution $f(x,\omega)$ of Eq. (4.52), is $(\Omega(\lambda) \cap \mathfrak{A})$-measurable.

We now show that a Fredholm equation with a random continuous kernel can be reduced to a Fredholm equation with degenerate kernel. Consider a random continuous Fredholm kernel $K(x,y,\omega)$, $x,y \in [a,b]$, and assume that it can be written in the form

$$K(x,y,\omega) = M(x,y) + N(x,y,\omega), \tag{4.68}$$

where the deterministic kernel $M(x,y)$ satisfies the condition

$$|b-a| \max |M(x,y| < 1$$

[that is, $M(x,y)$ is "small"] and $N(x,y,\omega)$ is a random degenerate kernel. Hence the kernel $K(x,y,\omega)$ can be regarded as a random kernel which results from the perturbation of a "small" kernel by a random degenerate kernel. Hence the random Fredholm equation

$$f - L(\omega) f = g, \tag{4.69}$$

where $L(\omega)$ denotes the Fredholm operator with kernel $K(x,y,\omega)$, can be written in the form

$$f - L_M f = L_N(\omega) f + g. \tag{4.70}$$

In (4.70) we have put

$$L_M[f] = \int_a^b M(x,y) f(y)\, dy, \qquad L_N(\omega)[f] = \int_a^b N(x,y,\omega) f(y)\, dy,$$

where $N(x,y,\omega)$ is of the form (4.61). If we put

$$f - L_M f = h, \tag{4.71}$$

then (cf. Sect. 4.2B)

$$f(x) = h(x) - \int_a^b \Gamma(x,y) h(y)\, dy, \tag{4.72}$$

where $\Gamma(x,y)$ is the resolvent kernel associated with $M(x,y)$ (cf. (4.28)). Inserting (4.72) in (4.70), we obtain

$$h = L_N(\omega) \left[h - \int_a^b \Gamma(x,y) h(y)\, dy \right] + g. \tag{4.73}$$

Put

$$T(\omega)[h] = L_N(\omega) \left[\int_a^b \Gamma(x,y) h(y)\, dy \right],$$

and consider the random operator

$$S(\omega)[h] = (L_N(\omega) - T(\omega))[h]. \tag{4.74}$$

Then, it is easily verified that the kernel, say $\tilde{K}(x,y,\omega)$, of the random integral operator $S(\omega)$ is degenerate, and that the integral equations (4.70) and (4.73) are equivalent.

In Chap. 5 we consider the asymptotic distribution of the eigenvalues of the Fredholm operators with random degenerate kernels (cf. Bharucha-Reid and Arnold [8]). Since the eigenvalues of the random Fredholm operator are the solutions of the random algebraic equation $A(\omega) - \lambda I = 0$, we investigate the asymptotic distribution of the random eigenvalues of $A(\omega)$. In the next section of this chapter we consider a Volterra equation with random degenerate kernel which arises in the study of ordinary differential equations with random coefficients.

C. Random Fredholm equations in the space of continuous functions

Consider the Fredholm equation

$$\int_a^b K(x,y) f(y)\, dy - \lambda f(x) = g(x) \tag{4.75}$$

in $C[a,b]$, the space of continuous functions defined on the interval $[a,b]$. With the norm $\|f\| = \max_{x\in[a,b]} |f(x)|$, the space $C[a,b]$ becomes a separable Banach space. Let L denote the Fredholm operator on C with kernel $K(x,y)$; that is,

$$L[f(x)] = \int_a^b K(x,y) f(y)\, dy, \qquad f \in C \tag{4.76}$$

In this section we consider the existence, uniqueness, and measurability of the solution of Eq. (4.75) when the kernel K is a random kernel $K(x,y,\omega)$. The first problem we consider is that of the measurability of the Fredholm operator

$$L(\omega)\,[f(x)] = \int_a^b K(x,y,\omega) f(y)\, dy. \tag{4.77}$$

With reference to the deterministic kernel $K(x,y)$, we assume that $K(x,y)$ is bounded for every $x,y \in [a,b]$, and is continuous except possibly at points on a finite number of continuous curves $y = \varphi_i(x)$, $x \in [a,b]$, $i = 1,2,\ldots,n$. With the above assumptions on $K(x,y)$, the Fredholm operator L is completely continuous on C (cf. Kolmogorov and Fomin [29, pp. 239–243]). We remark that kernels with the above properties are sometimes referred to as *mildly discontinuous*. As an example, a Volterra kernel on $[0,b] \times [0,b]$ which is continuous for $y < x$ and vanishes for $y \geqslant x$ is mildly discontinuous on $[0,b] \times [0,b]$. In this case we can take $n = 1$ and $\varphi_1(x) = x$.

Let $\mathfrak{K}$ denote the space of all mildly discontinuous kernels K defined on $[a,b] \times [a,b]$, and such that for every $x \in [a,b]$, $y \in [a,b]$ and every sequence of real numbers $b \geqslant \delta_1 > \delta_2 > \cdots > \delta_n \to 0$, (i) $K(x,0) = \lim_{n\to\infty} K(x,\delta_n)$ and (ii) $K(x,y) = \lim_{n\to\infty} K(x, y - \delta_n)$, provided $\delta_1 \leqslant y$ in (ii). $\mathfrak{K}$, as the space of all bounded functions on $[a,b] \times [a,b]$ with the above properties, is clearly a linear space; and with the norm $\|K\| = \sup |K(x,y)|$, where the supremum is taken over $x \in [a,b]$ and $y \in [a,b]$, $\mathfrak{K}$ becomes a separable normed linear space. Let $\mathfrak{B}(\mathfrak{K})$ denote the σ-algebra of subsets of $\mathfrak{K}$. We can now define a *random kernel* as a $\mathfrak{A}$-measurable mapping K of Ω into $\mathfrak{K}$.

The relationship between the measurability of the Fredholm operator with kernel $K(x,y,\omega)$ and the measurability of its kernel is established by the following result:

THEOREM 4.6. *Let K be a mapping of Ω into $\mathfrak{K}$, and let the transformation $L(\omega)$ of $\Omega \times C$ into C be defined for every $\omega \in \Omega$ and every $f \in C$ by* (4.77). *Then $L(\omega)$*

is a completely continuous linear operator on C for every $\omega \in \Omega$. Moreover, the following statements are equivalent.

(i) *$L(\omega)$ is a random operator on C;*
(ii) *$\{\omega : K(x,y,\omega) < \xi\} \in \mathfrak{A}$ for every $x, y \in [a,b]$, and every $\xi \in R$;*
(iii) *$K(x,y,\omega)$ is a random kernel;*
(iv) *$L(\omega)$ is an operator-valued random variable.*

Proof. The complete continuity of $L(\omega)$ for every $\omega \in \Omega$ follows from the classical result (cf. Kolmogorov and Fomin [29, p. 241, Theorem 1]). Since statements (i)–(iv) are assertions of measurability, we will base the proof of their equivalence on the fact that a mapping $x(\omega)$ of Ω into a separable Banach space $\mathfrak{X}$ is a generalized random variable if and only if for every bounded linear functional x^* belonging to a set which is total on $\mathfrak{X}$ the mapping $x^*(x(\omega))$ is a real-valued random variable.

For every $x, y \in [a,b]$, $f \in C$, and $K \in \mathfrak{K}$, put

$$g_{x,y}(K) = K(x,y), \tag{4.78}$$

and

$$h_{x,f}(K) = \int_a^b K(x,y) f(y)\, dy. \tag{4.79}$$

Then it is clear that the sets $\{g_{x,y}(K) : x, y \in [a,b]\}$ and $\{h_{x,f}(K) : x \in [a,b], f \in C\}$ are total sets of bounded linear functionals on $\mathfrak{K}$. Further, if for every $x \in [a,b]$ and $f \in C$ we put

$$r_x(f) = f(x), \tag{4.80}$$

then the set $\{r_x(f) : x \in [a,b]\}$ is a total set of bounded linear functionals on C. Finally, let

$$E_0 = \left\{L : L[f] = \int_a^b K(x,y) f(y)\, dy, \qquad K \in \mathfrak{K}, \quad f \in C\right\}.$$

If, for every $x \in [a,b]$ and $f \in C$, we put

$$s_{x,f}(L) = r_x(L[f]), \tag{4.81}$$

then the set $\{s_{x,f}(L) : x \in [a,b], f \in C\}$ is a total set of bounded linear functionals on E_0. Since E_0 is a subspace of $\mathfrak{L}(C)$, the algebra of bounded linear operators on C, it is a normed linear space; and for every $L \in E_0$

$$\begin{aligned} \|L\| &= \sup \|L[f]\| = \sup \left\| \int_a^b K(x,y) f(y)\, dy \right\| \\ &= \sup \max \left\| \int_a^b K(x,y) f(y)\, dy \right\| \\ &\leqslant \sup \max\, (b-a)\, \|K\|\, \|f\| \\ &= (b-a)\, \|K\|, \end{aligned} \tag{4.82}$$

where sup is taken over $\{f: \|f\| = 1\}$ and max over $x \in [a,b]$. Hence the separability of $\mathfrak{K}$ implies the separability of E_0.

The equivalence of (i)–(iv) now follows from (a) Theorem 1.2 (the applicability of which follows from the separability of the spaces C, $\mathfrak{K}$, and E_0, and the fact that the sets of bounded linear functionals defined by (4.78)–(4.81) are total on K, C, and E_0, respectively), (b) the fact that $(E_0, \mathfrak{B}(E_0))$ is a separable measurable space, where $\mathfrak{B}(E_0) = E_0 \cap \mathfrak{B}(\mathfrak{L}(C))$, and (c) the fact that

$$h_{x,f}(K) = s_{x,f}(L) = r_x(L[f]) = L[f]$$

for every $x \in [a,b]$, $f \in C$, and every $K \in \mathfrak{K}$ and $L \in E_0$, where $L(\omega)[f]$ is defined by (4.77).

We now turn to the problems of existence, uniqueness, and measurability of the solution of the random integral equation

$$(L(\omega) - \lambda)f = g(\omega), \tag{4.83}$$

where $L(\omega)$ is the random Fredholm operator given by (4.77), and $g(\omega)$ is a $C[a,b]$-valued random variable. To establish the existence, uniqueness, and measurability of the solution of Eq. (4.83) when $g(\omega)$ is a given generalized random variable with values in $C[a,b]$ we can use fixed-point methods; and to consider the same problems when $g(\omega)$ is an arbitrary generalized random variable with values in $C[a,b]$ we can use results on the existence and measurability of the inverse of the random operator $(L(\omega) - \lambda I)$.

As in Chap. 2, let $\rho(L)$ denote the set of those pairs $(\omega, \lambda) \in \Omega \times R$ for which the linear random operator $(L(\omega) - \lambda I)$ has a linear bounded inverse; and recall that $\{\omega : (\omega, \lambda) \in \rho(L)\} \in \mathfrak{A}$.

We now state and prove a theorem which establishes a sufficient condition for the invertibility of $(L(\omega) - \lambda I)$.

THEOREM 4.7. *Let $L(\omega)$ be the random Fredholm operator on C as given by* (4.77), *and let the real number λ satisfy the inequality $(b-a)\|K(x,y,\omega)\| < |\lambda|$ with probability one. Then the linear random operator $(L(\omega) - \lambda I)$ is invertible, that is, $\mu(\{\omega : (\omega, \lambda) \in \rho(L)\}) = 1$.*

Proof. From Theorem 3.12 we have that the inverse of $(L(\omega) - \lambda I)$ exists for every $\lambda \neq 0$ such that $\mu(\{\omega : \|L(\omega)\| < |\lambda|\}) = 1$. By hypothesis

$$(b-a)\|K(x,y,\omega)\| < |\lambda|$$

almost surely; hence using (4.82) we have

$$\|L(\omega)\| \leqslant (b-a)\|K(x,y,\omega)\| < |\lambda|$$

almost surely.

In view of the above, it follows from Theorems 4.6 and 4.7 that the formal solution of the random integral equation (4.83) with Fredholm kernel is given by

$$\begin{aligned} f(x,\omega) &= (L(\omega)-\lambda I)^{-1}[g(x,\omega)] \\ &= R(L(\omega),\lambda)[g(x,\omega)]. \end{aligned} \tag{4.84}$$

Since a Volterra operator is a special case of a Fredholm operator, Theorem 4.6 is applicable to the Volterra operator

$$L(\omega)[f(x)] = \int_a^x K(x,y,\omega)f(y)\,dy. \tag{4.85}$$

We now state and prove the analogue of Theorem 4.7 for integral equations with random Volterra kernels.

THEOREM 4.8. *Let $L(\omega)$ denote a random Volterra operator on $C[a,b]$; and let the kernel $K(x,y,\omega)$ satisfy the condition $\mu(\{\omega : K(x,y,\omega)=0\})=1$ for every $a \leqslant x < y \leqslant b$. Then for every real number $\lambda \neq 0$ the linear random operator $(L(\omega)-\lambda I)$ is invertible.*

Proof. As in the deterministic theory of Volterra equations of the second kind (cf. Yosida [45, Chap. 4]), estimates obtained using the iterated kernels $K^{(n)}.(x,y,\omega)$ lead to the inequality

$$\|L^n(\omega)\| \leqslant |\lambda(b-a)|^n \, \|K(x,y,\omega)\|^n/n!, \tag{4.86}$$

which holds almost surely and for every $n=1,2,\ldots$. Hence it follows from Theorem 3.12 that $(L(\omega)-\lambda I)$ is invertible.

D. Random Fredholm equations in Orlicz spaces

Let D denote a bounded closed subset of an n-dimensional Euclidean space; and let $K(x,y)$, $x,y \in D$, be a continuous Fredholm kernel. In this section we consider the Fredholm integral equation

$$\int_{I_\omega} K(x,y)f(y)\,dy - \lambda f(x) = g(x), \tag{4.87}$$

where I_ω is for every $\omega \in \Omega$ a subset of D. Hence the case we consider here is that of a Fredholm operator defined over a random domain:

$$L(\omega)[f(x)] = \int_{I_\omega} K(x,y)f(y)\,dy. \tag{4.88}$$

We will study Eq. (4.87) in the separable Orlicz space $L_\Phi(D,m)$, where m denotes the Lebesgue measure of D. However, it is known (cf. Krasnosel'skiĭ and Rutickiĭ [30, p. 120]) that, when D is a bounded closed subset of an

n-dimensional Euclidean space, $L_\Phi(D,m)$ is isomorphic and isometric to the Orlicz space $\mathfrak{M}_\Phi$ of functions defined on $[0,m]$, provided that in both cases the same convex function Φ is employed. Hence we can restrict our attention to Eq. (4.87) in the one-dimensional case; that is, we will take $D=[0,m]$, and $I_\omega \subset [0,m]$ for every $\omega \in \Omega$; and consider Eq. (4.87) in the space $\mathfrak{M}_\Phi$. As pointed out in Sect. 1.2B, if we take $\Phi(u)=\alpha u^p$ ($\alpha > 0$, $1 \leqslant p < \infty$), then the Orlicz space $L_\Phi(D,m)$ (equivalently, $\mathfrak{M}_\Phi$) contains the same functions as the Lebesgue spaces $L_p(D)$. Hence the results we obtain in this section are valid for random Fredholm equations of the form (4.87) in the Lebesgue spaces L_p. The results presented in this section are based on the papers of Bharucha-Reid [4, 5] and Hanš [20].

In order to obtain results for Eq. (4.87) which are analogous to those obtained in Sect. 4.3C, it is advantageous to rewrite Eq. (4.87) as an integral equation with a random kernel. For every $\omega \in \Omega$, put

$$\begin{aligned} \tilde{K}(x,y,\omega) &= K(x,y) \qquad \text{for} \quad x \in [0,m], \quad y \in I_\omega \\ &= 0 \qquad \text{for} \quad x \in [0,m], \quad y \in [0,m]-I_\omega. \end{aligned} \tag{4.89}$$

Since $K(x,y)$ is a continuous Fredholm kernel, the random kernel $\tilde{K}(x,y,\omega)$ defined by (4.89) is a mapping of Ω into $\tilde{\mathfrak{K}}$, where $\tilde{\mathfrak{K}}$ denotes the space of all mildly discontinuous kernels $\tilde{\mathfrak{K}}$ defined on $[0,m]\times[0,m]$, which also satisfy the conditions for the kernels K which belong to the space $\mathfrak{K}$ introduced in Sect. 4.3C. We can now use Theorem 4.6 to state necessary and sufficient conditions for the measurability of the Fredholm operator defined by (4.88) with $f \in \mathfrak{M}_\Phi$.

THEOREM 4.9. *The mapping* $L(\omega)$ *of* $\Omega \times \mathfrak{M}_\Phi$ *into* $\mathfrak{M}_\Phi$ *defined by* (4.88) *is for every deterministic kernel* $K(x,y) \in \mathfrak{K}$ *a random operator if and only if* $\{\omega : y \in I_\omega\} \in \mathfrak{A}$ *for every* $y \in [0,m]$.

Proof. The assertion of the above theorem follows from Theorem 4.6 if in condition (ii) in the statement of that theorem we put $K(x,y,\omega)=\chi_{I_\omega}(y)K(x,y)$ for every $\omega \in \Omega$, $x,y \in [0,m]$, where χ_{I_ω} denotes the indicator function of the set I_ω.

It is of interest to note that in general nothing can be said about I_ω when a particular kernel $K \in \mathfrak{K}$ is considered. As an example, consider the case when $K(x,y)=0$ for all $x,y \in [0,m]$. In this case the integral operator defined by (4.88) is equal to the null operator, and is, therefore, measurable regardless of the nature of I_ω. Nevertheless, we can state the following result, the proof of which is an immediate consequence of Theorem 4.6.

THEOREM 4.10. *If $K \in \mathfrak{K}$, then the mapping $L(\omega)$ of $\Omega \times \mathfrak{M}_\Phi$ into $\mathfrak{M}_\Phi$ defined by (4.88) is a random operator if and only if $\chi_{I_\omega}(y)K(x,y)$ is a real-valued random variable for every $x, y \in [0,m]$.*

We now consider the measurability of the Fredholm operator (4.88) when the set I_ω is of the form $I_\omega = [0, b(\omega)]$. In this case we have the following result, the proof of which follows from Theorem 4.9.

THEOREM 4.11. *If $I_\omega = [0, b(\omega)]$ for every $\omega \in \Omega$, then the mapping $L(\omega)$ of $\Omega \times \mathfrak{M}_\Phi$ into $\mathfrak{M}_\Phi$ defined by (4.82) is for every $K \in \mathfrak{K}$ a random operator if and only if $b(\omega)$ is a real-valued random variable.*

Let us now consider the measurability of $L(\omega)$ in the case when I_ω is of the form $I_\omega = [a(\omega), b(\omega)]$. In this case the measurability question is no longer unambiguous. We have the following result.

THEOREM 4.12. *If $I_\omega = [a(\omega), b(\omega)]$ for $\omega \in \Omega$, then the mapping $L(\omega)$ of $\Omega \times \mathfrak{M}_\Phi$ into $\mathfrak{M}_\Phi$ defined by (4.88) is for every $K \in \mathfrak{K}$ a random operator if and only if $a(\omega)$ and $b(\omega)$ are both real-valued random variables.*

Proof. The above result follows from Theorem 4.9 and the relation

$$\{\omega : b(\omega) \geqslant r_0\} = \bigcup \{\omega : r \in [a(\omega), b(\omega)]\},$$

which holds for every rational number r_0 provided the union is taken over all rational numbers $r \geqslant r_0$.

We now consider an example which points out the relationship between the structure of the σ-algebra $\mathfrak{U}$ and the measurability of $a(\omega)$ and $b(\omega)$. Suppose there exists a set $F \in \mathfrak{U}$ with $\mu(F) > 0$ such that $F \subset \{\omega : a(\omega) = b(\omega)\}$. In this case, depending on the structure of $\mathfrak{U}$, $a(\omega)$ and $b(\omega)$ may not be random variables. Consider the probability measure space $(\Omega, \mathfrak{U}, \mu)$, where $\Omega = R$, $\mathfrak{U}$ is the σ-algebra of all at most denumerable sets of real numbers and their complements, and μ is a complete probability measure defined by $\mu(A) = 0$ if A is at most denumerable, and $\mu(A) = 1$ otherwise. Let $I_\omega = [a(\omega), b(\omega)]$ for every $\omega \in \Omega$, where $a(\omega) = b(\omega) = \omega - [\omega/m]m$ for every $\omega \in \Omega$. Here we assume that $[\omega/m]$ is defined by $(\omega/m) - 1 < [\omega/m] \leqslant (\omega/m)$. Then the indicator function $\chi_{I_\omega}(y)$ is a real-valued random variable for every $y \in [0,m]$, but

$$\{\omega : 2a(\omega) < m\} = \{\omega : 2b(\omega) < m\} \notin \mathfrak{U}.$$

Having established the measurability of the Fredholm operator $L(\omega)$, it follows from Theorem 2.15 that if for every real number $\lambda \neq 0$ the set $\{\omega : \|L(\omega)\| < |\lambda|\} \in \mathfrak{U}$, then the transformation $L(\omega) - \lambda I$ is invertible for every $\omega \in \Omega(\lambda)$. Hence, if $m(I_\omega)\|K(x,y)\| < |\lambda|$ almost surely (where $m(I_\omega)$ is the Lebesgue measure of the set I_ω), then $L(\omega) - \lambda I$ is invertible; and the solution of Eq. (4.87) is, for every $g(x,\omega) \in L_\Phi(D)$, given by

$$f(x,\omega) = R(L(\omega),\lambda)\,[g(x,\omega)] \in L_\Phi(D), \tag{4.90}$$

and $f(x,\omega)$ is $\Omega(\lambda) \cap \mathfrak{U}$-measurable. The reader is referred to Bharucha-Reid [5] for a discussion of other results on random Fredholm equations in Orlicz spaces.

E. Integral equation formulation of a class of random differential equations

1. *Introduction.* Consider the random operator equation

$$L(\omega)f(x,\omega) = g(x,\omega), \tag{4.91}$$

where $L(\omega)$ is an ordinary differential operator of order n with random coefficients; that is,

$$L(\omega)\,[f] = \sum_{k=0}^{n} a_k(x,\omega)\, d^k f/dx^k. \tag{4.92}$$

The study of ordinary differential equations of order n with random coefficients was initiated by Samuels and Eringen [39] (cf. also Samuels [37, 38]), who restricted their attention to equations with the following properties: (i) small randomly varying parameters, (ii) slowly varying coefficients, and (iii) only one random coefficient. Perturbation methods were utilized to obtain a formal solution of Eq. (4.91); and the methods developed were applied to an *RLC* circuit with random capacitance and to dynamical instability of an elastic bar subject to random time-dependent axial force. Caughey and Dienes [10] also considered Eq. (4.91), with the forcing function and the lowest order coefficient being white-noise processes. A number of other authors have studied differential equations of first- and second-order with random coefficients. We refer to Bharucha-Reid [7] and Saaty [36, pp. 419–433] for a discussion of other studies and additional references. In Chap. 7 we consider, utilizing Itô's theory of random differential and integral equations, the first-order equation

$$dx(t,\omega)/dt = a(t,\omega)\,x(t,\omega), \tag{4.93}$$

where $a(t,\omega) = \alpha + \beta(t,\omega)$, with α a constant and $\beta(t,\omega)$ white noise.

In this section we consider, following Sibul [40], Eq. (4.91) within the framework of the theory of random integral equations.

2. *The random Volterra equation equivalent to a differential equation of order n with random coefficients.* We assume that the random coefficients $a_k(x,\omega)$ are of the form

$$a_k(x,\omega) = \alpha_k(x) + \beta_k(x,\omega), \qquad k = 0, 1, \ldots, n; \tag{4.94}$$

that is, the coefficients admit an additive decomposition into a deterministic function $\alpha_k(x)$ and a random function $\beta_k(x,\omega)$. It follows from (4.94) that the random operator $L(\omega)$ given by (4.92) admits the decomposition

$$L(\omega)[f] = A[f] + B(\omega)[f], \tag{4.95}$$

where

$$A[f] = \sum_{k=0}^{n} \alpha_k(x)\, d^k f/dx^k \tag{4.96}$$

and

$$B(\omega)[f] = \sum_{k=0}^{n} \beta_k(x,\omega)\, d^k f/dx^k. \tag{4.97}$$

Hence Eq. (4.91) is of the form

$$(A + B(\omega))f(x,\omega) = g(x,\omega). \tag{4.98}$$

Let us now assume that the deterministic differential operator A is invertible; that is, the Green's function $G(x,y)$ for A is known or can be constructed. In this case Eq. (4.98) is equivalent to the random integral equation

$$f(x,\omega) + \int_0^x G(x,y)\,dy \sum_{k=0}^{n} \beta_k(x,\omega)\, d^k f/dx^k = \int_0^x g(y,\omega)\,G(x,y)\,dy. \tag{4.99}$$

If we put

$$\int_0^x G(x,y)g(y,\omega)\,dy = \tilde{g}(x,\omega) \tag{4.100}$$

and

$$K(x,y,\omega) = G(x,y) \sum_{k=0}^{n} \beta_k(x,\omega)\, d^k/dx^k, \tag{4.101}$$

then Eq. (4.99) can be rewritten as

$$f(x,\omega) - \int_0^x K(x,y,\omega)f(y,\omega)\,dy = \tilde{g}(x,\omega). \tag{4.102}$$

Equation (4.102) is a Volterra equation of the second kind with random kernel $K(x,y,\omega)$ and random forcing function $\tilde{g}(x,\omega)$. It is of interest to note that the random kernel $K(x,y,\omega)$ is of the form $G(x,y)B(\omega)$; hence it arises from the multiplicative perturbation of the Green's function $G(x,y)$ by the random differential operator $B(\omega)$ defined by (4.97).

The formal solution of Eq. (4.102) is of the form

$$f(x,\omega) = \tilde{g}(x,\omega) + \sum_{k=1}^{\infty} \lambda^k \int_0^x K^{(k)}(x,y,\omega)\tilde{g}(y,\omega)\,dy, \tag{4.103}$$

where, as in the deterministic case, the iterated kernels $K^{(k)}(x,y,\omega)$ are defined by the recurrence relation

$$K^{(k)}(x,y,\omega) = \int_0^x K(x,\xi,\omega)K^{(k-1)}(\xi,y,\omega)\,d\xi, \qquad (4.104)$$

$k=2,3,\ldots$, and $K^{(1)}(x,y,\omega) = K(x,y,\omega)$. If the sum in (4.103) converges uniformly, the order of summation and integration can be interchanged. Therefore, assuming this to be the case, (4.103) can be rewritten as

$$f(x,\omega) = \tilde{g}(x,\omega) + \lambda\int_0^x \left\{\sum_{k=0}^{\infty} \lambda^k K^{(k+1)}(x,y,\omega)\,\tilde{g}(y,\omega)\,dy\right\}. \qquad (4.105)$$

Let

$$\Gamma(x,y,\lambda,\omega) = \sum_{k=0}^{\infty} \lambda^k K^{(k+1)}(x,y,\omega) \qquad (4.106)$$

denote the *resolvent kernel* of $K(x,y,\omega)$. Then the formal solution of Eq. (4.102) can be expressed in the form

$$f(x,\omega) = \tilde{g}(x,\omega) + \int_0^x \Gamma(x,y,\omega)\,\tilde{g}(y,\omega)\,dy. \qquad (4.107)$$

It follows from (4.95) that if $B(\omega) = \Theta$ almost surely (that is, the coefficients are deterministic, and $L(\omega) = A$ almost surely), then the random differential equation $L(\omega)f = g$ reduces to the deterministic differential equation $Af = g$, the solution of which is well known (cf. Coddington and Levinson [11, pp. 87–88]). The determination of solutions of Eq. (4.102), requires, in concrete problems, that the convergence of the Neumann series (4.106) be established, and the resolvent kernel $\Gamma(x,y,\omega)$ calculated. The latter, in particular, can present considerable difficulties.

We now restrict our attention to the case when only the coefficient of lowest order, β_0, is a random function; hence $\beta_k(x,\omega) = 0$ almost surely, $k = 1,2,\ldots,n$. In this case the convergence of the Neumann series can be established, since sufficient conditions for the uniform convergence of the Neumann series are the hypotheses of Theorems 4.6 and 4.8. Hence the hypotheses of Theorem 4.6 require that (i) the kernel $\tilde{K}(x,y,\omega) = G(x,y)\beta_0(x,\omega)$ be a $\mathfrak{U}$-measurable mapping of Ω into $\mathfrak{K}$, where $\mathfrak{K}$ denotes the space of kernels defined on $[0,m] \times [0,m]$, say, with $m > 0$; and (ii) the integral operator

$$\tilde{L}(\omega)[f] = \int_0^x G(x,y)\,\beta_0(y,\omega)f(y,\omega)\,dy, \qquad f \in C[0,m] \qquad (4.108)$$

be a random operator on $C[0,m]$. The hypotheses of Theorem 4.8 simply state that the kernel be a Volterra kernel almost surely. Hence, under the above conditions the random integral operator defined by (4.108), is invertible, and the Neumann series converges uniformly.

As is well known, a differential equation of order n is equivalent to a system of n differential equations of first order. This representation leads to a vector integral equation of Volterra type, the solution of which is analogous to the scalar solution (4.107). We refer to Sibul [40] for a detailed discussion of the solution and its properties.

To return to the solution of Eq. (4.102) in the general case, we now compute the *mean* and *covariance function* of the solution $f(x,\omega)$ given by (4.107). We now assume that the random coefficients $\beta_k(x,\omega)$ and the forcing function $g(x,\omega)$ are independent; hence the resolvent kernel $\Gamma(x,y,\omega)$ and the function $\tilde{g}(x,\omega)$ are independent. Under these assumptions we have

$$\mathscr{E}\{f(x,\omega)\} = \mathscr{E}\{\tilde{g}(x,\omega)\} - \int_0^x \mathscr{E}\{\Gamma(x,y,\omega)\}\,\mathscr{E}\{\tilde{g}(y,\omega)\}\,dy, \tag{4.109}$$

where

$$\mathscr{E}\{\tilde{g}(x,\omega)\} = \int_0^x G(x,y)\,\mathscr{E}\{g(y,\omega)\}\,dy. \tag{4.110}$$

Hence, in this case the *stochastic Green's function* is simply $\mathscr{E}\{\Gamma(x,y,\omega)\}$ (cf. Adomian [1]).

Before calculating the covariance of $f(x,\omega)$, we recall that second-order random functions can be defined as those random functions having covariances. Conversely, if a random function is of second order, its covariance exists and is finite. In this connection it is of interest to state conditions for which $f(x,\omega)$ is a second-order random function. Firstly, the coefficients $\beta_k(x,\omega)$ should be second-order random functions. Secondly, the forcing function $g(x,\omega)$ should be in the class C^∞, where C^∞ denotes the class of functions (mean-square) continuously differentiable an infinite number of times. Since $G(x,y)$ is continuous, if $g(x,\omega)\in C^\infty$ then $\tilde{g}(x,\omega)\in C^\infty$. That $\tilde{g}(x,\omega)$ must be in C^∞ is clear from the Neumann series for the resolvent kernel which occurs in (4.107). If the above conditions are satisfied the formal solution as given by (4.107) is meaningful, and $f(x,\omega)$ is a second-order random function.

The calculation of the covariance function of $f(x,\omega)$ is straightforward. Let

$$R_f(x_1,x_2) = \mathscr{E}\{f(x_1,\omega)f(x_2,\omega)\} \tag{4.111}$$

and

$$R_{\tilde{g}}(x_1,x_2) = \mathscr{E}\{\tilde{g}(x_1,\omega)\,\tilde{g}(x_2,\omega)\}. \tag{4.112}$$

Hence we are assuming that $f(x,\omega)$ and $\tilde{g}(x,\omega)$ are centered at their expectations. Then, from (4.107), we have

$$\begin{aligned} R_f(x_1,x_2) = {} & \mathscr{E}\left\{\tilde{g}(x_1,\omega) - \int_0^{x_1} \Gamma(x_1,y_1,\omega)\,\tilde{g}(y_1,\omega)\,dy_1\right\} \\ & \times \mathscr{E}\left\{\overline{\tilde{g}(x_2,\omega)} - \int_0^{x_2} \overline{\Gamma(x_2,y_2,\omega)}\,\overline{\tilde{g}(y_2,\omega)}\,dy_2\right\} \end{aligned}$$

$$R_f(x_1, x_2) = R_{\tilde{g}}(x_1, x_2) - \int_0^{x_1} \mathscr{E}\{\Gamma(x_1, y_1, \omega)\} R_{\tilde{g}}(y_1, x_2)\, dy_1$$
$$- \int_0^{x_2} \mathscr{E}\{\Gamma(x_2, y_2, \omega)\} R_{\tilde{g}}(x_1, y_2)\, dy_2$$
$$+ \int_0^{x_1}\int_0^{x_2} \mathscr{E}\{\Gamma(x_1, y_1, \omega)\, \Gamma(x_2, y_2, \omega)\} R_{\tilde{g}}(y_1, y_2)\, dy_1\, dy_2. \tag{4.113}$$

$R_{\tilde{g}}(x_1, x_2)$ can be calculated explicitly from (4.100). We have

$$R_{\tilde{g}}(x_1, x_2) = \int_0^{x_1}\int_0^{x_2} G(x_1, y_1)\, \overline{G(x_2, y_2)}\, \mathscr{E}\{g(x_1, \omega)\, \overline{g(x_2, \omega)}\}\, dx_1\, dx_2$$
$$= \int_0^{x_1}\int_0^{x_2} G(x_1, y_1)\, \overline{G(x_2, y_2)}\, R_g(x_1, x_2)\, dx_1\, dx_2, \tag{4.114}$$

where $R_g(x_1, x_2)$ denotes the covariance function of the random forcing function $g(x, \omega)$.

It is clear from (4.113), that in the special case when the coefficients of the differential operator are deterministic, then the last three terms of (4.113) vanish; and $R_f(x_1, x_2) = R_{\tilde{g}}(x_1, x_2)$.

Finally, we compute the *cross-correlation* between the forcing function $g(x, \omega)$ and the solution $f(x, \omega)$. We have

$$R_{fg}(x_1, x_2) = \mathscr{E}\{f(x_1, \omega)\, \overline{g(x_2, \omega)}\}$$
$$= \int_0^{x_1} G(x_1, y_1)\, R_g(y_1, x_2)\, dy_1$$
$$+ \int_0^{x_1} \mathscr{E}\{\Gamma(x_1, y_1, \omega)\}\Big[\int_0^{y_1} G(y_1, z)\, R_g(z, x_2)\, dz\Big] dy_1. \tag{4.115}$$

3. *A Volterra equation with random degenerate kernel.* We now consider a special case of Eq. (4.91) (equivalently, Eq. (4.102)) which leads to a Volterra equation with random degenerate kernel. Consider the case in which the kernel $K(x, y, \omega)$, as given by (4.101) is of the form

$$K(x, y, \omega) = \sum_{k=1}^{n} u_k(x)\, v_k(y, \omega). \tag{4.116}$$

A representation of the above form will obtain, for example, if the Green's function $G(x, y)$ of the deterministic differential operator A, as given by (4.96) is a sum of exponential function, and the only random coefficient is β_0 (that is, $\beta_0 \neq 0$ almost surely). In this case

$$u_k(x) = \nu_k \exp\{\gamma_k x\} \tag{4.117}$$

and

$$v_k(y, \omega) = \beta_0(y, \omega) \exp\{-\gamma_k y\}, \tag{4.118}$$

where ν_k and γ_k are complex numbers.

For a detailed discussion of the use of degenerate kernel techniques for solving random Volterra equations we refer to Sibul [40].

4.4 Some Random Linear Integral Equations Which Arise in Applied Problems

A. Introduction

In the first two subsections we consider some random Fredholm equations that arise in the study of wave propagation in random media and in connection with the numerical analysis of solutions of Fredholm equations of the first kind. Random Fredholm equations also arise in the theory of differential equations with random boundary conditions; and Boyce and his students (cf. Boyce [9]) have studied the random Fredholm equations which arise in connection with several concrete problems in random vibration theory. Eigenvalue problems for some random Fredholm equations are considered in Chap. 5. In the third subsection we consider a system of random Volterra integral equations which arise in hereditary mechanics.

B. Wave propagation in random media

The mathematical formulation of wave propagation in random media leads to linear partial differential equations whose coefficients are random functions of space and time; hence the mathematical theory of wave propagation in random media is a special case of the general theory of random differential equations. Mathematically, a wave motion is described by a function $\psi(t, x)$, where x denotes the vector of space variables and t denotes time. The transmission medium, say M, is characterized by a function $n(t,x)$ which enters the coefficients of the partial differential equation for $\psi(t, x)$. In the formulation of problems for random media, a single nonhomogeneous medium M is replaced by a family or ensemble of media $\{M(\omega), \omega \in \Omega\}$, where Ω is a probability space. If for every fixed $\omega \in \Omega$ the properties of the transmission medium are characterized by a function $n(t, x, \omega)$ (for example, the index of refraction) $\{M(\omega)\}$ is said to be a *random medium* if for every fixed x and t, $n(t, x, \omega)$ is a random variable.

The introduction of random Fredholm equations in the study of wave propagation in random media is due to Hoffman [22–24]; these equations arising as integral equations equivalent to Helmholtz equations with random coefficients. Our discussion in this section is based on Hoffman's formulation (cf. also Frisch [15]). For authoritative expositions on random equations and wave propagation in random media we refer to the articles of Frisch [15] and Keller [27, 28].

For initial value problems or propagation in time-dependent media, the *random wave propagation equation* is of the form

$$\partial\psi/\partial t = (A + B(\omega))\,\psi, \tag{4.119}$$

where the unknown function ψ (scalar or vector valued) is called the *wave* (or *field*), A is a deterministic linear partial differential operator (usually with constant coefficients), and $B(\omega)$ is a linear partial differential operator with random coefficients. Without loss of generality, we can assume the coefficients are centered random functions. Problems which involve the radiation of harmonic time-dependent waves in time-independent random media lead to the random differential equation

$$(A + B(\omega))\,\psi = j. \tag{4.120}$$

In Eq. (4.120), the operators A and $B(\omega)$ are as defined earlier, but are time independent; and j, called the *source term*, is a given deterministic function or distribution.

We now restrict our attention to the formulation of the problem of radiation of scalar waves by a harmonic point-source in a lossless, homogeneous, isotropic, time-independent random medium $\{M(\omega)\}$. In this case Eq. (4.120) is the *random Helmholtz equation*

$$\Delta\psi(x, \omega) + k_0^2 n^2(x, \omega)\,\psi(x, \omega) = \delta(x). \tag{4.121}$$

In Eq. (4.121), $k_0 > 0$ is the *free-space wave number*, $n(x,\omega)$ is the *index of refraction*, which we assume is a real-valued, homogeneous and isotropic random function of the form

$$n^2(x, \omega) = 1 + \mu(x, \omega). \tag{4.122}$$

The function $\mu(x,\omega)$ is assumed to be a centered homogeneous and isotropic random function; hence $\mathscr{E}\{n^2(x,\omega)\} = 1$. Together with Eq. (4.121) we consider the Sommerfeld radiation condition

$$\lim_{|x|\to\infty} |x| \left[\frac{\partial\psi}{\partial|x|} - ik_0(x, \omega)\,\psi\right] = 0, \tag{4.123}$$

which asserts that the solution ψ describes outwardly propagating waves.

We first consider Eq. (4.121) in the one-dimensional case. In this case we write Eq. (4.121) as

$$\frac{d^2\psi}{dx^2} + k_0^2(1 + \mu(x, \omega))\,\psi = 0. \tag{4.124}$$

If $\mu(x,\omega)=0$ outside a finite interval $[0,S]$, called the *scattering region*, we can assume a solution of the form

$$\psi=\psi_{\text{inc}}+\psi_{\text{sc}}, \tag{4.125}$$

where ψ_{inc} is a deterministic solution of the free-wave equation (that is, $\mu=0$ almost surely), and ψ_{sc} satisfies the radiation condition

$$\lim_{|x|\to\infty}\left[\frac{d\psi_{\text{sc}}}{d|x|}-ik_0\psi_{\text{sc}}\right]=0. \tag{4.126}$$

The Green's function associated with the free-space version of Eq. (4.124) is

$$G_0(x,y)=(1/2ik_0)\exp\{ik_0|x-y|\}; \tag{4.127}$$

hence Eq. (4.124) is equivalent to the random Fredholm integral equation

$$k_0^2\int_0^S K(x,y,\omega)\psi(y,\omega)\,dy+\psi(x,\omega)=\psi_{\text{inc}}(x) \tag{4.128}$$

The random kernel $K(x,y,\omega)$ is of the form

$$K(x,y,\omega)=G_0(x,y)\,\mu(y,\omega); \tag{4.129}$$

hence the random kernel arises from a multiplicative perturbation of the free-space Green's function by the random component of the index of refraction.

No exact solution of Eq. (4.128) has been given; however special cases have been studied by Bazer [3] and Kay and Silverman [26]. In the general cases, the theorems of Sect. 3.3 can be used to establish conditions for the existence, uniqueness and measurability of the solution $\psi(x,\omega)$. We also remark that as in Sect. 4.3D we can replace the random kernel $K(x,y,\omega)$ by a deterministic kernel, and consider the integral over a random scattering region $[0,S(\omega)]$.

We now consider Eq. (4.121), together with (4.123) in the vector-valued case. In this case the free-space Green's function is

$$G_0(x,y)=\frac{\exp\{ik_0|x-y|\}}{-4\pi|x-y|}; \tag{4.130}$$

hence the equivalent random Fredholm integral equation is of the form

$$k_0^2\int K(x,y,\omega)\psi(y,\omega)\,dy+\psi(x,\omega)=G_0(x,0), \tag{4.131}$$

where $K(x,y,\omega)=G_0(x,y)\mu(y,\omega)$.

To solve Eq. (4.131), we first assume that the random medium is not homogeneous; but that the randomness is restricted to a bounded domain D. Hence the random function $\mu(x,\omega)$ has compact support contained in D. Let

$$\varphi_1(x,\omega)=\mu(x,\omega)G_0(x,0)$$

and

$$\varphi_{n+1}(x,\omega) = k_0^2\,\mu(x,\omega) \int_D G_0(x,y)\,\varphi_n(y,\omega)\,dy,$$

$n = 1, 2, \ldots$. Then the formal solution of Eq. (4.131) is of the form

$$\psi(x,\omega) = G_0(x,0) - k_0^2 \int_D G_0(x,y) \sum_{n=1}^{\infty} \varphi_n(y,\omega)\,dy. \tag{4.132}$$

We now establish the convergence of the series $\sum_{n=1}^{\infty} \varphi_n(x,\omega)$ (cf. Frisch [15]).

THEOREM 4.13. *If $|\mu(x,\omega)| < M$ almost surely, a sufficient condition for the almost sure convergence of the series $\sum_{n=1}^{\infty} \varphi_n(x,\omega)$ in $L_2(D)$ is $\frac{1}{2}Mk_0^2 d^2 < 1$, where d denotes the diameter*† *of D.*

Proof. An application of Minkowski's inequality yields

$$\begin{aligned}
\|\varphi_{n+1}(x,\omega)\|_2 &\leqslant k_0^2\, M \left\| \int_D G_0(x,y)\,\varphi_n(y,\omega)\,dy \right\|_2 \\
&\leqslant k_0^2\, M\,\|\varphi_n(x,\omega)\|_2 \sup_{x,y\in D} \int |G_0(x,y)|\,dy \\
&\leqslant k_0^2\, M\,\|\varphi_n(x,\omega)\|_2 \sup_{x,y\in D} \int \frac{dy}{4\pi(x-y)} \\
&\leqslant k_0^2\, M\,\|\varphi_n(x,\omega)\|_2 \int_0^d \frac{4\pi x^2}{4\pi x}\,dx \\
&\leqslant \tfrac{1}{2} k_0^2\, M d^2 \|\varphi_n(x,\omega)\|_2 .
\end{aligned}$$

Since $L_2(D)$ is complete, the almost sure convergence of the series $\sum_{n=1}^{\infty} \varphi_n(x,\omega)$ follows from the convergence of the series $\sum_{n=1}^{\infty} \|\varphi_n(x,\omega)\|_2$.

Theorem 4.13 establishes the validity of (4.132) as the solution of Eq. (4.131).

Random integral equations have been utilized by Sibul [40] to study wave propagation in a randomly time and space varying medium. Consider the random wave equation

$$\nabla^2 \psi(t,x,\omega) - \frac{\partial^2}{\partial t^2}\left[\frac{1}{c^2} + a(t,x,\omega)\right] \psi(t,x,\omega) = g(t,x,\omega), \tag{4.133}$$

where the coefficient $a(t,x,\omega)$ and the source function $g(t,x,\omega)$ are independent, wide-sense, centered, stationary random functions, the wide-sense stationarity restricted to the time variable t. Let

$$a(t,x,\omega) = \int_{-\infty}^{\infty} A(\tau,x,\omega)\,e^{i\tau t}\,d\tau \tag{4.134}$$

† The *diameter* $d(D)$ is the supremum of the distances $\rho(x,y)$, $x,y \in D$.

$$g(t, x, \omega) = \int_{-\infty}^{\infty} G(\xi, x, \omega)\, e^{i\xi t}\, d\xi, \tag{4.135}$$

where $\tau, \xi \in R$. And, under the assumption that the solution $\psi(t, x, \omega)$ is a wide-sense stationary random function, put

$$\psi(t, x, \omega) = \int_{-\infty}^{\infty} \Psi(\lambda, x, \omega)\, e^{i\lambda t}\, d\lambda. \tag{4.136}$$

Sibul has shown that the spectral representation of the wave function $\psi(t, x, \omega)$ satisfies a random Fredholm equation of the form

$$\Psi(\xi, x, \omega) + L^{-1}\xi^2 \int_{-\infty}^{\infty} \Psi(\tau, x, \omega)\, A(\xi - \tau, x, \omega)\, d\tau = F(\xi, x, \omega), \tag{4.137}$$

where L^{-1} is the inverse Helmholtz operator (for appropriate boundary conditions), and $F(\xi, x, \omega) = L^{-1}[G(\xi, x, \omega)]$. The solution of Eq. (4.137) was obtained using both degenerate kernel approximation methods and the Neumann series expansion.

C. *Numerical solution of Fredholm equations of the first kind*

Consider the Fredholm integral equation of the first kind

$$\int_a^b K(x, y) f(y)\, dy = g(x), \tag{4.138}$$

where the kernel $K(x, y)$ is a continuous function of $x, y \in [a, b]$, and the known function $g(x) \in L_2[a, b]$, say. For a discussion of Fredholm equations of the first kind we refer to Pogorzelski [34, Chap. VI]. It has been pointed out by several authors that in studies concerned with numerical solutions of Eq. (4.138), the equation should be written in the form

$$\int_a^b K(x, y) f(y)\, dy = g(x) + \epsilon(x), \tag{4.139}$$

where $\epsilon(x)$ is an error term. If we make the reasonable assumption that $\epsilon(x)$ is a random function, then the right-hand side of Eq. (4.139) can be expressed as

$$g(x) + \epsilon(x, \omega) = g(x, \omega); \tag{4.140}$$

and Eq. (4.138) can be written as the random Fredholm equation of the first kind

$$\int_a^b K(x, y) f(y, \omega)\, dy = g(x, \omega). \tag{4.141}$$

Strand and Westwater [43] have considered the problem of estimating the solution $f(x, \omega)$ from observations on $g(x, \omega)$ at a prescribed set of points, say $x = x_i$ $(i = 1, 2, \ldots, n)$.

Let $f(x, \omega)$ be a second-order random function with continuous realizations. Since the kernel $K(x, y)$ is continuous, the integral in (4.141) is well defined. Consider a quadrature rule of the form

$$\int_a^b \varphi(y)\, dy = \sum_{j=1}^{m} \lambda_j \varphi(y_j). \tag{4.142}$$

Application of (4.142) to (4.141) yields, as the quadrature approximation to the random Fredholm equation the (random) matrix equation

$$Af(\omega) = g(\omega). \tag{4.143}$$

In (4.143) A is the $n \times m$ matrix

$$A = (a_{ij}) = (\lambda_j K(x_i, y_j)),$$

(where the λ_j are the weights associated with the quadrature abscissas y_j, $f(\omega)$ is the m-vector $(f(y_1, \omega), f(y_2, \omega), \ldots, f(y_m, \omega))$, and $g(\omega)$ is the n-vector $(g(x_1, \omega), g(x_2, \omega), \ldots, g(x_n, \omega))$. Assume that (i) $\mathscr{E}\{f(\omega)\}$ is known, and (ii) that the covariance matrix of $f(\omega)$, say R_f, is known. Then $\mathscr{E}\{g(\omega)\} = A\mathscr{E}\{f(\omega)\}$, and the covariance matrix of $g(\omega)$, say R_g, is of the form $R_g = AR_f A'$, where A' denotes the transpose of A.

It follows from (4.140) that in any practical situation observation made on a realization of $g(x, \omega)$ instead of $g(x)$, where $g(x)$ is the vector of measurements subject to error. The following assumptions are made:

(i) the components $\epsilon_i(\omega)$ $(i = 1, 2, \ldots, n)$ of the error vector are independent of $f(\omega)$, hence independent of $g(x)$;
(ii) the errors have a multivariate normal distribution with mean zero and known covariance matrix R_ϵ;
(iii) the quadrature errors are negligible with respect to ϵ;
(iv) the covariance matrices R_ϵ and R_f are both nonsingular.

Under the above assumptions, a solution of Eq. (4.141) is derived by Strand and Westwater [43], for a general set of basis vectors, which has the minimum expected mean-square error for a linear unbiased estimator. In particular, it is shown that this error is a monotone increasing function of the number of basis vectors; hence there is no computational advantage in using a basis with dimension less than m, the number of quadrature abscissas.

D. Hereditary mechanics

Distefano [13] has studied a pair of random Volterra integral equations which arise in the probabilistic analysis of the behavior of hereditary mechanical systems. In certain problems in the theory of hereditary systems the forcing term depends on the deviation of the system from a natural position of equilibrium as well as on an external source of excitation. Consider the following example: Forces α and η are applied to two hinged bars, and the deflection of the bars is prevented by a viscoelastic spring reacting with an upward force S. Here η is the axial load, and α is the resulting downward force on the spring. The bars are deflected a certain amount w which is, in general, a nonlinear function of η. The forces and deflection are functions of time $t \geqslant 0$; and $w(t)$, the displacement at time $t > 0$, is a functional of the force $S(\tau)$ exerted by the spring for $\tau \leqslant t$. The hereditary effects of the system are reflected in the fact that $S(\tau)$ is a function of $\alpha(\tau)$, $\eta(\tau)$, and $w(\tau)$ for each $\tau \leqslant t$.

A linearized version of the above problem leads to relations of the form

$$S(t) = \alpha(t) + \eta(t)\, w(t) \tag{4.144}$$

and

$$w(t) = S(t) + \int_0^t K(t,\tau)\, S(\tau)\, d\tau. \tag{4.145}$$

From (4.144) and (4.145) we obtain the integral equation

$$(1 - \eta(t))\, w(t) - \int_0^t K(t,\tau)\, \eta(\tau)\, w(\tau)\, d\tau = g(t), \tag{4.146}$$

where $K(t,\tau)$ is the "memory function" for the hereditary phenomenon, and

$$g(t) = \beta(t) + \int_0^t K(t,\tau)\, \beta(\tau)\, d\tau. \tag{4.147}$$

In (4.147) $\beta = \alpha$ when we assume the condition of initial straightness of the bar. Eq. (4.146) can be replaced by the pair of Volterra integral equations

$$u(t) - \int_0^t \varphi(\tau)\, K(t,\tau)\, u(\tau)\, d\tau = g(t) \tag{4.148}$$

$$v(t) - \varphi(t) \int_0^t K(t,\tau)\, v(\tau)\, d\tau = h(t), \tag{4.149}$$

when it is assumed that $0 < \eta(t) < 1$ for all $t \in [0, \infty]$, and

$$h(t) = g(t)\,\varphi(t), \qquad \varphi(t) = \frac{\eta(t)}{1 - \eta(t)},$$

$$u(t) = (1 - \eta(t))\, w(t), \qquad v(t) = \eta(t)\, w(t).$$

Then $u(t) + v(t) = w(t)$ and $\varphi(t) = v(t)/u(t)$.

Let us now assume that the axial load η is a random function $\eta(t,\omega)$. With this assumption, the functions φ, h, u and v, defined above, are also random functions; and Eqs. (4.148) and (4.149) become random Volterra equations of the form

$$u(t,\omega) - \int_0^t \varphi(\tau,\omega)\, K(t,\tau)\, u(\tau,\omega)\, d\tau = g(t,\omega) \tag{4.150}$$

$$v(t,\omega) - \varphi(t,\omega)\int_0^t K(t,\tau)\,v(\tau,\omega)\,d\tau = h(t,\omega). \tag{4.151}$$

Put

$$K_1(t,\tau,\omega) = \begin{cases} \varphi(\tau,\omega)\,K(\tau,\omega), & 0 \leqslant \tau \leqslant t < \infty \\ 0, & \text{otherwise,} \end{cases}$$

and

$$K_2(t,\tau,\omega) = \begin{cases} \varphi(t,\omega)\,K(t,\tau), & 0 \leqslant \tau \leqslant t < \infty \\ 0, & \text{otherwise.} \end{cases}$$

Then Eqs. (4.150) and (4.151) are clearly Volterra integral equations with random kernels of the form

$$u(t,\omega) - \int_0^t K_1(t,\tau,\omega)\,u(\tau,\omega)\,d\tau = g(t,\omega) \tag{4.152}$$

$$v(t,\omega) - \int_0^t K_2(t,\tau,\omega)\,v(\tau,\omega)\,d\tau = h(t,\omega). \tag{4.153}$$

The results of Sect. 4.3C and the results of Tsokos, presented in Sect. 6.4, can be used to study the existence, uniqueness and measurability of the solutions $u(t,\omega)$ and $v(t,\omega)$ of the above equations. Results of Tsokos on the stability of solutions of random Volterra equations are also applicable to the analysis of the asymptotic behavior of the solutions. Distefano [13] studied Eq. (4.150) using the method of truncated hierarchies, and Eq. (4.151) was studied using a method of successive approximations.

References

1. Adomian, G., Theory of random systems. *Trans. 4th Prague Conf. Information Theory, Statist. Decision Processes, and Random Processes (1965)*, pp. 205–222, 1967.
2. Anderson, M. W., Stochastic integral equations. Ph.D. Dissertation, Univ. of Tennessee, Knoxville, Tennessee, 1966.
3. Bazer, J., Multiple scattering in one dimension. *J. Soc. Indust. Appl. Math.* **12** (1964), 539–579.
4. Bharucha-Reid, A. T., On random solutions of Fredholm integral equations. *Bull. Amer. Math. Soc.* **66** (1960), 104–109.
5. Bharucha-Reid, A. T., On random solutions of integral equations in Banach spaces. *Trans. 2nd Prague Conf. Information Theory, Statist. Decision Functions, and Random Processes (1959)*, pp. 27–48, 1960.
6. Bharucha-Reid, A. T., Sur les équations intégrales aléatoires de Fredholm à noyaux séparables. *C. R. Acad. Sci. Paris* **250** (1960), 454–456, 657–658.
7. Bharucha-Reid, A. T., On the theory of random equations. *Proc. Symp. Appl. Math., 16th*, pp. 40–69. Amer. Math. Soc., Providence, Rhode Island, 1964.
8. Bharucha-Reid, A. T., and Arnold, L., On Fredholm integral equations with random degenerate kernels. *Zastos. Mat.* (Steinhaus Jubilee Volume) **10** (1969), 85–90.

9. Boyce, W. E., Random eigenvalue problems. *In* "Probabilistic Methods in Applied Mathematics" (A. T. Bharucha-Reid, ed.), Vol. 1, pp. 1–73. Academic Press, New York, 1968.
10. Caughey, T. K., and Dienes, J. K., The behavior of linear systems with random parametric excitation. *J. Math. and Phys.* **41** (1962), 300–318.
11. Coddington, E. A., and Levinson, N., "Theory of Ordinary Differential Equations." McGraw-Hill, New York, 1955.
11a. Corduneanu, C., "Principles of Differential and Integral Equations." Allyn & Bacon, Rockleigh, New Jersey, 1971.
12. Courant, R., and Hilbert, D., "Methods of Mathematical Physics," Vol. 1. Wiley (Interscience), New York, 1953.
13. Distefano, N., A Volterra integral equation in the stability of some linear hereditary phenomena. *J. Math. Anal. Appl.* **23** (1968), 365–383.
14. Dunford, N., and Schwartz, J. T., "Linear Operators, Pt. 1: General Theory." Wiley (Interscience), New York, 1958.
15. Frisch, U., Wave propagation in random media. *In* "Probabilistic Methods in Applied Mathematics" (A. T. Bharucha-Reid, ed.), Vol. 1, pp. 75–198. Academic Press, New York, 1968.
16. Goodwin, B. E., and Boyce, W. E., The vibration of a random elastic string: The method ot integral equations. *Quart. Appl. Math.* **22** (1964), 261–266.
17. Green, C. D., "Integral Equation Methods." Nelson, London, 1969.
18. Hanš, O., Reduzierende zulfällige Transformationen, *Czechoslovak Math. J.* **7** (1957), 58.
19. Hanš, O., Generalized random variables. *Trans. 1st Prague Conf. Information Theory, Statist. Decision Functions, and Random Processes* (*1956*), pp. 61–103, 1957.
20. Hanš, O., Random operator equations. *Proc. 4th Berkeley Symp. Math. Statist. Probability* (*1960*), Vol. II, pp. 185–202, 1961.
21. Hildebrand, F. B., "Methods of Applied Mathematics." Prentice-Hall, Englewood Cliffs, New Jersey, 1952.
22. Hoffman, W. C., The electromagnetic field in a randomly inhomogeneous medium. *IEEE Trans. Special Suppl.*, **AP-7** (1959), 301–306.
23. Hoffman, W. C., Electromagnetic wave propagation in a random medium. *Radio Sci.* **68D**(1964), 455–459.
24. Hoffman, W. C., Wave propagation in a general random continuous medium. *Proc. Symp. Appl. Math. 16th*, pp. 117–144. Amer. Math. Soc., Providence, Rhode Island, 1964.
25. Kantorovich, L. V., and Krylov, V. I., "Approximate Methods of Higher Analysis," translated from the Russian. Wiley (Interscience), New York, 1958.
25a. Kanwal, R. P., "Linear Integral Equations: Theory and Technique." Academic Press, New York, 1971.
26. Kay, I., and Silverman, R. A., Multiple scattering by a random stack of dielectric slabs. *Nuovo Cimento* (10) 9 Suppl. (1958), 626–645.
27. Keller, J. B., Wave propagation in random media. *Proc. Symp. Appl. Math., 13th*, pp. 227–246. Amer. Math. Soc., Providence, Rhode Island, 1962.
28. Keller, J. B., Stochastic equations and wave propagation in random media. *Proc. Symp. Appl. Math., 16th*, pp. 145–170. Amer. Math. Soc., Providence, Rhode Island, 1964.
29. Kolmogorov, A. N., and Fomin, S. V., "Introductory Real Analysis," rev. Engl. ed., translated from the Russian. Prentice-Hall, Englewood Cliffs, New Jersey, 1970.
30. Krasnosel'skiĭ, M. A., and Rutickiĭ, Ya. B., "Convex Functions and Orlicz Spaces," in Russian, Gosudarstv. Izdat. Fiz.-Mat. Lit., Moscow, 1958.

31. Loève, M., "Probability Theory," 3rd ed. Van Nostrand-Reinhold, Princeton, New Jersey, 1963.
32. Mihlin, S. G., "Integral Equations," translated from the Russian. Pergamon, Oxford, 1957.
33. Petrovskiĭ, I. G., "Lectures on the Theory of Integral Equations," translated from the Russian. Graylock Press, Albany, New York, 1957.
34. Pogorzelski, W., "Integral Equations and Their Applications," Vol. I, translated from the Polish. Pergamon, Oxford, 1967.
35. Riesz, F., and Sz.-Nagy, B., "Leçons d'analyse fonctionnelle." Akadémiai Kiadó, Budapest, 1953.
36. Saaty, T. L., "Modern Nonlinear Equations." McGraw-Hill, New York, 1967.
37. Samuels, J. C., On the stability of random systems and the stabilization of deterministic systems with random noise." *J. Acoust. Soc. Amer.* **32** (1960), 594–601.
38. Samuels, J. C., Theory of stochastic linear systems with Gaussian parameter variations." *J. Acoust. Soc. Amer.* **33** (1961), 1782–1786.
39. Samuels, J. C., and Eringen, A. C., On stochastic linear systems. *J. Math. and Phys.* **38** (1959), 83–103.
40. Sibul, L. H., Application of linear stochastic operator theory. Ph.D. Dissertation, Pennsylvania State Univ., University Park, Pennsylvania, 1968.
41. Špaček, A., Zufällige Gleichungen. *Czechoslovak Math. J.* **5** (1955), 462–466.
42. Stakgold, I., "Boundary Value Problems of Mathematical Physics," Vol. I. Macmillan, New York, 1967.
43. Strand, O. N., and Westwater, E. R., Minimum-rms estimation of the numerical solution of a Fredholm integral equation of the first kind. *SIAM J. Numer. Anal.* **5** (1968), 287–295.
44. Tricomi, F. G., "Integral Equations." Wiley (Interscience), New York, 1957.
45. Yosida, K., "Lectures on Differential and Integral Equations." Wiley (Interscience), New York, 1960.
46. Zaanen, A. C., "Linear Analysis." North-Holland Publ., Amsterdam, 1953.

CHAPTER 5

Eigenvalue Problems for Random Fredholm Integral Equations

5.1 Introduction

As shown in Chap. 3, eigenvalue problems for random operators lead to random operator equations of the form

$$L(\omega)f = \lambda f, \tag{5.1}$$

or

$$\lambda L(\omega)f = f. \tag{5.2}$$

In this chapter we consider some eigenvalue problems for random Fredholm operators which lead to equations of the form (5.1) and (5.2). Since a concrete eigenvalue problem requires the determination of the spectrum of a given operator, the results presented in this chapter can be considered as belonging to the spectral theory of random Fredholm operators.

The eigenvalue problem for a Fredholm integral equation with random degenerate kernel is considered in Sect. 5.2; and we show that the eigenvalue problem in this case can be reduced to an eigenvalue problem for random matrices. This reduction enables us to investigate the asymptotic distribution of the eigenvalues of a Fredholm operator with random degenerate kernel. The results presented in this section are due to Bharucha-Reid and Arnold [2]. In Sect. 5.3 we study some eigenvalue problems for ordinary differential equations with random coefficients or random boundary conditions, and show that in certain cases the differential equation is equivalent to a Fredholm integral equation with a random symmetric kernel. Kernel trace methods can then be used to estimate the random eigenvalues and their moments. The results of this section are due to Boyce and Goodwin (cf. Boyce [4]).

It is of interest to remark that the random Fredholm equations considered in this chapter have almost surely symmetric kernels. The kernel of the equation considered in Sect. 5.2 is degenerate, hence it has a finite number of eigenvalues. The equation studied in Sect. 5.3 has a nondegenerate kernel. Therefore the operator has a countable number of eigenvalues.

We close this introductory section by remarking that it should be possible to utilize certain results of Kagiwada *et al.* [8], and Wing [15, 16] to study eigenvalue problems for random integral equations.

5.2 Fredholm Integral Equations with Random Degenerate Kernels

In Sect. 4.3B we studied the random Fredholm equation

$$\int_0^1 K(x,y,\omega)f(y)\,dy - \lambda f(x) = g(x) \tag{5.3}$$

with degenerate kernel of the form

$$K(x,y,\omega) = \sum_{i=1}^{n} \alpha_i(x,\omega)\beta_i(y), \tag{5.4}$$

and showed that Eq. (5.3) with kernel (5.4) is equivalent to a system of random linear algebraic equations of the form

$$(A(\omega) - \lambda I)\,\xi = b. \tag{5.5}$$

In (5.5) $A(\omega)$ is an $n \times n$ random matrix with elements

$$a_{ij}(\omega) = \int_0^1 \alpha_j(x,\omega)\beta_i(x)\,dx, \qquad i,j = 1,2,\ldots,n; \tag{5.6}$$

and b and ξ are n-vectors with components

$$b_i = \int_0^1 \beta_i(x)g(x)\,dx, \qquad i = 1,2,\ldots,n \tag{5.7}$$

$$\xi_i = \int_0^1 \beta_i(x)f(x)\,dx, \qquad i = 1,2,\ldots,n. \tag{5.8}$$

In this section we utilize the fact that the eigenvalues $\lambda_1, \lambda_2, \ldots, \lambda_n$ of a Fredholm operator with degenerate kernel are the roots of the associated algebraic equation [10, p. 73]. Hence in the probabilistic case the random eigenvalues $\lambda_1(\omega), \lambda_2(\omega), \ldots, \lambda_n(\omega)$ of the Fredholm operator with degenerate kernel (5.4) are the roots of the random algebraic equation $A(\omega) - \lambda I = 0$; that is the eigenvalues of the random matrix $A(\omega)$.

Consider the random kernel

$$K(x, y, \omega) = \sum_{i=1}^{n} \alpha_i(x, \omega)\, \alpha_i(y, \omega) \tag{5.9}$$

(that is, (5.4) with $\alpha_i = \beta_i$), where the α_i's are almost surely independent $L_2[0,1]$-random functions having the same finite-dimensional probability distributions. For the sake of simplicity, we make the following assumptions:

(i) $m_k(x) = \mathscr{E}\{|\alpha_i(x, \omega)|^k\} > \infty$ for all $x \in [0, 1]$, and

(ii) $\int_0^1 m_k(x)\, dx > \infty$ for every $k = 1, 2, \ldots$.

Put

$$R(x, y) = \mathscr{E}\{\alpha_i(x, \omega)\, \alpha_i(y, \omega)\}; \tag{5.10}$$

and assume

$$\mathscr{E}\{\alpha_i(x, \omega)\} = 0.$$

In the case being considered, the kernel (5.9) is almost surely symmetric and positive definite, since

$$\int_0^1\int_0^1 K(x, y, \omega)\, h(x)\, \overline{h(y)}\, dx\, dy = \sum_{i=1}^{n} \left| \int_0^1 \alpha_i(x, \omega)\, h(x)\, dx \right|^2$$

for any continuous function $h(x)$. Therefore the eigenvalues of a Fredholm operator with kernel (5.9) are real and, moreover, nonnegative random variables. We remark that the *mean kernel* (that is, the expectation of K)

$$\tilde{K}(x, y) = \mathscr{E}\{K(x, y, \omega)\} = nR(x, y) \tag{5.11}$$

is also a symmetric and positive-definite kernel, but in general it is not degenerate.

From (5.6) and (5.9) it follows that the elements $a_{ij}(\omega)$ of the random matrix $A(\omega)$ are given by

$$a_{ij}(\omega) = \int_0^1 \alpha_i(x, \omega)\, \alpha_j(x, \omega)\, dx. \tag{5.12}$$

Since $a_{ij}(\omega) = a_{ji}(\omega)$ almost surely, the random matrix $A(\omega)$ is symmetric. The diagonal elements $a_{ii}(\omega)$ are independent, have the same distribution, and

$$\mathscr{E}\{a_{ii}(\omega)\} = \int_0^1 R(x, x)\, dx = \operatorname{Tr}(R). \tag{5.13}$$

The off-diagonal elements $a_{ij}(\omega)$, $i \neq j$, also have the same distribution

$$\mathscr{E}\{a_{ij}(\omega)\} = 0 \tag{5.14}$$

and

$$\mathscr{E}\{a_{ij}^2(\omega)\} = \int_0^1\int_0^1 R^2(x, y)\, dx\, dy. \tag{5.15}$$

We remark that the only difference between the random matrices considered by Arnold [1] and Wigner [13, 14] and the matrix $A(\omega)$ with elements (5.12) is that in addition to the symmetry condition there are other relations among the matrix elements; hence they are not independent. This difference is not trivial, and completely changes the type of result that can be obtained concerning the asymptotic distribution of the random eigenvalues. However, the elements of $A(\omega)$ are independent if they are not in the same row or column.

Define

$$R_1(x,y) = R(x,y), \qquad R_k(x,y) = \int_0^1 R(x,s)\, R_{k-1}(x,y)\, ds; \tag{5.16}$$

and

$$\mathrm{Tr}(R_k) = \int_0^1 R_k(x,x)\, dx. \tag{5.17}$$

Then, if the indices $i_1, i_2, \ldots, i_k$ are all different

$$\begin{aligned} &\mathscr{E}\{a_{i_1 i_2}(\omega)\, a_{i_2 i_3}(\omega) \cdots a_{i_k i_1}(\omega)\} \\ &= \underbrace{\int_0^1 \cdots \int_0^1}_{k\text{-fold}} R(x_1, x_2)\, R(x_2, x_3) \ldots R(x_k, x_1)\, dx_1\, dx_2 \cdots dx_k \\ &= \mathrm{Tr}(R_k). \end{aligned} \tag{5.18}$$

We now use the fact that for the eigenvalues $\lambda_1, \lambda_2, \ldots, \lambda_n$ of any $n \times n$ matrix $W = (w_{ij})$

$$\begin{aligned} \sum_{i=1}^{n} \lambda_i^k &= \mathrm{Tr}(W^k) \\ &= \sum_{i_1=1}^{n} \cdots \sum_{i_k=1}^{n} w_{i_1 i_2}\, w_{i_2 i_3}\, w_{i_k i_1} \end{aligned} \tag{5.19}$$

for $k = 1, 2, \ldots$. Hence from (5.18) and (5.19) we have the asymptotic result

$$\lim_{n\to\infty} (1/n^k) \sum_{i=1}^{n} \mathscr{E}\{\lambda_i^k(\omega)\} = \mathrm{Tr}(R_k) \tag{5.20}$$

for all $k = 1, 2, \ldots$.

Let $N_n(x, \omega)$ denote the number of eigenvalues of the random matrix $A(\omega)$ (equivalently the random Fredholm integral operator $L(\omega)$ with kernel (5.9)) which are less than x. Clearly $N_n(x, \omega)$ is a real-valued random variable for every fixed x. Since $A(\omega)$ is positive, we have $N_n(0, \omega) = 0$ and $N_n(\infty, \omega) = n$ almost surely; hence $\sigma(A(\omega))$, the spectrum of $A(\omega)$ (equivalently, $\sigma(L(\omega))$, the spectrum of $L(\omega)$), is a finite set with not more than n points. Furthermore,

$$\int_0^1 x^k\, dN_n(x, \omega) = \sum_{i=1}^{n} \lambda_i^k(\omega). \tag{5.21}$$

Relation (5.20) can now be rewritten as

$$\lim_{n\to\infty} \int_0^1 x^k \, d(\mathscr{E}\{N_n(nx,\omega)\}) = \operatorname{Tr}(R_k). \tag{5.22}$$

The convergence must be due to the fact that there are, on the average, very many small eigenvalues, but only a few large eigenvalues.

As an example, consider the case when

$$\alpha_i(x,\omega) = z_i(\omega). \tag{5.23}$$

In this case

$$R(x,y) = \mathscr{E}\{z_i^2(\omega)\} = \sigma^2; \tag{5.24}$$

and the eigenvalues of the random kernel (5.9) are

$$\lambda_1(\omega) = \lambda_2(\omega) = \cdots = \lambda_{n-1}(\omega) = 0$$

$$\lambda_n(\omega) = \sum_{i=1}^{n} z_i^2(\omega). \tag{5.25}$$

From the strong law of large numbers it follows that

$$\int_0^1 x^k \, dN_n(nx,\omega) = \left((1/n)\sum_{i=1}^{n} z_i^2(\omega)\right)^k \to \operatorname{Tr}(R_k) \tag{5.26}$$

almost surely for every $k = 1, 2, \ldots$.

5.3 Fredholm Integral Equations with Random Symmetric Kernels

A. Integral equation formulation of a class of boundary-value problems

Consider the boundary-value problem of (1) the linear differential equation in $L_2[a,b]$

$$(L - \lambda M)f = 0, \tag{5.27}$$

where L and M are ordinary differential operators of order $2n$ and $2m$, respectively, $(n > m \geqslant 0)$ and (2) $2n$ linear homogeneous boundary conditions

$$B_i(f) = 0, \qquad i = 1, 2, \ldots, 2n \tag{5.28}$$

at $x = a$ and $x = b$. We can, for example, take L and M to be of the form

$$L[f] = \sum_{k=0}^{n} (-1)^k [\alpha_k(x) f^{(k)}(x)]^{(k)} \tag{5.29}$$

$$M[f] = \sum_{k=0}^{m} (-1)^k [\beta_k(x) f^{(k)}(x)]^{(k)}, \tag{5.30}$$

where $\alpha_k(x)$ and $\beta_k(x)$ are real-valued functions having at least k continuous derivatives in $[a,b]$. We also assume $\alpha_k(x) > 0$ on $[a,b]$ and $\beta_k(x) \neq 0$ on $[a,b]$.

The eigenvalue problem for Eq. (5.27) is to determine values of the parameter λ for which the solution of boundary-value problem (that is, Eq. (5.27) together with boundary conditions (5.28)) exists. In theory there is no difficulty, for the eigenvalues can be found as the roots of an algebraic (determinantal) equation obtained by substitution of the general solution of Eq. (5.27) into the boundary conditions (5.28).

Differential boundary-value problems of the form (5.27), (5.28) admit an equivalent formulation as integral or integrodifferential equations by utilizing the Green's function $G(x,y)$ associated with the differential operator L subject to the boundary conditions (5.28) (cf. Courant and Hilbert [6, Chap. V, Sect. 14], Yosida [17, Chap. 2]). For (5.27), (5.28), the equivalent integrodifferential equation is

$$f(x) = \lambda \int_a^b G(x,y)\, Mf(y)\, dy. \tag{5.31}$$

We now restrict our attention to an important special case of Eq. (5.27), equivalently Eq. (5.31), in which the operator M is given by

$$M[f] = \beta_0(x) f \tag{5.32}$$

If we now assume that $\beta_0(x) > 0$ on $[a,b]$ and put

$$\varphi(x) = [\beta_0(x)]^{1/2} f(x), \tag{5.33}$$

Eq. (5.31) becomes

$$\varphi(x) = \lambda \int_a^b K(x,y)\, \varphi(y)\, dy, \tag{5.34}$$

where

$$K(x,y) = G(x,y)\, [\beta_0(x)\, \beta_0(y)]^{1/2}. \tag{5.35}$$

Hence we see that the integral equation equivalent to the differential system (5.27), (5.28), with M given by (5.32), is a homogeneous Fredholm integral equation of the second kind.† If the operator L together with the boundary conditions (5.28) are self-adjoint, then the Green's function $G(x,y)$ is a symmetric function of its arguments; and, in turn, the kernel $K(x,y)$ of the Fredholm operator is symmetric. Moreover, $K(x,y)$ is a nondegenerate kernel; hence the Fredholm operator has a countable number of real eigenvalues $\lambda_1, \lambda_2, \ldots$, and their sequence tends to infinity (cf. Pogorzelski [11, Chap. V]).

The eigenvalues λ_k of Eq. (5.31) are the same as those of Eq. (5.34); and since Eqs. (5.27) and (5.31) are equivalent, the eigenvalue problems for the ordinary differential equation with M given by (5.32) and the Fredholm integral equation (5.34) are equivalent. Furthermore, the eigenfunctions φ_k

† Equation (5.31) can also be reduced to a Fredholm integral equation in the general case (cf. Maass [9]).

and f_k of the two problems are related by (5.33). We will assume, as usual, that the eigenfunctions are orthonormal.

A standard procedure for approximating the eigenvalues of a Fredholm equation with symmetric kernel is the so-called *kernel trace method* (cf. Mihlin [10, Chap. II], Pogorzelski [11, Chap. V], Tricomi [12, Chap. III]). The iterated kernels $K_n(x,y)$ generated by $K(x,y)$ are defined recursively by

$$\begin{aligned} K_1(x,y) &= K(x,y) \\ K_2(x,y) &= \int_a^b K(x,\xi)\,K(\xi,y)\,d\xi \\ &\vdots \\ K_n(x,y) &= \int K(x,\xi)\,K_{n-1}(\xi,y)\,d\xi, \qquad n=2,3,\ldots \end{aligned} \tag{5.36}$$

and admit the expansions

$$K_n(x,y) = \sum_{k=1}^{\infty} \frac{\varphi_k(x)\,\varphi_k(y)}{\lambda_k^n}, \qquad n=1,2,\ldots \tag{5.37}$$

all of which converge absolutely and uniformly both in x and in y. If we now put $x=y$ in (5.37), and then integrate over the interval $[a,b]$ we obtain the *trace relations*

$$\sum_{k=1}^{\infty} \lambda_k^{-n} = \int_a^b K_n(x,x)\,dx, \qquad n=1,2,\ldots. \tag{5.38}$$

We remark that (5.37) and (5.38) are valid for $n \geqslant 2$ as long as the kernel K is square integrable (that is, an L_2-kernel). If K is continuous and positive,† then (5.37) and (5.38) also hold for $n=1$. If we restrict our attention to the eigenvalue λ_1, then from (5.38) we have

$$\lambda_1^{-n} = \int_a^b K_n(x,x)\,dx - \sum_{k=2}^{\infty} \lambda_k^{-n}. \tag{5.39}$$

If all eigenvalues are positive, which will be the case if the kernel is positive, we can obtain the following upper bound for λ_1^{-2} by dropping the series on the right-hand side of (5.39):

$$\lambda_1^{-n} \leqslant \int_a^b K_n(x,x)\,dx, \qquad n=1,2,\ldots. \tag{5.40}$$

B. Kernel trace estimates for the moments of random eigenvalues

In this section we consider two random eigenvalue problems of the form (5.27), (5.28). These problems will be formulated as Fredholm integral

† A symmetric kernel $K(x,y)$, $x,y \in [a,b]$ is said to be *positive* if for every function $h \in L_2[a,b]$

$$\int_a^b \int_a^b K(x,y)\,h(x)\,h(y)\,dx\,dy \geqslant 0.$$

equations with random symmetric kernels; and kernel trace methods employed to estimate the moments of the random eigenvalues. Hence the method employed is an "honest" method in the sense of Keller (cf. Sect. 3.4C); that is, the kernel trace relations (5.38) will be applied to the random eigenvalues as in the deterministic case, and then the statistical properties of the random eigenvalues investigated.

1. ***A differential equation with a random coefficient.*** Consider a differential equation of the form (5.27) with boundary conditions (5.28); and assume

1. the operator L is deterministic,
2. the operator M is of the form

$$M[f] = \beta_0(x, \omega) f \tag{5.41}$$

where $\beta_0(x, \omega)$ is a positive real-valued random function.

If we now put

$$\varphi(x, \omega) = [\beta_0(x, \omega)]^{1/2} f(x) \tag{5.42}$$

and proceed as in the deterministic case, we obtain a random Fredholm integral equation

$$\varphi(x, \omega) = \lambda \int_a^b K(x, y, \omega)\, \varphi(y, \omega)\, dy. \tag{5.43}$$

In Eq. (5.43) the random kernel is of the form

$$K(x, y, \omega) = G(x, y)\, [\beta_0(x, \omega)\, \beta_0(y, \omega)]^{1/2}; \tag{5.44}$$

where, as before, $G(x,y)$ is the Green's function associated with the deterministic operator L and the boundary conditions (5.28). We now assume that L together with the boundary conditions are self-adjoint; hence $G(x,y)$ is symmetric, and $K(x,y,\omega)$ is almost surely symmetric. We remark that the existence, uniqueness and measurability of the solution of Eq. (5.43) can be established by the methods used in Sect. 4.3.

Let us now assume that $\beta_0(x, \omega)$ is of the form

$$\beta_0(x, \omega) = 1 + b(x, \omega), \tag{5.45}$$

where $b(x, \omega)$ is a positive real-valued random function with $\mathscr{E}\{b(x, \omega)\} = 0$; hence

$$\mathscr{E}\{\beta_0(x, \omega)\} = 1. \tag{5.46}$$

In this case the random differential equation is of the form

$$Lf = \lambda \beta_0(x, \omega) f; \tag{5.47}$$

and the deterministic differential equation obtained by replacing $\beta_0(x, \omega)$ by its mean value is of the form

$$L\tilde{f} = \lambda \tilde{f}, \tag{5.48}$$

with boundary conditions (5.28). Equation (5.48) is, of course, equivalent to the deterministic integral equation

$$\tilde{f}(x) = \lambda \int_a^b G(x, y)\tilde{f}(y)\, dy, \tag{5.49}$$

since, in this case it follows from (5.33) that $\tilde{\varphi}(x) = \tilde{f}(x)$.

Returning to the random integral equation (5.43), an application of the kernel trace method yields the relations

$$\sum_{k=1}^{\infty} \lambda_k^{-1}(\omega) = \int_a^b K(x, x, \omega)\, dx = \int_a^b G(x, x)\,[1 + b(x, \omega)]\, dx, \tag{5.50}$$

$$\begin{aligned}\sum_{k=1}^{\infty} \lambda_k^{-2}(\omega) &= \int_a^b K_2(x, x, \omega)\, dx \\ &= \int_a^b\int_a^b G(x, \xi)\, G(\xi, x)\,[1 + b(x, \omega)]\,[1 + b(\xi, \omega)]\, dx\, d\xi,\end{aligned} \tag{5.51}$$

and so on for values of $n > 2$. From (5.50) and (5.51) we obtain the following kernel trace relations for the expected values of the random eigenvalues:

$$\sum_{k=1}^{\infty} \mathscr{E}\{\lambda_k^{-1}(\omega)\} = \int_a^b G(x, x)\, dx \tag{5.52}$$

$$\sum_{k=1}^{\infty} \mathscr{E}\{\lambda_k^{-2}(\omega)\} = \int_a^b\int_a^b G^2(x, \xi)\,[1 + \mathscr{E}\{b(x, \omega)\, b(\xi, \omega)\}]\, dx\, d\xi. \tag{5.53}$$

If the series in (5.52) and (5.53) converge fast enough, so that the terms after the first can be neglected, these equations provide estimates for $\mathscr{E}\{\lambda_1^{-1}(\omega)\}$ and $\mathscr{E}\{\lambda_1^{-2}(\omega)\}$. From these expressions an estimate for $\operatorname{Var}\{\lambda_1^{-1}(\omega)\}$ can be obtained.

We refer to Boyce [4] for a discussion, in a special case, of the relationship between the eigenvalues of the random problem (5.43) and the mean problem (5.48).

2. ***A differential equation with random boundary conditions.*** Consider the differential equation

$$\frac{d^2 f}{dx^2} + \lambda f = 0, \tag{5.54}$$

with boundary conditions

$$f(0) = 0, \qquad f'(1) + \xi f(1) = 0. \tag{5.55}$$

Boundary-value problems of the above form arise, for example, in the study of the transverse vibrations of an elastic string or the longitudinal vibration of an elastic bar. The boundary conditions (5.55) can be interpreted as follows:

The string or bar is fixed at $x = 0$; however, depending on the value of the *support coefficient* ξ (assumed to be nonnegative), the string or bar is either free ($\xi = 0$) or fixed ($\xi > 0$) at $x = 1$.

We now assume that the support coefficient ξ is a nonnegative real-valued random variable; hence the differential equation (5.54) is deterministic, but the boundary conditions (5.55) are random. Therefore the solution of Eq. (5.54) will be a random function with values in $L_2[0,1]$.

The above random boundary-value problem, which is a problem of Sturm–Liouville type, can be formulated as the Fredholm integral equation with random kernel

$$f(x) = \lambda \int_0^1 K(x, y, \omega) f(y)\, dy. \tag{5.56}$$

In Eq. (5.56) the random kernel is of the form

$$K(x, y, \omega) = G(x, y) - \left(\frac{\xi(\omega)}{1 + \xi(\omega)}\right) xy, \tag{5.57}$$

where

$$G(x, y) = \begin{cases} x, & x \leqslant y \\ y, & x > y, \end{cases} \tag{5.58}$$

is the Green's function for the boundary-value problem with $\xi = 0$. Put $\xi_0(\omega) = \xi(\omega)/(1 + \xi(\omega))$. Then $0 \leqslant \xi_0(\omega) < 1$ almost surely, since $\xi(\omega) \geqslant 0$. Since the Green's function is known, it is possible in this case to obtain concrete kernel trace estimates. Hence the first trace equation is

$$\sum_{k=1}^{\infty} \lambda_k^{-1}(\omega) = \int_0^1 K(x, x, \omega)\, dx = \int_0^1 (x - \xi_0(\omega)\, x^2)\, dx.$$

Therefore

$$\sum_{k=1}^{\infty} \lambda_k^{-1}(\omega) = \tfrac{1}{2} - \tfrac{1}{3}\xi_0(\omega). \tag{5.59}$$

Similarly

$$\sum_{k=1}^{\infty} \lambda_k^{-2}(\omega) = \int_0^1 K_2(x, x, \omega)\, dx = \tfrac{1}{6} - \tfrac{4}{15}\,\xi_0(\omega) + \tfrac{1}{9}\xi_0^2(\omega). \tag{5.60}$$

It follows from the maximum–minimum property of eigenvalues (cf. Courant and Hilbert [6, Chap. VI]) that

$$\lambda_k^* \leqslant \lambda_k(\omega) \leqslant \lambda_k^{**}, \tag{5.61}$$

where

$$\lambda_k^* = (k - \tfrac{1}{2})^2\, \pi^2, \qquad \lambda_k^{**} = k^2\, \pi^2. \tag{5.62}$$

In (5.62) λ_k^* denotes the kth eigenvalue of the boundary-value problem with $\xi > 0$, and λ_k^{**} denotes the kth eigenvalue of the boundary-value problem with

$\xi = 0$. From (5.59), (5.61), and (5.62) we obtain the following bounds for $\lambda_1^{-1}(\omega)$:

$$\tfrac{1}{2} - \tfrac{1}{3}\xi_0(\omega) - \sum_{k=2}^{\infty} (\lambda_k^*)^{-1} \leqslant \lambda_1^{-1}(\omega) \leqslant \tfrac{1}{2} - \tfrac{1}{3}\xi_0(\omega) - \sum_{k=2}^{\infty} (\lambda_k^{**})^{-1}. \tag{5.63}$$

Since

$$\sum_{k=2}^{\infty} (\lambda_k^*)^{-1} = \frac{4}{\pi^2}\left(\frac{\pi^2}{8} - 1\right), \qquad \sum_{k=2}^{\infty} (\lambda_k^{**})^{-1} = \frac{1}{\pi^2}\left(\frac{\pi^2}{6} - 1\right),$$

(5.63) can be rewritten as

$$\frac{4}{\pi^2} - \frac{1}{3}\xi_0(\omega) \leqslant \lambda^{-1}(\omega) \leqslant \frac{1}{3} + \frac{1}{\pi^2} - \frac{1}{3}\xi_0(\omega). \tag{5.64}$$

(5.64) gives explicit bounds on $\lambda_1^{-1}(\omega)$ as functions of the random support coefficient $\xi(\omega)$. The expectation operation applied to (5.64) yields the following bounds for $\mathscr{E}\{\lambda_1^{-1}(\omega)\}$:

$$\frac{4}{\pi^2} - \frac{1}{3}\mathscr{E}\{\xi_0(\omega)\} \leqslant \mathscr{E}\{\lambda^{-1}(\omega)\} \leqslant \frac{1}{3} + \frac{1}{\pi^2} - \frac{1}{3}\mathscr{E}\{\xi_0(\omega)\}. \tag{5.65}$$

Similar calculations based on (5.60) give the following bounds for the second moment $\mathscr{E}\{\lambda_1^{-2}(\omega)\}$:

$$\begin{aligned}\frac{16}{\pi^4} - \frac{4}{15}\mathscr{E}\{\xi_0(\omega)\} + \tfrac{1}{9}\mathscr{E}\{\xi_0^2(\omega)\} &\leqslant \mathscr{E}\{\lambda_1^{-2}(\omega)\} \\ &\leqslant \frac{7}{45} + \frac{1}{\pi^4} - \frac{4}{15}\mathscr{E}\{\xi_0(\omega)\} + \tfrac{1}{9}\mathscr{E}\{\xi_0^2(\omega)\}.\end{aligned} \tag{5.66}$$

Bounds for the higher moments $\mathscr{E}\{\lambda_1^{-n}\}$, $n = 3, 4, \ldots$, can be obtained by using the trace relations based upon higher iterated kernels.

We refer to the papers of Boyce [3, 4], Boyce and Goodwin [5], and Goodwin and Boyce [7] for the treatment of other random differential equations utilizing random Fredholm equations, and for discussions of the application of other methods (asymptotic, hierarchy, iteration, perturbation, and variational) to random eigenvalue problems.

References

1. Arnold, L., On the asymptotic distribution of the eigenvalues of random matrices. *J. Math. Anal. Appl.* **20** (1967), 262–268.
2. Bharucha-Reid, A. T., and Arnold, L., On Fredholm integral equations with random degenerate kernels. *Zastos. Mat.* (Steinhaus Jubilee Volume) **10** (1969), 85–90.
3. Boyce, W. E., Stochastic nonhomogeneous Sturm-Liouville problems. *J. Franklin Inst.* **282** (1966), 206–215.

4. Boyce, W. E., Random eigenvalue problems. *In* "Probabilistic Methods in Applied Mathematics" (A. T. Bharucha-Reid, ed.), Vol. 1, pp. 1–73. Academic Press, New York, 1968.
5. Boyce, W. E., and Goodwin, B. E., Random transverse vibrations of elastic beams. *J. Soc. Indust. Appl. Math.* **12** (1964), 613–629.
6. Courant, R., and Hilbert, D., "Methods of Mathematical Physics," Vol. 1. Wiley (Interscience), New York, 1953.
7. Goodwin, B. E., and Boyce, W. E., The vibration of a random elastic string: The method of integral equations. *Quart. Appl. Math.* **22** (1964), 261–266.
8. Kagiwada, H., Kalaba, R., and Schumitzky, A., Differential systems for eigenvalues of Fredholm integral equations. *J. Math. Anal. Appl.* **23** (1968), 227–234.
9. Maass, H., Über die Zurückführung der Eigenwertprobleme bei gewöhnlichen Differentialgleichungen auf Integralgleichungen. *Math. Z.* **58** (1953), 385–390.
10. Mihlin, S. G., "Integral Equations," translated from the Russian. Pergamon, Oxford, 1957.
11. Pogorzelski, W., "Integral Equations and Their Applications," translated from the Polish, Vol. I. Pergamon, Oxford, 1967.
12. Tricomi, F. G., "Integral Equations." Wiley (Interscience), New York, 1957.
13. Wigner, E. P., Characteristic vectors of bordered matrices with infinite dimensions. *Ann. of Math.* **62** (1955), 548–564.
14. Wigner, E. P., On the distribution of the roots of certain symmetric matrices. *Ann. of Math.* **67** (1958), 325–326.
15. Wing, G. M., On a method of obtaining bounds on the eigenvalues of certain integral equations. *J. Math. Anal. Appl.* **11** (1965), 160–175.
16. Wing, G. M., Some convexity theorems for eigenvalues of Fredholm integral equations. *J. Math. Anal. Appl.* **19** (1967), 330–338.
17. Yosida, K., "Lectures on Differential and Integral Equations." Wiley (Interscience), New York, 1960.

CHAPTER 6

Random Nonlinear Integral Equations

6.1 Introduction

In Chaps. 4 and 5 we studied random *linear* integral equations, that is, random equations of integral type which were linear with respect to the unknown function. In this chapter we will consider random nonlinear integral equations. Following the terminology employed in the study of deterministic integral equations, we define a random *nonlinear* integral equation as a random integral equation that is not linear.

Nonlinear integral equations are of great theoretical interest and are of importance in many branches of science, engineering, and technology. As in the case of linear integral equations, nonlinear integral equations arise as mathematical models of concrete physical phenomena and as integral equation formulations of nonlinear differential equations. For example, initial-value problems can lead to nonlinear Volterra integral equations of the form

$$x(t) = x(a) + \int_a^t f(\tau, x(\tau))\, d\tau; \tag{6.1}$$

and boundary-value problems can lead to Hammerstein integral equations of the form

$$x(t) + \int_a^b K(t, \tau) f(\tau, x(\tau))\, d\tau = 0. \tag{6.2}$$

Also, nonlinear Fredholm integral equations can be deduced, for example, from linear Fredholm integral equations with difference kernels, or Wiener–Hopf integral equations with finite ranges (cf. Anselone [2, pp. 299–308]).

We refer to the books of Anselone [2], Krasnosel'skiĭ [18], Pogorzelski [31],

Saaty [32], and Tricomi [35] for expositions of the theory of nonlinear integral equations and their applications.

In Sect. 6.2 we consider the integral equation formulation of some random nonlinear differential equations. The results of this section, which contains two main subsections, are due to Strand [33, 34] and Cameron [8]. A nonlinear integral equation with random right-hand side is studied in Sect. 6.3. The results of this section are due to Buche [6a], based on a paper of Beneš [5]. Section 6.4, based on a paper of Tsokos [36], is concerned with integral equations of Volterra type with random kernels and random right-hand sides. Section 6.5 is devoted to random integral equations of Uryson type, which include random Hammerstein equations as an important special case. The results of this section are due to Bharucha-Reid [6] and Mukherjea [23, 24]. A measure-theoretic problem associated with a nonlinear integral equation with random right-hand side is considered in Sect. 6.5. Finally, in Sect. 6.6 we present a brief discussion of some random integral equations arising in applied fields.

Random nonlinear Fredholm integral equations are not considered in this chapter, since the methods employed to study the nonlinear random equations of Volterra and Hammerstein type can also be used to study nonlinear random Fredholm equations. We refer to the recent book of Tsokos and Padgett [37a] for a detailed study of rather general classes of random nonlinear Fredholm and Volterra equations.

6.2 Integral Equation Formulation of Some Random Nonlinear Differential Equations

A. Introduction

The probabilistic formulation of many problems in mathematical physics, as well as in other branches of science and engineering, leads to the following problem: *Given an n-dimensional vector random function (field) $f(t,x,\omega)$, find an n-dimensional random vector $x(t,\omega)$ which satisfies the system of n random differential equations*

$$dx(t,\omega)/dt = f(t, x(t,\omega), \omega), \tag{6.3}$$

and satisfies a given initial value $x(0,\omega) = x_0$ or a random initial value $x(0,\omega) = x_0(\omega)$. If $n = 3$ and the function f represents a random velocity field, the above problem is the correct mathematical formulation of the diffusion problem arising in the turbulent flow of a fluid (cf. Kampé de Fériet [17]).

As in the case of deterministic ordinary differential equations (cf. Yosida [39]), a random differential equation of the form (6.3) can also be written as a random nonlinear integral equation of Volterra type:

$$x(t,\omega) = x_0(\omega) + \int_0^t f(\tau, x(\tau,\omega), \omega)\, d\tau. \tag{6.4}$$

The differential equation formulation requires the differentiability (in some sense) of a random function, while the integral equation formulation requires, in the main, that the function f be integrable in some appropriate sense.

In Sect. 6.2B we consider the integral equation formulation of random differential equations of the form (6.3). Three different problems for differential equations of the form (6.3) are considered, and in each case the integral equation formulation is employed. Section 6.2C is devoted to a brief summary of a result concerning a random nonlinear differential equation in the Banach space $C_0[0,1]$.

B. Integral equation formulation of a class of random nonlinear differential equations

Consider the differential equation

$$dx(t,\omega)/dt = f(t, x(t,\omega), \omega), \qquad x(t_0, \omega) = x_0(\omega). \tag{6.5}$$

We first state three problems for equations of the form (6.5). Throughout T will denote the closed interval $[a,b]$ or the semiopen interval $[a,\infty)$.

Problem* 1. *The sample function (SF), or realization, problem. Assume the function $f: T \times R_n \times \Omega \to R_n$ has the property that if $x\colon T \to R_n$ is absolutely continuous, then for almost all $\omega \in \Omega$, $f(t, x(t,\omega), \omega)$ is integrable on T. *A function $x\colon T \times \Omega \to R_n$ is said to solve the SF-problem*

$$x'(t,\omega) = f(t, x(t,\omega), \omega), \qquad x(a,\omega) = x_0(\omega)$$

if and only if for almost every $\omega \in \Omega$, the following conditions are satisfied:

(1.1) $x(t,\omega)$ is absolutely continuous on T.
(1.2) $x(a,\omega) = x_0(\omega)$.
(1.3) $x'(t,\omega) = f(t, x(t,\omega), \omega)$ for almost every $t \in T$.

In order to formulate the next two problems we consider two Banach spaces of functions on Ω and concepts of differentiability for functions with values in these spaces. Let $L_p(\Omega) = L_p(\Omega, \mathfrak{A}, \mu)$, and let $L_p^n(\Omega)$ denote the direct product of $L_p(\Omega)$ with itself n times. The norm of an element of $L_p^n(\Omega)$ is given by $\|x\| = \max(\|x_1\|, \|x_2\|, \ldots, \|x_n\|)$. The *$L_p$-derivative of a mapping $x\colon R \to L_p^n(\Omega)$ at t* is an element $x' \in L_p^n(\Omega)$ such that

$$\lim_{h\to 0} \frac{x(t+h) - x(t)}{h} = x'$$

in the norm topology of $L_p^n(\Omega)$. If the above limit exists in the weak topology of $L_p^n(\Omega)$, x' is called the *W_p-derivative of x at t*. The mapping x is said to be *W_p-pseudodifferentiable* if for every continuous linear functional $x^*\colon L_p^n(\Omega) \to R$, $x^*(x(t))$ is differentiable almost everywhere.

We also need to introduce mappings $g\colon T \times L_p^n(\Omega) \to L_p^n(\Omega)$ such that the domain of $g(t,x)$ is permitted to vary with t. When we write $g\colon T \times D_p^n(t) \to L_p^n(\Omega)$, we mean that the domain of g is the set $\{(t,x)\colon t \in T,\ x \in D(^n_p t)\}$, where $D_p^n(\cdot)$ maps T into subsets of $L_p^n(\Omega)$.

Problem* 2. *The L_p-problem. Let $g\colon T \times D_p^n(t) \to L_p^n(\Omega)$ and $x_0 \in D_p^n(a)$ be given. *A function $x\colon T \to L_p^n(\Omega)$ is said to solve the L_p-problem* if and only if the following conditions are satisfied:

(2.1) $x(t) \in D_p^n(t)$ for all $t \in T$.
(2.2) $x(t)$ is strongly absolutely continuous.
(2.3) $x(a) = x_0$.
(2.4) $g(t, x(t))$ is Bochner integrable on T.
(2.5) The L_p-derivative of x exists for almost all $t \in T$ and satisfies $x'(t) = g(t, x(t))$.

Problem* 3. *The W_p-problem. Let $g\colon T \times D_p^n(t) \to L_p^n(\Omega)$ and $x_0 \in D_p^n(a)$ be given. *A function $x\colon T \to L_p^n(\Omega)$ is said to solve the W_p-problem* if the following conditions are satisfied:

(3.1) $x(t) \in D_p^n(t)$ for all $t \in T$.
(3.2) $x(t)$ is absolutely continuous.
(3.3) $x(a) = x_0$.
(3.4) $g(t, x(t))$ is Bochner integrable on T.
(3.5) The W_p-pseudoderivative of x exists for almost all $t \in T$ and satisfies $x'(t) = g(t, x(t))$.

We now state and prove three theorems which establish the integral equation formulations for each of the above problems.

THEOREM 6.1. *A function $x(t,\omega)\colon T \times \Omega \to R_n$ is a solution of the SF-problem if and only if for all $t \in T$*

$$x(t,\omega) = x_0(\omega) + \int_a^t f(\tau, x(\tau,\omega), \omega)\, d\tau \tag{6.6}$$

with probability one. (The integral in (6.6) is, for every fixed ω, the usual Lebesgue integral.)

Proof. Since the sample functions are assumed to be absolutely continuous, for fixed $\omega \in \Omega$, Eq. (6.6) is a well-known characterization of absolutely continuous functions (cf. Hewitt and Stromberg [15, Sect. 18]).

THEOREM 6.2. *A function $x(t)\colon T \to L_p^n(\Omega)$ is a solution of the L_p-problem if and only if for all $t \in T$*

$$x(t) = x_0 + \int_a^t g(\tau, x(\tau))\, d\tau. \tag{6.7}$$

(The integral in (6.7) is the Bochner integral.)

Proof. The sufficiency follows from the fact that if $x(t)$ is weakly absolutely continuous, of strong bounded variation, and almost surely weakly differentiable with derivative $x'(t)$, then $x(t)$ is strongly differentiable, $x'(t)$ is Bochner integrable, and $x(t) = x_0 + \int_0^t x'(\tau)\, d\tau$ (cf. Hille and Phillips [16, p. 88, Theorem 3.8.6]); and the necessity follows from the fact that (i) $x(t)$: $T \to L_p^n(\Omega)$ is strongly absolutely continuous and (ii) $x'(t)$ exists almost surely if and only if $x(t)$ is the Bochner integral of $x'(t)$.

THEOREM 6.3. *A function $x(t)$: $T \to L_p^n(\Omega)$, $p > 1$, is a solution of the W_p-problem if and only if for all $t \in T$*

$$x(t) = x_0 + \int_a^t g(\tau, x(\tau))\, d\tau. \tag{6.8}$$

(The integral in (6.8) is the Pettis integral.)

Proof. The proof is based on results of Phillips [30, pp. 132–133, 137–138] and Strand [33]; from which it follows that $x(t) = x_0 + \int_a^t x'(\tau)\, d\tau$ if and only if (i) $x'(t)$ is Pettis integrable and (ii) $x(t)$ is absolutely continuous.

The next theorems establish the relationships that obtain between the three problems and their integral equation formulations. The following notation will be employed: If $x(\omega)$ is a random variable, then $\tilde{x}(\omega)$ denotes the class of random variables equivalent to $x(\omega)$. For example, if $x(\omega)$: $\Omega \to R$, then $\tilde{x}(\omega) \in L_p(\Omega)$ denotes its equivalence class. We will consider the L_p-problem

$$x'(t) = g(t, x(t)), \qquad x(a) = x_0, \tag{6.9}$$

and the SF-problem

$$dy(\omega, t)/dt = f(t, y(t, \omega), \omega), \qquad y(a, \omega) = x_0(\omega), \tag{6.10}$$

where g, f, and x_0 satisfy the assumptions made in the statements of the above problems. We also assume that if $x(\omega)$: $\Omega \to R_n$ has $\tilde{x}(\omega) \in D_p^n(t)$, then

$$g(t, \tilde{x}) = \tilde{f}(t, x(t, \omega), \omega). \tag{6.11}$$

The following theorem expresses the relationship between the L_p-, and W_p-problems.

THEOREM 6.4. (1) *If x is a solution of the L_p-problem, then it is also a solution of the W_p-problem.* (2) *If x is a solution of the W_p-problem, then it is a solution of the L_p-problem if and only if $g(t, x(t))$ is Bochner integrable.*

Proof. The above theorem follows from the integral equation formulations of the two problems (Theorems 6.2 and 6.3), together with the fact that the Bochner integral is a special case of the Pettis integral. The necessity of the condition in part (2) of the theorem is condition (2.4) as in the statement of the L_p-problem. The next theorem relates the W_p- and SF-problems. The functions g and f are as in (6.9)–(6.11).

THEOREM 6.5. *Let y: $T \times \Omega \to R_n$, and define $x(t)$ by $x(t) = \tilde{y}(t, \omega)$; and assume $f(t, y(t, \omega), \omega)$ is integrable with respect to the product measure on $T \times \Omega$. Then* (1) *if x solves the W_p-problem, then y solves the SF-problem; and* (2) *if y solves the SF-problem and $g(t, x(t))$ is Pettis integrable, then x solves the W_p-problem.*

Proof. We omit the proof since it is analogous to that of Theorem 6.4. However, we remark that the proof utilizes the following result: Let $p > 1$. If $x(t, \omega)$ is Pettis integrable and also integrable with respect to the product measure on $R \times \Omega$, then

$$\int x(t, \omega)\, dt = \int x(t)\, dt,$$

where the integral on the left is a Lebesgue integral (obtained by fixing ω) and the integral on the right is a Pettis integral.

Finally, we state, without proof, the following result which establishes the relationship between the L_p- and SF-solutions.

THEOREM 6.6. *Assume the function g in the L_p-problem* (6.9) *and the function f in the SF-problem* (6.10) *are related by* (6.11). *Then,* (1) *if $x(t)$ is a solution of the L_p-problem on T, it is also a solution of the SF-problem on T; and* (2) *if $y(t, \omega)$ is a solution of the SF-problem, then $x(t) = \tilde{y}(t, \omega)$ is a solution of the L_p-problem if and only if $g(t, x(t))$ is Bochner integrable.*

In the above theorem the statement that a solution of the L_p-problem is a solution of the SF-problem means that for each t an element in the equivalence class $x(t) \in L_p^n(\Omega)$ is chosen so that the statement is true.

We now state and prove two theorems of Picard type which establish the existence and uniqueness of solutions of the SF- and L_p-problems. These theorems are generalizations of the classical Picard theorem for deterministic ordinary differential equations (cf. Saaty [32, pp. 90–94], Yosida [39, Chap. 1]). As in the classical case, these theorems establish the existence of a solution by successive approximation procedures; hence they establish the existence of a fixed point of an integral operator. For example, in the case of the SF-problem we introduce the transformation on R_n

$$y(t,\omega) = T(\omega)\,[x(t,\omega)] \tag{6.12}$$

$$T(\omega)\,[x] = x_0(\omega) + \int_a^t f(\tau, x(\tau,\omega), \omega)\,d\tau. \tag{6.13}$$

If ξ is a fixed point of the random transformation $T(\omega)$ defined by (6.13) (that is, $T(\omega)\xi = \xi$), then ξ is a solution of the SF-problem.

The uniqueness in both problems is based on the following lemma:

LEMMA 6.1. (A generalized Gronwall inequality.) *Assume the functions $K(t), x(t), y(t), K(t)x(t)$ and $x(t)y(t)$ are integrable on $[a,b]$, and that $x(t) \geqslant 0$. If*

$$y(t) \leqslant K(t) + \int_a^t x(\tau)\,y(\tau)\,d\tau$$

for almost every t, then for almost every $t \in [a,b]$

$$y(t) \leqslant K(t) + \int_a^t K(\tau)\,x(\tau)\exp\left\{\int_\tau^t x(\xi)\,d\xi\right\} d\tau.$$

A proof of the above result in the special case of continuous $K(t)$, $x(t)$, and $y(t)$ is given in Coppel [9, p. 19], and the same proof is applicable to the more general situation as stated in Lemma 6.1.

In Theorems 6.7 and 6.8 the parameter set $T = [a,b]$ is a finite or infinite interval.

THEOREM 6.7. *The SF-problem has a unique solution $x(t,\omega)$ if the following conditions are satisfied:*

i. *There exists a finite function $k: T \times \Omega \to R$, integrable on T for almost all $\omega \in \Omega$ such that for $\xi_1, \xi_2 \in R_n$ the function f: $T \times R_n \times \Omega \to R_n$ satisfies the Lipschitz condition*

$$\|f(t,\xi_1,\omega) - f(t,\xi_2,\omega)\| \leqslant K(t,\omega)\,\|\xi_1 - \xi_2\| \tag{6.14}$$

for almost all ω;

ii. $$\int_a^b \|f(\tau, x_0(\omega), \omega)\|\,d\tau < M(\omega) < \infty \tag{6.15}$$

for almost all $\omega \in \Omega$.

Proof. We define a sequence of random functions as follows:

$$\begin{aligned} x_0(t,\omega) &= x_0(\omega) \\ x_1(t,\omega) &= x_0(\omega) + \int_a^t f(\tau, x_0(\tau,\omega), \omega)\,d\tau \\ &\;\;\vdots \\ x_n(t,\omega) &= x_0(\omega) + \int_a^t f(\tau, x_{n-1}(\tau,\omega), \omega)\,d\tau. \end{aligned} \tag{6.16}$$

The existence of the integrals in the definition of $x_n(t,\omega)$, $n = 1, 2, \ldots$ is an immediate consequence of the assumptions on the function f as indicated in the statement of the SF-problem.

From (6.16), (6.14), and (6.15) we obtain

$$\|x_{n+1}(t,\omega) - x_n(t,\omega)\| \leqslant M(\omega)\,[K(t,\omega)]^n/n!,$$

where

$$K(t,\omega) = \int_a^b k(\tau,\omega)\,d\tau.$$

It follows from the above that the sequence $\{x_n(t, \omega)\}$ converges uniformly on T for almost all $\omega \in \Omega$ to some random function $x(t,\omega)$; and since the convergence is uniform, $x(t,\omega)$ has absolutely continuous sample functions. We also have

$$\left\| x(t,\omega) - x_0(\omega) - \int_a^t f(\tau, x(\tau,\omega),\omega)\,d\tau \right\|$$

$$\leqslant \lim_{n\to\infty} \int_a^t K(\tau,\omega)\,\|x_n(\tau,\omega) - x(\tau,\omega)\|\,d\tau = 0,$$

so $x(t,\omega)$ is the desired solution of the SF-problem.

The uniqueness of the solution $x(t,\omega)$ follows from Lemma 6.1.

We now consider the analogous result for the L_p-problem.

THEOREM 6.8. *Suppose* (i) *the function* $g\colon T \times L_p^n(\Omega) \to L_p^n(\Omega)$ *satisfies the Lipschitz condition*

$$\|g(t,\xi_1) - g(t,\xi_2)\| \leqslant k(t)\,\|\xi_1 - \xi_2\| \tag{6.17}$$

for $\xi_1, \xi_2 \in L_p^n(\Omega)$, *where* $k(t)$ *is integrable on* T; *and* (ii) *if* $x\colon T \to L_n^p(\Omega)$ *is (norm) absolutely continuous, then* $g(t,x)$ *is Bochner integrable. Then there exists a unique function* $x(t)\colon T \to L_p^n(\Omega)$ *which is (norm) absolutely continuous and satisfies the* L_p*-problem.*

Proof. We define a sequence of random functions as follows:

$$\begin{aligned} x_0(t) &= x(a) = x_0 \\ x_1(t) &= x_0 + \int_a^t g(\tau, x_0(\tau))\,d\tau \\ &\;\;\vdots \\ x_n(t) &= x_0 + \int_a^t g(\tau, x_{n-1}(\tau))\,d\tau; \end{aligned} \tag{6.18}$$

and put

$$M = \int_a^b g(\tau, x_0)\,d\tau \tag{6.19}$$

$$K(t) = \int_a^t k(\tau)\,d\tau. \tag{6.20}$$

From (6.18), (6.17), (6.19), and (6.20) it follows that

$$\|x_{n+1}(t) - x_n(t)\| \leqslant M[K(t)]^n/n!.$$

The remainder of the existence proof is analogous to that of Theorem 6.7, and is, therefore, omitted. Similarly, we omit the proof of uniqueness, since it follows from Lemma 6.1.

We close this subsection with an example which demonstrates the (unfortunately) limited applicability of Theorem 6.8. Consider the L_p-problem for the random *linear* differential equation

$$x'(t) = A(\omega)\,x(t), \qquad x(0) = 1, \tag{6.21}$$

where $A(\omega)$ is a real-valued random variable. The use of Theorem 6.8 to establish existence of Eq. (6.21) requires the existence of a constant K such that

$$\|A(\omega)\,x(\omega)\| \leqslant K\|x(\omega)\| \tag{6.22}$$

for all $x \in L_p(\Omega)$. However, $x(\omega)$ satisfies (6.22) if and only if $A(\omega)$ is bounded; that is, if there exists an M such that $|A(\omega)| < M$ for almost all $\omega \in \Omega$. The sufficiency of the boundedness condition is clear; for we have only to let $K = M$. Now suppose $A(\omega)$ is not bounded; put $x_n(\omega) = (A(\omega))^n$, and define

$$m(\xi) = \int |A(\omega)|^{\xi}\,d\mu. \tag{6.23}$$

Then inequality (6.22) applied to $x_n(\omega)$ takes the form

$$\|A^{n+1}(\omega)\| \leqslant K\|A^n(\omega)\|;$$

or, using (6.23)

$$m(p(n+1)) \leqslant Km(pn), \qquad n = 1, 2, \ldots,$$

for some constant K. Now suppose $x(\omega)$ is not essentially bounded; then for any M

$$\lim_{\xi\to\infty}\left[m(\xi)\Big/\int_{|x(\omega)|>M} |x(\omega)|^{\xi}\,d\mu\right] = 1.$$

But then

$$\lim_{n\to\infty}\frac{m(p(n+1))}{m(pn)}$$

$$= \lim_{n\to\infty}\left[\int_{|x(\omega)|>M} |x(\omega)|^{p(n+1)}\,d\mu\Big/\int_{|x(\omega)|>M} |x(\omega)|^{pn}\,d\mu\right] \geqslant M^p.$$

Hence there is no constant K such that (6.22) holds.

The above example shows that even for linear differential equations with random coefficients that have Gaussian or Poisson distributions we cannot expect a Lipschitz condition to be satisfied.

C. A random nonlinear differential equation in the space of continuous functions

Let $C_0[0,1]$ denote the space of functions continuous on the interval $T=[0,1]$ and vanishing at 0. Consider the measure space $(C_0, \mathfrak{B}, w)$, where $\mathfrak{B}$ is the σ-algebra of Borel subsets of $C_0[0,1]$, and w is Wiener measure.

Consider the random nonlinear differential equation

$$dy(t,\omega)/dt = f(t, y(t,\omega) + w(t,\omega)), \qquad y(0,\omega)=0, \tag{6.24}$$

where $w(t,\omega)$ is a Wiener process and $y(t,\omega)$ is, for every fixed $\omega \in \Omega$, an element of $C_0[0,1]$. The function $f(t,u)$: $T \times R \to R$ is a real-valued continuous function of t. Put $x(t,\omega) = y(t,\omega) + w(t,\omega)$; then Eq. (6.24) is equivalent to the random nonlinear Volterra equation

$$x(t,\omega) - \int_0^t f(\tau, x(\tau,\omega))\, d\tau = w(t,\omega). \tag{6.25}$$

Cameron [8] considered the following problem: Find conditions on the function $f(t,\omega)$ such that Eq. (6.24) has a solution $y(t,\omega)$ for almost all sample functions of $w(t,\omega)$, that is, all sample functions of $w(t,\omega)$ except those belonging to a set of Wiener measure zero. The following theorem establishes conditions on $f(t,u)$ that are sufficient for the existence of an almost sure solution of Eq. (6.24).

THEOREM 6.9. (i) *Let $f(t,u)$ have continuous partial derivatives of first order f_t and f_u in the region $\{(t,u): t \in [0,1],\ u \in R\}$. Put*

$$g(t,u) = \int_0^u f(t,\xi)\, d\xi, \qquad u \in R.$$

(ii) *Let the following order of growth conditions be satisfied: for every $t \in T$ and $u \in R$*

a. $f(t,u)\operatorname{sgn} u \geqslant -A_1 \exp\{Bu^2\}$

b. $f_u(t,u) + 4g_t(t,u) \leqslant 2\alpha^2 u^2 + A_2$

c. $g(1,u) \geqslant -\frac{1}{2}\alpha u^2 \cot\beta - A_3,\ u \in R,$

where A_1, A_2, A_3, α, β and B are positive constants, $\alpha < \beta < \pi$ and $B < 1$. Then Eq. (6.24) has a unique solution $y(t,\omega) \in C_0[0,1]$ for almost all $w(t,\omega) \in C_0[0,1]$; equivalently, Eq. (6.25) has a solution $x(t,\omega) \in C_0[0,1]$ for almost all $w(t,\omega) \in C_0[0,1]$.

The importance of the above theorem is due to the fact that it can be utilized in cases where the classical (that is, deterministic) theorems are not applicable. Cameron [8] shows that there exists at least one $w \in C_0$ for which Eq. (4.24) has no solution; hence Theorem 6.9 cannot be contained in any of the classical theorems. We also remark that for $f(t,u) = -u^2$ the conditions of Theorem 6.9 are not satisfied; and Woodward [38] has shown that for almost all $\omega \in \Omega$, $y(t,\omega)$ does not exist.

6.3 A Nonlinear Integral Equation with Random Right-Hand Side

The study of certain nonlinear problems in circuit and control theory leads to an integral equation of the form

$$x(t) - \int_{-\infty}^{\infty} K(t-\tau)\psi(x(\tau),\tau)\,d\tau = y(t). \tag{6.26}$$

In Eq. (6.26) the function $\psi(x,t)$ represents a time-variable nonlinear element. $K(t)$ is an impulse response characteristic of a linear system, and the right-hand side $y(t)$ is an input signal. Beneš [5] has studied Eq. (6.26) in the real Marcinkiewicz space M_2. The function space $M_2(-\infty,\infty)$ (cf. Marcinkiewicz [20]) is the class of all measurable, locally integrable, real-valued functions $x(t)$, $t \in T = (-\infty,\infty)$ for which

$$\|x\|^2 = \limsup_{A\to\infty}(1/2A)\int_{-A}^{A} |x(t)|^2\,dt < \infty. \tag{6.27}$$

(6.27) can be referred to as a weak "finite power" condition. Let M_0 denote the subspace of functions of zero power, that is, those $x(t) \in M_2$ with $\|x\| = 0$; and consider the quotient space M_2/M_0 consisting of all cosets $\{x + M_0\}$, where $x \in M_2$. With the norm $\|x\| = \|\{x + M_0\}\|$ the quotient space M_2/M_0 is a Banach space. There is a natural homomorphism λ of M_2 onto M_2/M_0 defined by λ: $x \to \{\xi : \|\xi - x\| = 0\}$, $x \in M_2$. Operators mapping M_2 into itself can be extended to M_2/M_0 according to λ; that is, if T: $M_2 \to M_2$, then $T[\lambda x] = \lambda T x$ for $x \in M_2$.

Random solutions of Eq. (6.26) in M_2 are of interest in those cases in which the right-hand side, or input, is a random function $y(t,\omega)$, $t \in T$, $\omega \in \Omega$. If the input has finite power and is stationary, then the averages

$$\lim_{A\to\infty}(1/2A)\int_{-A}^{A} |y(t,\omega)|^2\,dt$$

will exist, and will be finite for almost all $\omega \in \Omega$; hence almost all sample functions of $y(t,\omega)$ will belong to M_2. In these cases it is of interest to seek solutions $x(t,\omega)$ of Eq. (6.26) which also belongs to M_2.

Using the Banach contraction mapping theorem, Beneš proved the following existence and uniqueness theorem for Eq. (6.26):

THEOREM 6.10. *Suppose* (i) $\psi(x,t)$ *satisfies the condition*

$$\alpha(x_1 - x_2) \leqslant \psi(x_1, t) - \psi(x_2, t) \leqslant \beta(x_1 - x_2) \tag{6.28}$$

for all t, *and all* x_1, x_2 *with* $x_1 \geqslant x_2$, *and some constants* α, β $(\beta > 0)$, *and* (ii) $K(t)$ *is an* L_2-*function such that*

$$\int_{-\infty}^{\infty} t^2 |K(t)|^2 \, dt < \infty \tag{6.29}$$

$$\left| \tfrac{1}{2}(\alpha + \beta) - 1 \Big/ \int_{-\infty}^{\infty} e^{-i\mu t} K(t) \, dt \right| > \tfrac{1}{2}(\beta - \alpha). \tag{6.30}$$

Let $y(t)$ *be any function in* M_2. *Then there exists a solution* $x(t) \in M_2$ *of Eq.* (6.26), *and* λx *is unique in* M_2/M_0; *that is, the solution is unique to within a function belonging to* M_0.

We now consider Eq. (6.26) with input a random function $y(t, \omega)$. In this case it is necessary to consider the space M_2 of all random functions $x(t, \omega)$ such that (1) $x(t, \omega)$ is a measurable random function, (2) $x(t, \omega)$ is locally integrable for every fixed $\omega \in \Omega$, and (3) the weak finite power condition

$$\|x(\omega)\| = \limsup_{A \to \infty} (1/2A) \int_{-A}^{A} |x(t, \omega)|^2 \, dt < \infty \tag{6.31}$$

is satisfied for almost all $\omega \in \Omega$. As before, we denote by M_0 the subspace of functions of zero power, that is, those $x(t, \omega)$ with $\|x(\omega)\| = 0$ almost surely.

Consider the random nonlinear integral equation

$$x(t, \omega) = y(t, \omega) + \int_{-\infty}^{\infty} K(t - \tau) \psi(x(\tau, \omega) \tau) \, d\tau, \tag{6.32}$$

where $y(t, \omega)$: $T \times \Omega \to M_2$, $\psi(x(\tau, \omega) \tau)$: $M_2 \times T \to M_2$, and $K(t)$: $T \to T$ satisfies (6.29). The following theorem, which establishes the existence and uniqueness of the random solution $x(t, \omega)$ of Eq. (6.32), extends Theorem 6.10 to the case of random inputs.

THEOREM 6.11. *Suppose* (i) $\psi(x,t)$ *satisfies* (6.28) *with* $x_1(t, \omega) \in M_2$, $x_2(t, \omega) \in M_2$ *where* $x_1(t, \omega) \geqslant x_2(t, \omega)$ *almost surely, and* (ii) $(K(t))$ *is an* L_2-*function satisfying* (6.29) *and* (6.30). *Let* $Y(t, \omega)$ *be an arbitrary measurable random function in* M_2. *Then there exists a random solution* $x(t, \omega) \in M_2$ *of Eq.* (6.32), *and the solution is unique in* M_2/M_0.

We state without proof two lemmas which will be used in the proof of Theorem 6.11.

LEMMA 6.2. (Lemma 4 in Beneš [5].) *Let* $\tilde{K}(\mu) = \mathfrak{F}\{K(t)\}$ *denote the Fourier transform of* $K(t)$. *If* (i) $K(t) \in L_2$, (ii) (6.29) *is satisfied,* (iii) $\tilde{K}(\mu) = 1$ *for all* μ, *and* (iv) $\tilde{H}(\mu) = \tilde{K}(\mu)[1 - \tilde{K}(\mu)]^{-1}$, *then* $H(t) = \mathfrak{F}^{-1}\{H(\mu)\} \in L_2$ *and satisfies* (6.29).

LEMMA 6.3. (Lemma 5 in Beneš [5].) *Let* $F(t)$ *be a function such that* $(1 + t^2)|F(t)|^2 \in L_1$, *and let* $\tilde{F}(\mu) = \mathfrak{F}\{F(t)\}$. *Then* $F(t) \in L_1$. *If* $\tilde{F}(\mu) \neq 1$ *for all* μ, *and* $G(t) = \mathfrak{F}^{-1}\{\tilde{F}(\mu)[1 - \tilde{F}(\mu)^{-1}]\}$, *then the operator* $I - \tilde{F}$ *on* M_2/M_0 *has a bounded inverse representable as*

$$(I - \tilde{F})^{-1}\lambda x = (I + \tilde{G})\lambda x.$$

***Proof of Theorem* 6.11.** We first rewrite Eq. (6.32) in the form

$$x(t, \omega) = y(t, \omega) + Ax(t, \omega). \tag{6.33}$$

If we define W as the operator

$$W[x(t, \omega)] = \psi(x(t, \omega), t) \tag{6.34}$$

and V as the linear operator

$$V[x(t, \omega)] = \int_{-\infty}^{\infty} K(t - \tau)\, x(\tau, \omega)\, d\tau, \qquad x(\tau, \omega) \in M_2, \tag{6.35}$$

then it is clear that the operator A in Eq. (6.33) is of the form $A = VW$; and that the existence and uniqueness of the solution of Eq. (6.33) can be established by finding a fixed point of the operator $A = VW$. It follows from (6.29) that V maps M_2 into itself. In order that A be well defined W must map $x(t, \omega)$ into the domain of V. It follows from (i) that ψ is an almost surely continuous function of $x(t, \omega)$, and that

$$|\psi(x(t, \omega), t)| \leqslant \max\{|\alpha|\, |\beta|\}\, |x(t, \omega)|$$

almost surely. Hence $\psi(x(t, \omega), t) \in M_2$; and A maps M_2 into itself. We now show that A is a contraction operator on M_2. Since $\psi(x, t)$ and $K(t)$ are deterministic functions, the operators W and V are deterministic; hence A is a deterministic operator, and the classical Banach contraction mapping theorem can be used to establish the existence of a fixed point of A.

Using the definition of the operator V, Eq. (6.32) can be written as

$$(I - \tfrac{1}{2}(\alpha + \beta)\, V)\, x(t, \omega) = y(t, \omega) + \int_{-\infty}^{\infty} K(t - \tau)\, [\psi(x(\tau, \omega), \tau)$$

$$- \tfrac{1}{2}(\alpha + \beta)\, x(\tau, \omega)]\, d\tau. \tag{6.36}$$

It follows from (6.29) that $K(t) \in L_1$, and from (6.30) that $\tfrac{1}{2}(\alpha + \beta)\tilde{F}(\mu) \neq 1$ for all μ; hence by Lemmas 6.2 and 6.3 the operator $I - \tfrac{1}{2}(\alpha + \beta)\, V$ has a bounded inverse on M_2/M_0 represented by the identity minus a convolution.

Hence we can rewrite (6.36) as

$$\begin{aligned} x(t,\omega) &= (I - \tfrac{1}{2}(\alpha+\beta)V)^{-1} y(t,\omega) \\ &\quad + \int_{-\infty}^{\infty} H(t-\tau)\,[\psi(x(\tau,\omega),\omega) - \tfrac{1}{2}(\alpha+\beta)\,x(\tau,\omega)]\,d\tau \\ &= y_1(t,\omega) + Sx(t,\omega), \end{aligned} \tag{6.37}$$

where $H(t)$ is the L_1-function with

$$\tilde{H}(\mu) = \mathfrak{F}\{H(t)\} = \frac{\tilde{K}(\mu)}{1 - \tfrac{1}{2}(\alpha+\beta)\,\tilde{K}(\mu)},$$

and the second equality defines the function $y_1(t,\omega)$ and operator S in an obvious way.

From (i) it follows that

$$|\psi(x_1,t) - \psi(x_2,t) - \tfrac{1}{2}(\alpha+\beta)(x_1-x_2)| \leqslant \tfrac{1}{2}(\beta-\alpha)(x_1-x_2)$$

for almost all $\omega \in \Omega$. Hence for any $x_1(t,\omega) \in M_2$, $x_2(t,\omega) \in M_2$,

$$\|S(x_1-x_2)\| \leqslant \sup_{\mu} \left| \frac{\tilde{K}(\mu)}{1 - \tfrac{1}{2}(\alpha+\beta)\,\tilde{K}(\mu)} \right| \tfrac{1}{2}(\beta-\alpha)\,\|x_1-x_2\| \tag{6.38}$$

for almost all $\omega \in \Omega$. From (6.30) it follows that the constant on the right-hand side of (6.38) is less than one. Hence, rewriting Eq. (6.37) as

$$(I-S)\,x(t,\omega) = y(t,\omega),$$

and observing that

$$\|(I-S)(x_1-x_2)\| = \|S(x_1-x_2)\|,$$

it follows from the completeness of M_2, and the fact that M_2/M_0 is a metric space, that the Banach contraction mapping theorem is applicable; and this establishes the existence and uniqueness of the random solution of Eq. (6.32).

6.4 Nonlinear Integral Equations of Volterra Type with Random Kernels and Random Right-Hand Sides

A. Introduction

Random linear integral equations of Volterra type were considered in Sects. 4.2 and 4.3, and in Sect. 6.2 we considered some nonlinear integral equations of Volterra type which arose as integral equation formulations of some random nonlinear differential equations. In this section, which is based on a paper of Tsokos [36], we study a rather general class of nonlinear integral equations of Volterra type with random kernels and random right-hand sides. The equation to be considered is of the form

$$x(t, \omega) - \int_0^t K(t, \tau, \omega) f(\tau, x(\tau, \omega))\, d\tau = y(t, \omega), \tag{6.39}$$

where $\omega \in \Omega$, $t \in R^+$. We assume that the unknown function $x(t, \omega)$ and the known function $y(t, \omega)$ are functions of $t \in R^+$ with values in $L_2(\Omega) = L_2(\Omega, \mathfrak{A}, \mu)$. The function $f(t, x(t, \omega))$ will, under appropriate conditions, also be a function of $t \in R^+$ with values in $L_2(\Omega)$. The random kernel $K(t, \tau, \omega)$ is assumed to be a μ-essentially bounded function for every t and τ $(0 \leqslant \tau \leqslant t < \infty)$, with values in $L_\infty(\Omega)$ for every fixed t and τ. Hence the product $K(t, \tau, \omega) f(t, x(t, \omega))$ will be in $L_2(\Omega)$. We also assume that the mapping $(t, \tau) \to K(t, \tau, \omega)$ from the set $\{(t, \tau): 0 \leqslant \tau \leqslant t < \infty\}$ into $L_\infty(\Omega)$ is continuous; that is,

$$\mu - \operatorname*{ess\,sup}_{\omega} |K(t_n, \tau_n, \omega) - K(t, \tau, \omega)| \to \infty \quad \text{a.s.}$$

as $n \to \infty$ whenever $(t_n, \tau_n) \to (t, \tau)$ as $n \to \infty$.

Section 6.4B is devoted to the existence and uniqueness of a random solution $x(t, \omega)$ of Eq. (6.39). In this section the notion of a pair of Banach spaces being admissible with respect to an operator is utilized.† In order to introduce this notion we need to define several spaces: (1) The space $C_c = C_c(R^+, L_2(\Omega))$ is defined as the space of all continuous functions from R^+ into $L_2(\Omega)$, with the topology of uniform convergence on every interval $[0, b]$, $b > 0$. The space C_c is a locally convex space, whose topology is defined by means of the family of seminorms

$$\|x(t, \omega)\|_n = \sup_{t \in [0, n]} \left\{ \int_\Omega |x(t, \omega)|^2 \, d\mu \right\}^{1/2}. \tag{6.40}$$

$n = 1, 2, \ldots.$ (2) The space $C_g = C_g(R^+, L_2(\Omega))$ is defined as the space of all continuous functions from R^+ into $L_2(\Omega)$ such that

$$\left\{ \int_\Omega |x(t, \omega)|^2 \, d\mu \right\}^{1/2} \leqslant M g(t), \qquad t \in R^+, \tag{6.41}$$

where M is a positive number and $g(t)$, $t \in R^+$, is a positive continuous function. The norm in C_g is defined by

$$\|x(t, \omega)\|_g = \sup_{t \in R^+} \left\{ \frac{1}{g(t)} \|x(t, \omega)\|_{L_2} \right\}. \tag{6.42}$$

(3) The space $C = C(R^+, L_2(\Omega))$ is defined as the space of all continuous and bounded functions on R^+ with values in $L_2(\Omega)$. (4) Finally, let $\mathfrak{X} = \mathfrak{X}(R^+, L_2(\Omega))$

† The concept of admissibility was introduced in the theory of differential equation by Massera and Schaffer (cf. [21, Chap. 5]), and in the theory of integral equations by Corduneanu [10, 11].

and $\mathfrak{Y} = \mathfrak{Y}(R^+, L_2(\Omega))$ be a pair of Banach spaces of continuous functions from R^+ into $L_2(\Omega)$ such that $\mathfrak{X}, \mathfrak{Y} \subset C_c$; and let L be a linear operator from C_c into itself.

***Definition* 6.1.** The pair of Banach spaces $(\mathfrak{X}, \mathfrak{Y})$ is said to be *admissible with respect to the operator* L: $C_c(R^+, L_2(\Omega)) \to C_c(R^+, L_2(\Omega))$ if and only if $L[\mathfrak{X}] \subset \mathfrak{Y}$.

We now state a fundamental lemma which will be used in Sect. 6.4B.

LEMMA 6.4. *Let L be a continuous operator from* $C_c(R^+, L_2(\Omega))$ *into itself. If* (*i*) $\mathfrak{X}$ *and* $\mathfrak{Y}$ *are Banach spaces with topologies stronger than the topology of* C_c, *and* (*ii*) *the pair* $(\mathfrak{X}, \mathfrak{Y})$ *is admissible with respect to L, then L is a continuous operator from* $\mathfrak{X}$ *to* $\mathfrak{Y}$.

We refer to Corduneanu [10] for a proof of the above lemma, which involves showing that L is a closed operator, and then using the closed graph theorem.

We remark that if L is a continuous operator it is also bounded; hence we can find a constant $M > 0$ such that

$$\|Lx(t, \omega)\|_{\mathfrak{Y}} \leqslant M \|x(t, \omega)\|_{\mathfrak{X}} . \tag{6.43}$$

In Sect. 6.4C we consider a random nonlinear differential equation of the form

$$dx(t, \omega)/dt = A(\omega)\, x(t, \omega) + f(t, x(t, \omega)), \qquad t \in R^+ \tag{6.44}$$

and show that it can be formulated as a random integral equation of the form (6.39).

Tsokos [37] has also studied random nonlinear Volterra integral equations of convolution type

$$x(t, \omega) - \int_0^t K(t - \tau, \omega)\, \Phi(x(\tau, \omega))\, d\tau = y(t, \omega), \tag{6.45}$$

which arise as integral equation formulations of certain random nonlinear differential systems.

B. Existence and uniqueness

We now consider the existence and uniqueness of a random solution of Eq. (6.39). A random function $x(t, \omega)$ will be called a random solution of Eq. (6.39) if for every fixed $t \in R^+$, $x(t, \omega) \in L_2(\Omega)$ and satisfies Eq. (6.39) with probability one.

We now state and prove the following existence and uniqueness theorem for Eq. (6.39).

THEOREM 6.12. *Assume*

i. $\mathfrak{X}$ *and* $\mathfrak{Y}$ *are Banach spaces of continuous functions from* R^+ *into* $L_2(\Omega)$ *with topologies stronger than the topology of* $C_c(R^+, L_2(\Omega))$, *and the pair* $(\mathfrak{X}, \mathfrak{Y})$ *is admissible with respect to the random integral operator*

$$L(\omega)\,x(t,\omega) = \int_0^t K(t,\tau,\omega)\,x(\tau,\omega)\,d\tau, \tag{6.46}$$

where the random kernel $K(t,\tau,\omega)$ *is continuous in the sense indicated earlier;*

ii. *The mapping* $x(t,\omega) \to f(t, x(t,\omega))$ *is an operator on the set*

$$S = \{x(t,\omega) : x(t,\omega) \in \mathfrak{Y}, \quad \|x(t,\omega)\|_{\mathfrak{Y}} \leqslant \rho\},$$

for some $\rho \geqslant 0$, *with values in* $\mathfrak{X}$, *satisfying the condition*

$$\|f(t, x_1(t,\omega)) - f(t, x_2(t,\omega))_{\mathfrak{X}} \leqslant k\|x_1(t,\omega) - x_2(t,\omega\|_{\mathfrak{Y}}, \tag{6.47}$$

for $x_1, x_2 \in S$ *and* k *a positive constant.*

iii. $y(t,\omega) \in \mathfrak{Y}$.

Then, there exists a unique random solution of Eq. (6.39) *whenever* (*a*) $k < N^{-1}$ *and* (*b*) $\|y(t,\omega)\|_{\mathfrak{Y}} + N\|f(t,0)\|_{\mathfrak{X}} \leqslant \rho(1 - kN)$, *where* N *is the norm of* $L(\omega)$.

Proof. We define a random operator $W(\omega)$ from S into $\mathfrak{Y}$ as follows:

$$W(\omega)\,[x(t,\omega)] = y(t,\omega) + \int_0^t K(t,\tau,\omega)\,f(\tau, x(\tau,\omega))\,d\tau. \tag{6.48}$$

We first show that under the hypothesis of the theorem $W(\omega)$ is a contraction operator. Let $x_1(t,\omega)$ and $x_2(t,\omega)$ be elements of S. Then from (6.48) we have

$$W(\omega)\,[x_1(t,\omega) - x_2(t,\omega)] = \int_0^t K(t,\tau,\omega)\,[f(t, x_1(\tau,\omega)) - f(t, x_2(\tau,\omega))]\,d\tau. \tag{6.49}$$

Since $W(\omega)\,[S] \subset \mathfrak{Y}$, and $\mathfrak{Y}$ is a Banach space, we have

$$W(\omega)\,[x_1(t,\omega) - x_2(t,\omega)] \in \mathfrak{Y}.$$

It follows from assumptions (i) and (ii) that $[f(t, x_1(t,\omega)) - f(t, x_2(t,\omega)] \in \mathfrak{X}$. Since, by Lemma 6.4, $L(\omega)$ is a continuous operator from $\mathfrak{X}$ into $\mathfrak{Y}$, there exists a constant $N > 0$ such that

$$\|L(\omega)\,x(t,\omega)\|_{\mathfrak{Y}} \leqslant N\|x(t,\omega)\|_{\mathfrak{X}}.$$

From (6.49) we have

$$\|W(\omega)\,[x_1(t,\omega) - x_2(t,\omega)\|_{\mathfrak{Y}} \leqslant N\|f(t, x_1(t,\omega)) - f(t, x_2(t,\omega))\|_{\mathfrak{X}}.$$

Applying the Lipschitz condition on $f(t,x)$ as given by (ii), we have

$$\|W(\omega)[x_1(t,\omega)-x_2(t,\omega)]\|_{\mathfrak{Y}} \leqslant kN\|x_1(t,\omega)-x_2(t,\omega)\|_{\mathfrak{X}}.$$

Using condition (a) (that is, $kN<1$), the above inequality implies that $W(\omega)$ is a contraction operator.

We now show that $W(\omega)[S]\subset S$. For every $x(t,\omega)\in S$ the operator $W(\omega)$ is well defined. Hence, it follows from assumption (iii) and an application of Lemma 6.4 that we can write

$$\|W(\omega)\,x(t,\omega)\|_{\mathfrak{Y}} \leqslant \|y(t,\omega)\|_{\mathfrak{Y}} + N\|f(t,x(t,\omega))\|_{\mathfrak{X}}. \tag{6.50}$$

The norm of $f(t,x)$ in (6.48) can be written as

$$\begin{aligned}\|f(t,x(t,\omega))\|_{\mathfrak{X}} &= \|f(t,x(t,\omega))-f(t,0)+f(t,0)\|_{\mathfrak{X}}\\ &\leqslant \|f(t,x(t,\omega))-f(t,0)\|_{\mathfrak{X}} + \|f(t,0)\|_{\mathfrak{X}}.\end{aligned}$$

Another application of the Lipschitz condition on $f(t,x)$ yields

$$\|f(t,x(t,\omega))\|_{\mathfrak{X}} \leqslant k\|x(t,\omega)-0\|_{\mathfrak{Y}} + \|f(t,0)\|_{\mathfrak{X}};$$

which enables us to rewrite (6.50) as follows:

$$\|W(\omega)\,x(t,\omega)\|_{\mathfrak{Y}} \leqslant \|y(t,\omega)\|_{\mathfrak{Y}} + kN\|x(t,\omega)\|_{\mathfrak{Y}} + N\|ft,0)\|_{\mathfrak{X}}.$$

Since $x(t,\omega)\in S$, $\|x(t,\omega)\|_{\mathfrak{Y}}\leqslant\rho$; therefore the above inequality becomes

$$\|W(\omega)\,x(t,\omega)\|_{\mathfrak{Y}} \leqslant \|y(t,\omega)\|_{\mathfrak{Y}} + kN\rho + N\|f(t,0)\|_{\mathfrak{X}}. \tag{6.51}$$

Using condition (b), that is, $\|y(t,\omega)\|_{\mathfrak{Y}} + N\|f(t,0)\|_{\mathfrak{X}} \leqslant \rho(1-kN)$, (6.51) becomes

$$\|W(\omega)\,x(t,\omega)\|_{\mathfrak{Y}} \leqslant \rho(1-kN)+kN\rho=\rho.$$

Hence, by definition of the set S, $W(\omega)x(t,\omega)\in S$ for every $x(t,\omega)\in S$, or $W(\omega)[S]\subset S$.

Finally, since we have shown that $W(\omega)$ is a contraction operator and $W(\omega)[S]\subset S$, it follows from the Špaček-Hanš fixed point theorem that the operator $W(\omega)$ has a unique fixed point $x(t,\omega)$ for every $t\in R^+$; that is $W(\omega)x(t,\omega)=x(t,\omega)$. Hence there exists a unique random solution $x(t,\omega)\in S\subset\mathfrak{X}$ of the random integral equation (6.39).

Padgett and Tsokos [26] (cf. also [37a]) have studied a *random integral equation of the mixed Volterra–Fredholm type* of the form

$$\begin{aligned}x(t,\omega) = y(t,\omega) &+ \int_0^t K_1(t,\tau,\omega)f(\tau,x(\tau,\omega))\,d\tau\\ &+\int_0^\infty K_2(t,\tau,\omega)g(\tau,x(\tau,\omega))\,d\tau, \qquad t\geqslant 0.\end{aligned} \tag{6.52}$$

Using the theory of admissibility of Hilbert spaces and the fixed point theorems of Banach, Schauder, and Krasnosel'skiĭ (cf. Sect. 3.3B), the existence and uniqueness of a random solution of Eq. (6.52) was established. Their results generalize those of Anderson (Sect. 4.2), as well as those of Tsokos presented above.

C. Integral equation formulation of a random nonlinear differential equation

Consider the random nonlinear differential equation

$$dx(t,\omega)/dt = A(\omega)\,x(t,\omega) + f(t, x(t,\omega)), \qquad t \in R^+, \tag{6.53}$$

where $x(t,\omega)$ is an n-dimensional random vector, $A(\omega)$ is an $n \times n$ random matrix, and $f(t,x)$ is an n-dimensional vector-valued function for $t \in R^+$, $x \in R$. Equation (6.53) can be reduced to the random integral equation

$$x(t,\omega) - \int_0^t \exp\{A(\omega)(t-\tau)\} f(\tau, x(\tau,\omega))\,d\tau = e^{A(\omega)t} x_0(\omega). \tag{6.54}$$

Put

$$y(t,\omega) = e^{A(\omega)t} x_0(\omega) \tag{6.55}$$

and

$$K(t,\tau,\omega) = e^{A(\omega)(t-\tau)}, \qquad 0 \leqslant \tau \leqslant t < \infty; \tag{6.56}$$

then Eq. (6.54) is of the form (6.39).

We now assume that (1) the random matrix $A(\omega)$ is *μ-stable*, that is there exists an $\alpha > 0$ such that

$$\mu\left\{\omega:\ \max_{\lambda\in\sigma(A(\omega))} \operatorname{Re}\lambda < -\alpha\right\} = 1,$$

where $\sigma(A(\omega))$ denotes the spectrum of $A(\omega)$; and (2) $f(t,x)$ is a continuous function from $R^+ \times R_n \to R$ satisfying the Lipschitz condition

$$|f(t,x_1) - f(t,x_2)| \leqslant k|x_1 - x_2|,$$

with $f(t,0) = 0$, and k sufficiently small. Under these conditions we will show that there exists a unique random solution $x(t,\omega)$ of Eq. (6.54).

Consider the pair of Banach spaces (C_g, C_g) with $g(t) = e^{-\beta t}$, where $\beta \in (0,\alpha)$. For $x(t,\omega) \in C_g$, we define the following random integral operator:

$$L(\omega)\,x(t,\omega) = \int_0^t K(t,\tau,\omega)\,x(\tau,\omega)\,d\tau. \tag{6.57}$$

From the definition of the norm in C_g (cf. (6.42)), it follows, using (6.56) and (6.57), that

$$\|L(\omega)\,x(t,\omega)\| \leqslant \int_0^t \exp\{A(\omega)(t-\tau)\}\,\|x(\tau,\omega)\|\,d\tau. \tag{6.58}$$

Now, it has been shown by Morozan [22] that there exists a subset E of Ω such that $\mu(E) = 1$ and

$$\|K(t, \tau, \omega)\| = \|e^{A(\omega)(t-\tau)}\| \leqslant Me^{-\alpha(t-\tau)}, \tag{6.59}$$

for $\omega \in E$, $M > 0$, and α as defined above. Applying (6.59) to (6.58) we obtain

$$\|L(\omega)\, x(t, \omega)\| \leqslant M \int_0^t e^{-\alpha(t-\tau)} \frac{\|x(\tau, \omega)\|}{g(\tau)}\, g(\tau)\, d\tau. \tag{6.60}$$

Since $g(t) = e^{-\beta t}$, $\beta \in (0, \alpha)$, (6.60) can be rewritten as

$$\begin{aligned}\|L(\omega)\, x(t, \omega)\| &\leqslant M \int_0^t e^{-\alpha t}\, e^{(\alpha-\beta)\tau} \|x(\tau, \omega)\|\, e^{\beta\tau}\, d\tau \\ &\leqslant M\|x(t, \omega)\|_{C_g} e^{-\alpha t} \int_0^t e^{(\alpha-\beta)\tau}\, d\tau \\ &\leqslant M\|x(t, \omega)\|_{C_g} (\alpha - \beta)^{-1}(e^{-\beta t} - e^{-\alpha t}), \qquad t \in R^+. \end{aligned} \tag{6.61}$$

Since $\beta \in (0, \alpha)$, the above inequality can be majorized to yield

$$\|L(\omega)\, x(t, \omega)\| \leqslant M\|x(t, \omega)\|_{C_g} (\alpha - \beta)^{-1}\, e^{-\beta t}.$$

Hence

$$\|L(\omega)\, x(t, \omega)\|_{C_g} \leqslant M(\alpha - \beta)^{-1}\|x(t, \omega)\|_{C_g}.$$

Therefore, $x(t, \omega) \in C_g$ implies $L(\omega)[C_g] \subset C_g$; which, in turn, implies that the pair of Banach spaces (C_g, C_g) with $g(t) = e^{-\beta t}$ is admissible with respect to the random integral operator $L(\omega)$.

From the above, namely the form of $y(t, \omega)$, the properties of the kernel $K(t, \tau, \omega)$, the Lipschitz condition on $f(t, x)$, and the admissibility of the pair (C_g, C_g) with respect to $L(\omega)$, it follows that the hypotheses of Theorem 6.11 are satisfied; hence there exists a unique random solution $x(t, \omega)$ of Eq. (6.54) (equivalently Eq. (6.53)), and $x(t, \omega)$ is exponentially μ-stable, that is,

$$\lim_{t\to\infty} \left\{\int_\Omega |x(t, \omega)|^2\, d\mu\right\}^{1/2} = 0.$$

D. Approximate solutions of Eq. (6.39)

Padgett and Tsokos [29] have defined a sequence of successive approximation, $x_n(t, \omega)$, which converges to the unique random solution of Eq. (6.39) at each $t \geqslant 0$ with probability one and in mean square under the conditions of Theorem 6.12. The sequence is defined as follows, assuming that the distribution of $y(t, \omega)$ is known or that a value of $y(t, \omega)$ can be observed for each $t \geqslant 0$:

$$\begin{aligned} x_0(t, \omega) &= y(t, \omega) \\ x_{n+1}(t, \omega) &= W(\omega)\,[x_n(t, \omega)], \qquad n \geqslant 0, \end{aligned} \tag{6.62}$$

where $W(\omega)$ is defined by (6.48).

Padgett and Tsokos [28] utilize a stochastic approximation procedure due to Burkholder [7] to study approximate solutions of Eq. (6.39). In particular, conditions on the functions $y(t, \omega)$, $K(t, \tau, \omega)$, and $f(t, x)$ are given for which the approximation procedure converges to a realization of the solution $x(t, \omega)$ with probability one. We refer to Tsokos and Padgett [37a] for a detailed discussion of approximate solutions of random Volterra equations.

6.5 Random Nonlinear Integral Equations of Uryson Type

A. Introduction

In this section we consider a probabilistic analogue of the *Uryson integral equation*

$$\lambda \int_G K(t, \tau, x(\tau))\, dm(\tau) - x(t) = y(t). \tag{6.63}$$

In Eq. (6.63) G is usually a closed and bounded subset of R_n, the kernel $K(t, \tau, \xi)$ is a continuous function of $t, \tau \in G$ and $\xi \in R$, and m is Lebesgue measure on G. Equation (6.63), which is a very general type of nonlinear integral equation, has been studied in the space of continuous functions $C(G)$, the Lebesgue spaces $L_p(G)$, and the Orlicz spaces $L_\Phi(G)$, We refer to the books of Anselone [2], Krasnosel'skiĭ [18], and Saaty [32] for authoritative discussions of the theory of Uryson integral equations, and for references to the relevant literature.

A special case of Eq. (6.63) which has been the subject of extensive study is the *Hammerstein integral equation*

$$\lambda \int_G K(t, \tau) f(\tau, x(\tau))\, dm(\tau) - x(t) = y(t). \tag{6.64}$$

A random integral equation of Hammerstein type was considered in [6], this equation being associated with a probabilistic analogue of Duffing's vibration problem. As is well known, the forced vibrations of finite amplitude of a pendulum are governed by the nonlinear differential equation

$$\frac{d^2 \varphi(t)}{dt^2} + \alpha^2 \sin \varphi(t) = F(t), \tag{6.65}$$

where $F(t)$ is a periodic driving function. If we consider Eq. (6.65) with the boundary conditions

$$\varphi(0) = \varphi(1) = 0, \tag{6.66}$$

then the boundary value problem (6.65)–(6.66) is equivalent to the nonlinear integral equation

$$\varphi(t) = -\int_0^1 K(t, \tau)\, [F(\tau) - \alpha^2 \sin \varphi(\tau)]\, d\tau \tag{6.67}$$

(cf. Saaty [32, p. 265], Tricomi [35, p. 215]). In Eq. (6.67) the kernel $K(t,\tau)$, which is the Green's function of the differential boundary-value problem, is of the form

$$K(t,\tau)=\begin{cases}\tau(1-t), & 0\leqslant\tau\leqslant t\leqslant 1\\ t(1-\tau), & 0\leqslant t\leqslant\tau\leqslant 1.\end{cases} \tag{6.68}$$

If we now assume that the driving function is a random function $F(t,\omega)$, then Eq. (6.65) becomes

$$\varphi(t,\omega)=-\int_0^1 K(t,\tau)\,[F(\tau,\omega)-\alpha^2\sin\varphi(\tau,\omega)]\,d\tau. \tag{6.69}$$

If we put

$$G(t,\omega)=\int_0^1 K(t,\tau)\,F(\tau,\omega)\,d\tau,\qquad\text{and}\qquad \psi(t,\omega)=\varphi(t,\omega)+G(t,\omega),$$

then Eq. (6.67) becomes the homogeneous random Hammerstein equation

$$\psi(t,\omega)+\int_0^1 K(t,\tau)\,f(\tau,\psi(\tau,\omega))\,d\tau=0, \tag{6.70}$$

where

$$f(t,\xi,\omega)=\alpha^2\sin(\xi-G(t,\omega)).$$

With appropriate assumptions on the random function $F(t,\omega)$ the integral defining $G(t,\omega)$ is well defined, and $G(t,\omega)$ is a random function.

In [6] the existence, uniqueness, and measurability of the random solution of the random Uryson equation was established using the Špaček–Hanš analogue of the Banach contraction mapping theorem. The Uryson integral operator was defined over a random domain; but, as we have shown in the case of Fredholm equations, this case is equivalent to a Uryson operator with a random kernel.

In Sect. 6.5B we state and prove an existence theorem for a random solution of a Uryson equation using Mukherjea's fixed point theorem (Theorem 3.7).

B. An existence theorem

Consider the random Uryson equation

$$\int_{I(\omega)} K(t,\tau,x(\tau,\omega),\omega)\,d\tau-x(t,\omega)=y(t,\omega), \tag{6.71}$$

where $I(\omega)=[a(\omega),b(\omega)]\subset[0,1]$; that is $I(\omega)$ is a subset of the interval $[0,1]$, the endpoints $a(\omega)$ and $b(\omega)$ being real-valued random variables with $0\leqslant a(\omega)<b(\omega)\leqslant 1$ almost surely. We now state and prove the following existence theorem.

THEOREM 6.13. *If (a) $(\Omega,\mathfrak{A},\mu)$ is atomic, (b) the kernel $K(t,\tau,\xi,\omega)$, where $t,\tau\in[0,1]$, $\xi\in(-\infty,\infty)$, is such that* (i) *for every ξ, K is a measurable mapping* *from Ω into $\mathfrak{F}$, a separable Banach space of bounded functions on* $[0,1] \times [0,1]$ *with the sup norm,* (ii) *for each* $\omega \in \Omega$, *K is continuous with respect to ξ uniformly in t and τ,* (iii) *for every triple* (t, ξ, ω), *K is a measurable function of* τ, (iv) *$y(t, \omega)$ is an element of $C[0,1]$ for every* $\omega \in \Omega$, (v)

$$\int_0^1 \sup_{|\xi|<1} |K(t, \tau, \xi, \omega)| \, d\tau + |y(t, \omega)| \leqslant 1$$

for each ω and each t, and (vi)

$$\lim_{\substack{h \to 0 \\ t+h \in [0,1]}} \int_0^1 \sup_{|\xi|<1} |K(t+h, \tau, \xi, \omega) - K(t, \tau, \xi, \omega)| \, d\tau = 0,$$

then there exists a random solution $x(t, \omega)$ of Eq. (6.70) *in* $C[0,1]$.

Proof. Let $\mathfrak{K}$ be the closure (in the sup norm) of the space of all functions of the form

$$\sum \chi_{[r,s]}(\tau) \, h(t, \tau),$$

where (i) the summation is taken over a finite set, (ii) r and s are rational points in $[0,1]$, and (iii) $h(t, \tau) \in \mathfrak{F}$. Then $\mathfrak{K}$ is a separable Banach space under the sup norm.

Now, if $x_n(\tau)$ is a sequence of step functions such that

$$m(\{\tau : |x_n(\tau) - x(\tau)| < 1/n\}) > 1 - 1/n,$$

where m is Lebesgue measure on the line and the $x_n(\tau)$ are of the form

$$\sum \chi_{[r_{in}, s_{in}]}(\tau) \, h_{in}$$

(where r_{in} and s_{in} are rational numbers), then $K(t, \tau, x_n(\tau), \omega) \in \mathfrak{K}$. Let $a_n(\omega)$ and $b_n(\omega)$ be simple functions such that $a_n(\omega)$ and $b_n(\omega)$ converges uniformly to $a(\omega)$ and $b(\omega)$, respectively, and $0 \leqslant a_n(\omega) \leqslant a(\omega) < b(\omega) \leqslant b_n(\omega) \leqslant 1$ for every ω. Then

$$\chi_{[a_n(\omega), b_n(\omega)]}(\tau) \, K(t, \tau, x_n(\tau), \omega)$$

is a $\mathfrak{K}$-valued random variable, since for every $t, \tau \in [0,1]$ it is a real-valued random variable. Therefore, its integral with respect to τ (being a bounded linear functional on $\mathfrak{K}$ for every fixed $t \in [0,1]$) is a real-valued random variable; and passing to the limit we have

$$\int_{a(\omega)}^{b(\omega)} K(t, \tau, x(\tau), \omega) \, d\tau = \lim_{n \to \infty} \int_{a_n(\omega)}^{b_n(\omega)} K(t, \tau, x_n(\tau), \omega) \, d\tau.$$

Put

$$T(\omega)[x(t)] = \int_{I(\omega)} K(t, \tau, x(\tau), \omega) \, d\tau - y(t, \omega). \tag{6.72}$$

Then (6.71) defines a real-valued random variable for every fixed $t \in [0, 1]$. It now follows from the assumptions on the kernel K and a theorem of Ladyzhenskiĭ (cf. [18, p. 34]) that for every fixed ω the operator $T(\omega)$ defined by (6.72) is a completely continuous operator on $C_1[0, 1]$. Since (6.72) defines a real-valued random variable for every $t \in [0, 1]$, $T(\omega)[x(t)]$ is a $C[0, 1]$-valued random variable; and it follows from Definition 2.21 that $T(\omega$)is a random integral operator on $C_1[0, 1]$ into $C[0, 1]$. The statement of the theorem now follows from Theorem 3.7.

6.6 A Measure-Theoretic Problem Associated with a Nonlinear Integrodifferential Equation with Random Right-Hand Side

Consider a random equation of the form

$$Tf(x, \omega) = g(x, \omega) \tag{6.73}$$

where the right-hand side $g(x, \omega)$ is a random function with values in a separable Hilbert space $\mathfrak{H}$. Let μ be a probability measure on the measurable space $(\mathfrak{H}, \mathfrak{B})$, where $\mathfrak{B}$ is the σ-algebra of Borel sets of $\mathfrak{H}$. That is,

$$\mu(B) = \mathscr{P}\{g \in B\}, \qquad B \in \mathfrak{B}. \tag{6.74}$$

Now let T be a deterministic nonlinear operator which transforms a Borel set B in a one-to-one manner into a Borel set TB. In this way we can define another measure ν on $(\mathfrak{H}, \mathfrak{B})$ by putting

$$\nu(B) = \mu(TB) = \mathscr{P}\{f \in B\} \tag{6.75}$$

As pointed out in Sect. 1.5E, if the measure ν corresponding to the solution process $f(x, \omega)$ is absolutely continuous with respect to the measure μ corresponding to the input process $g(x, \omega)$, then $f(x, \omega)$ will satisfy almost surely all those properties that the input process satisfies almost surely.

In this section we consider the case when T is a nonlinear operator of the form $T = I + S$, and state a theorem due to Baklan and Šatašvili [4] which gives sufficient conditions on T in order that the measures μ and ν be absolutely continuous with respect to one another. As a concrete example, a nonlinear integrodifferential equation with random right-hand side is considered; and the density of the solution measure ν with respect to the input measure μ is calculated.

Let R denote the correlation operator associated with the random function $g(x, \omega)$. We now assume that the nonlinear operator T is of the form $T = I + S$ and satisfies the following conditions:

i. There exists a bounded and continuous operator S^* such that

$$Sz = RS^* z. \tag{6.76}$$

for all $z \in R[\mathfrak{H}]$.

ii. The operator S^* at each point $z \in R[\mathfrak{H}]$ has a uniformly variational derivative $U^*(z)$.

iii. For each $z \in R[\mathfrak{H}]$ the determinant

$$D(z) = 1 + \sum_{k=1}^{\infty} \frac{1}{k!} \epsilon(\sigma_k) \prod_{j=1}^{k} [\mathrm{Tr}(U^j(z))]^{l_j(\sigma_k)} \neq 0, \tag{6.77}$$

where

$$U(z) = RU^*(z) \tag{6.78}$$

is a variational derivative operator of S. In (6.77), Tr denotes trace,

$$\sigma_k = \begin{pmatrix} 1, 2, \ldots, k \\ \alpha_1, \alpha_2, \ldots, \alpha_k \end{pmatrix}$$

is a permutation of the numbers $1, 2, \ldots, k$; $l_j(\sigma_k)$ denotes the number of cycles of length j in the permutation σ_k, and

$$\epsilon(\sigma_k) = \begin{cases} 1 & \text{when } \sigma_k \text{ is even} \\ -1 & \text{when } \sigma_k \text{ is odd.} \end{cases}$$

We state the following theorem.

THEOREM 6.14. *Consider the nonlinear operator equation $Tf(x,\omega) = g(x,\omega)$ in a separable Hilbert space $\mathfrak{H}$, where $g(x,\omega)$ is a Gaussian random function with values in $\mathfrak{H}$. Let μ denote the probability measure associated with $g(x,\omega)$, and let ν denote the probability measure associated with the solution process $f(x,\omega)$. If the operator T is such that conditions* (i)–(iii) *are satisfied, then μ and ν are absolutely continuous with respect to one another, and*

$$\frac{d\nu}{d\mu}[g] = D(g) \exp\{-\tfrac{1}{2}(Sg, S^* g) - (g, S^* g)\}. \tag{6.79}$$

As a concrete application of Theorem 6.14, we now consider a boundary-value problem for a nonlinear integrodifferential equation with random right-hand side. Let M be a bounded region in Euclidean space R_3 with a sufficiently regular boundary Γ; and let $L_2^0(M)$ denote the Hilbert space of functions which are square-integrable on M and vanish on the boundary Γ. Consider the boundary-value problem

$$\begin{aligned} \Delta f(x,\omega) + \lambda \int_M K(x,y)\, h(y, f(y,\omega))\, dy &= w(dx)/dx \\ f(x,\omega)|_{x \in \Gamma} &= 0. \end{aligned} \tag{6.80}$$

In (6.80) Δ is a Laplacian operator, $K: M \to M$, $h: M \times L_2^0 \to L_2^0$ are arbitrary functions, and w denotes Wiener measure on M.

By the solution of the random equation (6.80) we mean the random function $f(x, \omega)$ which satisfies the random integrodifferential equation

$$f(x, \omega) + \lambda \int_M d\xi \int G(x, \xi) K(\xi, y) h(y, f(y, \omega)) \, dy = g(x, \omega), \qquad (6.81)$$

where $G(x, y)$ is the Green's function of the Laplacian operator Δ, and

$$g(x, \omega) = \int_M G(x, y) \, w(dy). \qquad (6.82)$$

Since $G(x, y)$ is a deterministic function, the integral in (6.82) is well defined, and $g(x, \omega)$ is a Gaussian random function with values in $L_2^0(M)$. It is clear that Eq. (6.81) is of the form $(I + S)f(x, \omega) = g(x, \omega)$.

In [3, pp. 25, 26] it was shown that the correlation operator R of $g(x, \omega)$ is Δ^{-2}; therefore from (6.76) and (6.81)

$$S^* f = R^{-1} Sf = \lambda \int_M \Delta K(x, y) h(y, f(y, \omega)) \, dy. \qquad (6.83)$$

If we now assume that there exists constants C_1 and C_2 such that

$$\int_M \int_M (\Delta K(x, y))^2 \, dx \, dy \leqslant C_1,$$

$$|h_f'(y, f)| \leqslant C_2,$$

and

$$|\lambda| C_2 \left\{ \int_M \int_M \left| \int_M G(x, \xi) K(\xi, y) \, d\xi \right|^2 dx \, dy \right\}^{1/2} < 1,$$

then it can be verified that conditions (i)–(iii) of Theorem 6.14 are satisfied; and the density of the probability measure ν (solution process) with respect to the probability measure μ (input process) is of the form

$$\frac{d\nu}{d\mu}[g] = D(g) \exp\left\{ -\tfrac{1}{2}\lambda^2 \int_M \left| \int_M K(x, y) h(y, g(y, \omega)) \, dy \right|^2 dx \right.$$
$$\left. - \lambda \int_M \int_M K(x, y) h(y, g(y, \omega)) \Delta g(x, \omega) \, dx \, dy \right\}. \qquad (6.84)$$

6.7 Some Random Nonlinear Integral Equations Which Arise in Applied Problems

A. Introduction

In the following four subsections we consider briefly four random nonlinear integral equations which arise in the study of certain concrete problems in telephone traffic theory, the statistical theory of turbulence, wave propagation in random media, and control theory.

B. A random nonlinear integral equation in telephone traffic theory: The Fortet equation

Consider a telephone exchange with m (m finite) available trunks or channels; and assume that the calls arrive at the exchange at times t_n; and denote by $G(t)$ the distribution of arrival times. Let the random function $x(t, \omega)$ denote the total number of conversations held at time $t > 0$, with $x(0, \omega) = 0$; and let

$$y(t, \omega) = \begin{cases} 1 & \text{if} \quad x(t, \omega) < m \\ 0 & \text{if} \quad x(t, \omega) \geqslant m. \end{cases} \tag{6.85}$$

If $V(\xi)$ is any function such that

$$V(\xi) = \begin{cases} 1 & \text{for} \quad \xi = 0, 1, \ldots, m-1 \\ 0 & \text{for} \quad \xi \geqslant m, \end{cases} \tag{6.86}$$

then we can put

$$y(t, \omega) = V(x(t, \omega)). \tag{6.87}$$

In order to determine $x(t, \omega)$, Fortet [12] (cf. also Saaty [32, pp. 409, 410]) introduced the random nonlinear integral equation

$$x(t, \omega) = \int_0^t y(\tau, \omega)\, K(t, \tau, \omega)\, dG(\tau), \tag{6.88}$$

which, using (6.87), can be rewritten as

$$x(t, \omega) = \int_0^t V(x(\tau, \omega))\, K(t, \tau, \omega)\, dG(\tau). \tag{6.89}$$

In Eq. (6.89) $K(t, \tau, \omega)$ is a random kernel which assumes only the values 0 or 1, and whose first and second moments exist. If $G(t)$ is a Poisson distribution, then the integral in Eq. (6.89) exists in the mean-square sense. We will refer to Eq. (6.89) as the *Fortet integral equation.*

We now use Picard's method to establish the existence of a solution to the Fortet equation. Let

$$V(\xi) = \begin{cases} 1 & \text{if} \quad \xi \leqslant m-1 \\ m - \xi & \text{if} \quad m-1 \leqslant \xi. \end{cases} \tag{6.90}$$

Hence $V(\xi)$ as defined by (6.90) satisfies the Lipschitz condition

$$|V(\xi_1) - V(\xi_2)| \leqslant |\xi_1 - \xi_2| \tag{6.91}$$

for any $\xi_1, \xi_2 \in R$. We can define a sequence of random functions $\{x_n(t, \omega)\}$ as follows:

$$
\begin{aligned}
x_0(t, \omega) &= \int_0^t K(t, \tau, \omega)\, dG(\tau) \\
x_1(t, \omega) &= \int_0^t V(x_0(\tau, \omega))\, K(t, \tau, \omega)\, dG(\tau) \\
x_2(t, \omega) &= \int_0^t V(x_1(\tau, \omega))\, K(t, \tau, \omega)\, dG(\tau) \\
&\vdots \\
x_n(t, \omega) &= \int_0^t V(x_{n-1}(\tau, \omega))\, K(t, \tau, \omega)\, dG(\tau).
\end{aligned}
\tag{6.92}
$$

We have $x_0(t,\omega) \geqslant x(t,\omega)$ for every t; and we note that, since $V(\xi)$ is a non-decreasing function, $x_n(t,\omega) \leqslant x(t,\omega)$ if n is odd and $x_n(t,\omega) \geqslant x(t,\omega)$ if n is even.

If we put

$$d_n(t, \omega) = |x_n(t, \omega) - x_{n-1}(t, \omega)|, \tag{6.93}$$

with $d_0(t,\omega) \leqslant 1$, and use (6.93) we obtain

$$d_{n+1}(t, \omega) \leqslant \int_0^1 d_n(t, \omega)\, K(t, \tau, \omega)\, dG(\tau). \tag{6.94}$$

Since $K(t, \tau, \omega) \leqslant 1$, iteration of (6.94) yields

$$d_n(t, \omega) \leqslant G^n(t)/n!, \qquad n = 1, 2, \ldots.$$

Hence, it follows that the sequence $\{x_n(t,\omega)\}$ defined by (6.92) converges uniformly in t on every finite interval for almost all $\omega \in \Omega$ to the random function $x(t,\omega)$; and since the convergence is uniform, $x(t,\omega)$ has absolutely continuous sample functions.

Padgett and Tsokos [25] have established the existence and uniqueness of a solution of Eq. (6.89) in $L_2(\Omega)$ using the results presented in Sect. 6.4; and have indicated the applicability of Eq. (6.89) in queueing theory.

C. *A random integral equation in turbulence theory*

The motion of a tagged point in a continuous fluid in turbulent motion leads to a random nonlinear integral equation of the form

$$r(a, t, \omega) = a + \int_0^t u(r(a, \tau, \omega), \tau, \omega)\, d\tau. \tag{6.95}$$

In Eq. (6.95) $r(t,a,\omega)$ is the position at time t of a material point located at a at $t = 0$, and $u(t,x,\omega)$ is the Eulerian velocity field, specified in laboratory coordinates. The problem of interest in the statistical theory of turbulence is to determine the statistical properties of the random function $r(t,a,\omega)$ when the statistical properties of the random kernel $u(r,\tau,\omega)$ representing the velocity field are known.

Put $a=0$, and consider the random integral equation

$$r(t,\omega)=\int_0^t \varphi(r(\tau,\omega),t,\tau,\omega)\,d\tau. \tag{6.96}$$

For the motion of a tagged material point in a fluid (with $r(a,0,\omega)=a=0$) we can put

$$\varphi(r,t,\tau,\omega)=\begin{cases}u(r,\tau,\omega), & \tau\leqslant t\\ 0, & \tau>t.\end{cases} \tag{6.97}$$

In order to study Eq. (6.96), Lumley [19] considered the following discrete analogue:

$$\begin{aligned}r_n(t_i,\omega)&=\sum_{k=1}^{n}\int_{t_{k-1}}^{t_k}\varphi(r(\tau,\omega),t_i,\tau,\omega)\,d\tau\\ &=\sum_{k=1}^{n}\varphi(r(t_k,\omega),t_i,t_k,\omega)\,\Delta k,\end{aligned} \tag{6.98}$$

where $t_i\in(t_{i-1},t_{i+1})$, $i=1,2,\ldots,n$ and $\Delta_k=t_k-t_{k-1}$. Put $r_n(t_i,\omega)=x_i^{(n)}(\omega)$ $(i=1,2,\ldots,n)$, and

$$f_i^{(n)}(x_1^{(n)},x_2^{(n)},\ldots,x_n^{(n)})=x_i^{(n)}(\omega)-\sum_{k=1}^{n}\varphi(x_k^{(n)}(\omega),t_i,t_k,\omega)\,\Delta_k. \tag{6.99}$$

The system (6.99) is a set of n random vector functions (each having three components) of n random vector arguments (each having three components); and

$$f_i^{(n)}(x_1^{(n)}(\omega),x_2^{(n)}(\omega),\ldots,x_n^{(n)}(\omega))=0,\qquad i=1,2,\ldots,n, \tag{6.100}$$

is a system of random algebraic equations which constitute a finite-dimensional approximation to Eq. (6.96). Lumley did not consider the convergence of $r_n\to r$; but using a n-dimensional generalization of the Rice–Kac theorem for the probability density of the zeros of a random function, derived a formula for the joint probability density of the $x_i^{(n)}(\omega)$ which satisfy the system of random algebraic equations (6.100).

Padgett and Tsokos [26] have established the existence and uniqueness of the solution of Eq. (6.96) in $L_p(\Omega)$, $1\leqslant p<\infty$, using the methods of Sect. 6.4.

D. *A set of coupled random integral equations associated with nonlinear wave motion*

In Sect. 4.4B we showed that the Helmholtz equation for linear wave propagation in a random medium leads to a random Fredholm integral equation. In the case of nonlinear wave motion with random initial conditions in a deterministic medium Hasselmann [14] (cf. also Frisch [13, pp. 151–152]) obtained a set of coupled random nonlinear integral equations for the amplitudes of the different wave modes. Let $a_j(t,k)$ denote the amplitude of the jth wave mode with wave vector k, $\xi_j(k)$, the dispersion equation for the jth wave mode, and the C_{jkl} coupling coefficients (which are centered random functions); then the set of integral equations derived by Hasselmann is of the form

$$\frac{\partial}{\partial t} a_j(t,k,\omega) = i\xi_j(k)\, a_j(t,k,\omega) + \sum_{k,l=1}^{n} C_{jkl}(k,k',k'')\, a_k(t,k',\omega)\, a_l(t,k'',\omega)\, d^3k'\, d^3k'', \tag{6.101}$$

for $j = 1,2,\ldots,n$. Under appropriate assumptions for the initial values $a_j(0,k,\omega)$, Hasselmann calculated the mean energy densities using a diagram technique.

E. A random nonlinear integral equation in control theory

Ahmed [1] has studied the following class of random nonlinear integral equations in the Lebesgue spaces L_p, $p \geqslant 1$:

$$x = \lambda T(\omega)\, x + y, \tag{6.102}$$

where

$$T(\omega)[x(t)] = \sum_{n=1}^{\infty} \underbrace{\int \cdots \int}_{I \times \cdots \times I} K_n(t, \tau_1, \ldots, \tau_n, \omega) \prod_{i=1}^{n} x(\tau_i, \omega)\, d\tau_1 \cdots d\tau_n,$$

and $I = [t_0, T]$. Equations of the above form arise in the study of certain problems in optimal control theory. Using fixed point methods, Ahmed established the existence and uniqueness of a random solution of Eq. (6.102).

References

1. Ahmed, N. U., A class of stochastic nonlinear integral equations on L^p spaces and its application to optimal control. *Information and Control* **14** (1969), 512–523.
2. Anselone, P. M., ed., "Nonlinear Integral Equations." Univ. of Wisconsin Press, Madison, 1964.
3. Baklan, V. V., and Šatašvili, A. D., Conditions of absolute continuity of probability measures corresponding to Gaussian random variables in a Hilbert space (Ukrainian). *Dopovidi Akad. Nauk Ukrain. RSR* (1965), pp. 23–26.

4. Baklan, V. V., and Šatašvili, A. D., Transformation of Gaussian measures by nonlinear mappings in Hilbert space (Ukrainian). *Dopovidi Akad. Nauk Ukrain. RSR* (1965), pp. 1115–1117.
5. Beneš, V., A nonlinear integral equation in the Marcinkiewicz space M_2. *J. Math. and Phys.* **44** (1965), 24–35.
6. Bharucha-Reid, A. T., On random solutions of nonlinear integral equations. *In* "Nonlinear Problems in Engineering" (W. F. Ames, ed.), pp. 23–28. Academic Press, New York, 1964.
6a. Buche, A. B., Unpublished manuscript, 1969.
7. Burkholder, D. L., On a class of stochastic approximation procedures. *Ann. Math. Statist.* **27** (1956), 1044–1059.
8. Cameron, R. H., Differential equations involving a parametric function. *Proc. Amer. Math. Soc.* **8** (1957), 834–840.
9. Coppel, W. A., "Stability and Asymptotic Behavior of Differential Equations." Heath, Indianapolis, Indiana, 1965.
10. Corduneanu, C., Problèmes globaux dans la théorie des équations intégrales de Volterra. *Ann. Mat. Pura Appl.* **67** (1965), 349–363.
11. Corduneanu, C., Some perturbation problems in the theory of integral equations. *Math. Systems Theory* **1** (1967), 143–155.
12. Fortet, R., Random distributions with an application to telephone engineering. *Proc. 3rd Berkeley Symp. Math. Statist. Probability (1955)*, Vol. II, pp. 81–88, 1956.
13. Frisch, U., Wave propagation in random media. *In* "Probabilistic Methods in Applied Mathematics" (A. T. Bharucha-Reid, ed.), Vol. 1, pp. 75–198. Academic Press, New York, 1968.
14. Hasselmann, K., Feyman diagrams and interaction rule of wave-wave scattering processes. *Rev. Geophys.* **4** (1966), 1–32.
15. Hewitt, E., and Stromberg, K., "Real and Abstract Analysis." Springer-Verlag, Berlin and New York, 1969.
16. Hille, E., and Phillips, R. S., "Functional Analysis and Semi-Groups." Amer. Math. Soc., Providence, Rhode Island, 1957.
17. Kampé de Fériet, J., Random integrals of differential equations. *In* "Lectures on Modern Mathematics" (T. L. Saaty, ed.), Vol. III, pp. 277–321. Wiley, New York, 1965.
18. Krasnosel'skiĭ, M. A., "Topological Methods in the theory of Non-Linear Integral Equations," translated from the Russian. Macmillan, New York, 1964.
19. Lumley, J. L., An approach to the Eulerian-Lagrangian problem. *J. Math. Phys.* **3** (1962), 309–312.
20. Marcinkiewicz, J., Une remarque sur les espaces de M. Besikowitch. *C. R. Acad. Sci. Paris* **208** (1939), 157–159.
21. Massera, J. L., and Schaffer, J. J., "Linear Differential Equations and Function Spaces." Academic Press, New York, 1966.
22. Morozan, T., Stability of linear systems with random parameters. *J. Differential Equations* **3** (1967), 170–178.
23. Mukherjea, A., Random transformations on Banach spaces. Ph.D. Dissertation, Wayne State Univ., Detroit, Michigan, 1967.
24. Mukherjea, A., On a random nonlinear integral equation of Uryson type. To be published.
25. Padgett, W. J., and Tsokos, C. P., On a stochastic integral equation of the Volterra type in telephone traffic theory. *J. Appl. Probability* **8** (1971), 269–275.
26. Padgett, W. J., and Tsokos, C. P., Existence of a solution of a stochastic integral equation in turbulence theory. *J. Math. Phys.* **12** (1971), 210–212.

27. Padgett, W. J., and Tsokos, C. P., A random Fredholm integral equation. *Proc. Amer. Math. Soc.* **33** (1972), 534–542.
28. Padgett, W. J., and Tsokos, C. P., On the solution of a random integral equation by a method of successive approximations. *Z. Wahrscheinlichkeitstheorie und Verw. Gebiete.* In press.
29. Padgett, W. J., and Tsokos, C. P., Random solution of a stochastic integral equation: Almost sure and mean square convergence of successive approximations. To be published.
30. Phillips, R. S., Integration in a convex linear topological space. *Trans. Amer. Math. Soc.* **47** (1940), 114–145.
31. Pogorzelski, W., "Integral Equations and Their Applications," translated from the Polish, Vol. 1. Pergamon, Oxford, 1967.
32. Saaty, T. L., "Modern Nonlinear Equations." McGraw-Hill, New York, 1967.
33. Strand, J. L., Stochastic ordinary differential equations. Ph.D. Dissertation, Univ. of California, Berkeley, California, 1967.
34. Strand, J. L., Random ordinary differential equations. *J. Differential Equations* **7** (1970), 538–553.
35. Tricomi, F. G., "Integral Equations." Wiley (Interscience), New York, 1957.
36. Tsokos, C. P., On a stochastic integral equation of the Volterra type. *Math. Systems Theory* **3** (1969), 222–231.
37. Tsokos, C. P., On a nonlinear stochastic differential system. To be published.
37a. Tsokos, C. P., and Padgett, W. J., "Random Integral Equations with Applications to Stochastic Systems." Springer-Verlag, Berlin and New York, 1971.
38. Woodward, D. A., On a special integral equation. *Proc. Amer. Math. Soc.* **10** (1959), 853–854.
39. Yosida, K., "Lectures on Differential and Integral Equations." Wiley (Interscience), New, York, 1960.

CHAPTER 7

Itô Random Integral Equations

7.1 Introduction

The random integral equations studied in Chaps. 4, 5, and 6 were all probabilistic analogues of various types of well-known linear and nonlinear integral equations. In this chapter we consider an important class of random integral equations which were first introduced by Itô in his fundamental memoir on random differential equations [36]. The importance of these equations is due to the fact that (1) a large class of Markov processes in R_n can be represented as solutions of such equations, (2) they serve as models for physical processes and stochastic control systems, and (3) they are utilized in the probabilistic study of partial differential equations.

The integral equation referred to as the *Itô random integral equation* is of the form

$$x(t,\omega) = x_0(\omega) + \int_{t_0}^{t} a(\tau, x(\tau,\omega))\,d\tau + \int_{t_0}^{t} b(\tau, x(\tau,\omega))\,dw(\tau,\omega), \tag{7.1}$$

where $x(t,\omega) \in R$, the functions $a(t,x)$ and $b(t,x)$ satisfy certain conditions to be stated later, and $w(t,\omega)$ is a Wiener process. Equation (7.1) is the integral equation formulation of the *Itô random differential equation*

$$dx(t,\omega) = a(t,x)\,dt + b(t,x)\,dw(t,\omega) \tag{7.2}$$

with initial data $x(t_0,\omega) = x_0(\omega)$. Equation (7.2) can also be interpreted as a nonlinear transformation of a Wiener process; that is, the Wiener process $w(t,\omega)$, $t \geqslant 0$, is transformed into $x(t,\omega)$, $t \geqslant t_0$, such that $x(t_0,\omega) = x_0(\omega)$ and (7.2) obtains. In (7.2), $dx(t,\omega)$ denotes the infinitesimal increment in x during the interval $[t, t+dt]$, and $dw(t,\omega)$ denotes the corresponding increment in w.

We mention two special cases of Eq. (7.2). (1) If $b = 0$ in Eq. (7.2), it is natural to interpret the equation as the first-order deterministic differential equation

$$dx/dt = a(t, x). \tag{7.3}$$

In this case the solution will be a random function if Eq. (7.3) is considered together with a random initial condition, that is, $x(t_0) = x_0$ is a random variable. (2) If (i) b is a function of t, but is independent of x, and (ii) $a = 0$, then we have

$$x(t, \omega) = x_0(\omega) + \int_{t_0}^{t} b(\tau)\, dw(\tau, \omega). \tag{7.4}$$

We remark that in this case the solution process $x(t, \omega)$ is essentially the Wiener process with a change of variable in the time parameter. In fact, a Wiener process may be considered as the solution process of an Itô random integral equation with $a = 0$ and $b = 1$.

In physics and engineering a number of problems lead to random differential equations of so-called Langevin type. In physics these equations are encountered in the study of Bownian motion. Consider the Brownian motion of a particle of mass m immersed in a fluid. Let $x(t)$ denote the position of the particle at time t, and let $v(t)$ denote its velocity. The medium surrounding the particle can be considered as acting on the particle in two ways. Firstly, the medium offers resistance to the motion of the particle, this resistance being in the form of a frictional force equal to $\beta v(t)$, where β is the (mean) dynamic frictional coefficient. Secondly, the fluctuations in the number of collisions of molecules of the fluid with the particle appear as a random force $F(t)$. Hence the equation of motion of the particle is of the form

$$m\, dv/dt = -\beta v(t) + F(t). \tag{7.5}$$

Equation (7.5) is referred to as *Langevin's equation*, since it was first derived by Langevin in a fundamental paper [53] on the theory of Brownian motion. The proper probabilistic analogue of Eq. (7.5) is

$$dv(t, \omega) = -\alpha v(t, \omega) + dF(t, \omega), \tag{7.6}$$

since we do not expect $v(t)$ to be differentiable. In Eq. (7.6), $\alpha = \beta/m > 0$ and m^{-1} is considered as a constant factor of $dF(t, \omega)$. The random integral equation equivalent to Eq. (7.6) is

$$v(t, \omega) = v(0, \omega) - \beta \int_0^t v(\tau, \omega)\, d\tau + F(t, \omega). \tag{7.7}$$

The first rigorous mathematical treatment of the Langevin equation is due to Doob [18]. For detailed discussions of the Langevin equation and related topics in the theory of Brownian motion we refer to Chandrasekhar [14], Middleton [60, Chap. 10], and Papoulis [64, Chap. 15].

A generalized Langevin equation of the form

$$dv/dt = -\int_{t_0}^{t} \gamma(t-\tau)\, v(\tau)\, d\tau + F(t,\omega) \tag{7.8}$$

has been considered by several physicists. In Eq. (7.8) $\gamma(t)$ represents a retarded effect of the (time-dependent) frictional force, and $F(t,\omega)$ is the random force which is not correlated with the initial velocity. We refer to Kannan and Bharucha-Reid [43a] for the formulation of Eq. (7.8) as a random integral equation and an analysis of the solution process.

In engineering, a large class of dynamic systems studied in control theory lead to ordinary differential equations of the form

$$dx/dt = f(t,x), \qquad t \in [t_0, \infty), \tag{7.9}$$

where $x(t) \in R_n$. Since many dynamic systems are subject to random perturbations, a probabilistic generalization of Eq. (7.9) is the formal random differential equation

$$dx/dt = f(t,x) + \Phi(t,x)\, y(t,\omega), \tag{7.10}$$

where $\Phi(t,x)$ is an $n \times m$ matrix-valued function of t and x, and $y(t,\omega)$ is a m-dimensional Gaussian white noise process. Random equations of the form (7.10), which are referred to in the engineering literature as Langevin equations, were first proposed as a model for a randomly disturbed dynamic system by Pontryagin *et al.* [65]. The introduction of Langevin equations in stochastic control theory is due to Barrett [5], Chuang and Kazda [15], and Khazan [48]. A precise version of Eq. (7.10) is given by the Itô random differential equation

$$dx(t,\omega) = f(t,x)\, dt + \Phi(t,x)\, dw(t,\omega), \tag{7.11}$$

where $w(t,\omega)$ is a Wiener process in R_m; and Eq. (7.11) is interpreted as the Itô random integral equation

$$x(t,\omega) = x_0(\omega) + \int_{t_0}^{t} f(\tau, x(\tau,\omega))\, d\tau + \int_{t_0}^{t} \Phi(\tau, x(\tau,\omega))\, dw(\tau,\omega). \tag{7.12}$$

For an interesting discussion of the relation between the Langevin and Itô equations we refer to Wonham [78].

The purpose of this chapter is to give an introductory account of the theory of Itô random integral equations, and to give a brief survey of certain applications of this theory. In Sect. 7.2 we present the basic theory of Itô random integral equations, restricting our attention to the one-dimensional case; that is, we will study solutions of Itô random integral equations in R. Section 7.3 is devoted to Itô random integral equations in Hilbert spaces, these equations arising as the integral equation formulation of certain random differential equations in Hilbert spaces. Finally, in Sect. 7.4, we present a brief summary of some additional studies on Itô random integral equations and their applications.

Itô equations are of great importance in control theory and in the theory of partial differential equations; however, we will not include a survey of these applications in this book. The reader interested in Itô equations in control theory is referred to the books of Bucy and Joseph [11], Jazwinski [39], and Kushner [52, 52a], and the articles of Fleming [23], Wong and Zakai [76], and Wonham [78]. Readers interested in the probabilistic approach to partial differential equations based on Itô equations are referred to the book of Dynkin [21], and the papers of Daletskiĭ [16], Fleming [22], Freĭdlin [24–28], Khasminskiĭ [44–47], and Tanaka [71].

7.2 Itô Random Integral Equations: Basic Theory

A. Introduction

Consider the Itô random integral equation

$$x(t, \omega) = x_0(\omega) + \int_{t_0}^{t} a(\tau, x)\, d\tau + \int_{t_0}^{t} b(\tau, x)\, dw(\tau, \omega), \tag{7.13}$$

where $a(t,x)$ and $b(t,x)$ are assumed to be known, and $w(t, \omega)$ is a Wiener process. The second integral can not be interpreted as an ordinary Stieltjes integral, since the integrator function is a Wiener process; and, as is well known, the realizations of a Wiener process are, with probability one, of unbounded variation. Hence in order to study solutions of Eq. (7.13) the first thing that must be done is to define the "stochastic" integral in Eq. (7.13). In Sect. 7.2B we define the Itô stochastic integral and discuss its relationship to several other stochastic integrals. Once the Itô stochastic integral has been defined, the Itô random integral equation is well defined and meaningful, and the question of the existence and uniqueness of solutions of Eq. (7.13) can be investigated. Section 7.2C is devoted to the existence and uniqueness of solutions of Itô random integral equations. Finally, in Sect. 7.2D, we consider the class of Markov processes (referred to as Itô processes) which is determined by solutions of Itô random integral equations.

The reader interested in more detailed and/or sophisticated treatments of Itô random integral equations is referred to the books of Doob [19], Dynkin [21], Gihman and Skorohod [31], Itô and McKean [37], McKean [56], and Skorohod [67].

B. The Itô stochastic integral

The well-known *Wiener integral* is an integral of the form

$$\int_{a}^{b} f(t)\, dw(t, \omega), \tag{7.14}$$

where $f(t)$ is a given deterministic function of t, and $w(t,\omega)$ is a Wiener process in R defined on $[a,b]$. Hence the Wiener integral is a stochastic integral with respect to a Wiener process (equivalently, Wiener measure). We refer to Kac [40, Chap. IV], McKean [56, Chap. 2], Nelson [63, Chap. 7], Skorohod [67, Chap. 2], and Saaty [66, pp. 405–407] for discussions of the Wiener integral and its properties.

In order to study the nonlinear Itô random equations it is necessary to generalize the Wiener integral by assuming that the integrand f is a random function $f(t,\omega)$. The generalization of the Wiener integral which is considered in this section is called the Itô integral, since this integral was defined, and its basic properties established, by Itô [36].

Let $\{w(t,\omega), t \geqslant 0\}$ be a Wiener process in R, and let $\{f(t,\omega), t \geqslant 0\}$ be a real-valued random function. In many applications the dependence of f on w can be described as *nonanticipative*; that is, the random function $f(t,\omega)$ will depend, at most, on present and past values of $w(\tau,\omega)$, $\tau \leqslant t$, but not on values of $w(\tau,\omega)$ for $\tau > t$. To characterize this type of dependence, we introduce a family $\{\mathfrak{F}_t, t \geqslant 0\}$ of σ-algebras of subsets of Ω with the following properties:

i. $\mathfrak{F}_{t_1} \subset \mathfrak{F}_{t_2}$, for $t_1 < t_2$;
ii. for every t, $w(t,\omega)$ is $\mathfrak{F}_t$-measurable;
iii. for $t_1 > t_2 \geqslant t$, the increments $w(t_1,\omega) - w(t_2,\omega)$ are independent (in the probabilistic sense) of $\mathfrak{F}_t$;
iv. for each fixed t, the random variable $f(t,\omega)$ is $\mathfrak{F}_t$-measurable.

We now assume that $f(t,\omega)$ is a measurable random function, and that

$$\mu\left(\left\{\omega : \int_0^T |f(t,\omega)|^2 \, dt < \infty\right\}\right) = 1, \tag{7.15}$$

where $T < \infty$; and we denote by $M_2(\mathfrak{F}_t)$ the class of functions satisfying (7.15). Hence $M_2(\mathfrak{F}_t)$ is the class of functions $f(t,\omega)$ which are square-integrable almost surely as functions of t on the interval $[0,T]$.

We now proceed to define the *Itô stochastic integral*

$$\int_a^b f(t,\omega) \, dw(t,\omega) \tag{7.16}$$

for $f \in M_2$ and $0 \leqslant a < b < \infty$. As in the case of nonstochastic integrals, we first define (7.16) for f a step or simple function; and then use a completeness argument to obtain the Itô stochastic integral for all $f \in M_2$ as the limit of an approximating sequence. Hence we must first show that every $f \in M_2$ can be approximated in an appropriate sense.

The function $f = f(t,\omega)$ is said to be a *step or simple function* if there exists a partition of $[a,b]$, say $a = t_0 < t_1 < \cdots < t_n = b$, such that $f(t,\omega) = f(t_i,\omega)$ for $t \in [t_i, t_{i+1})$, $i = 1,2,\ldots,n$. We remark that the points t_i are independent of ω.

We now prove the following lemma.

LEMMA 7.1. *For every $f \in M_2$ there exist a sequence of step functions $f_n \in M_2$ such that*

$$\mu\left(\left\{\omega : \lim_{n\to\infty} \int_a^b |f(t,\omega) - f_n(t,\omega)|^2\, dt = 0\right\}\right) = 1.$$

Proof. If $f(t,\omega)$ is continuous almost surely, then the assertion of the lemma is obvious; since we can put

$$f_n(t,\omega) = f(k/n,\omega) \qquad \text{for} \quad \frac{k}{n}(b-a) \leqslant t - a \leqslant \frac{(k+1)}{n}(b-a), \quad k = 0, 1, \ldots, n.$$

If we now take $f(t,\omega)$ to be measurable and bounded (by a constant depending on ω, say $M(\omega)$), then there is a sequence g_n of continuous functions, correspondingly bounded, such that $g_n \to f$ almost surely for almost all t. Consequently.

$$\mu\left(\left\{\omega : \lim_{n\to\infty} \int_a^b |f - g_n|^2\, dt = 0\right\}\right) = 1.$$

Finally, an arbitrary $f \in M_2$ can be approximated in mean-square by bounded measurable functions in M_2. Hence, the step functions are dense in M_2, in the sense of L_2-convergence.

We now define the Itô stochastic integral for step functions in M_2 as follows:

$$\int_a^b f(t,\omega)\, dw(t,\omega) = \sum_{k=0}^{n} f(t_k,\omega)\,[w(t_{k+1},\omega) - w(t_k,\omega)]. \tag{7.17}$$

The Itô stochastic integral defined by (7.17) has the following properties:

1. If $f_1(t,\omega)$ and $f_2(t,\omega)$ are step functions in M_2, then

$$\int_a^b [\alpha f_1(t,\omega) + \beta f_2(t,\omega)]\, dw(t,\omega) = \alpha \int_a^b f_1(t,\omega)\, dw(t,\omega) + \beta \int_a^b f_2(t,\omega)\, dw(t,\omega). \tag{7.18}$$

2. If f is a step function in M_2 and $\mathscr{E}\{|f(t,\omega)| \,|\, \mathfrak{F}_a\} < \infty$ for $t \in [a,b]$, then

$$\mathscr{E}\left\{\int_a^b f(t,\omega)\, dw(t,\omega) \,\middle|\, \mathfrak{F}_a\right\} = 0 \tag{7.19}$$

almost surely.

3. If f is a step function in M_2 and $\mathfrak{E}\{|f(t,\omega)|^2 \,|\, \mathfrak{F}_a\} < \infty$ almost surely for $t \in [a,b]$

$$\mathscr{E}\left\{\left(\int_a^b f(t,\omega)\, dw(t,\omega)\right)^2 \,\middle|\, \mathfrak{F}_a\right\} = \int_a^b \mathscr{E}\left\{f^2(t,\omega) \,\middle|\, \mathfrak{F}_a\right\} dt \tag{7.20}$$

almost surely.

4. If f is a step function in M_2, then for every $N > 0$ and $\gamma > 0$,

$$\mu\left(\left\{\omega : \left|\int_a^b f(t,\omega)\,dw(t,\omega)\right| > \gamma\right\} \leqslant \frac{N}{\gamma^2} + \mu\left(\left\{\omega : \int_a^b |f(t,\omega)|^2\,dt > N\right\}\right). \tag{7.21}$$

Now suppose that (i) f_n is a sequence of step functions in M_2, (ii) $f \in M_2$, and (iii)

$$\lim_{n\to\infty} \int_a^b |f(t,\omega) - f_n(t,\omega)|^2\,dt = 0 \tag{7.22}$$

in probability. The existence of such a sequence of step functions is guaranteed by Lemma 7.1. Then, we have

$$\lim_{n,m\to\infty} \int_a^b |f_m(t,\omega) - f_n(t,\omega)|^2\,dt = 0$$

in probability; that is, for every $\epsilon > 0$,

$$\lim_{m,n\to\infty} \mu\left(\left\{\omega : \int_a^b |f_m(t,\omega) - f_n(t,\omega)|^2\,dt > \epsilon\right\}\right) = 0.$$

It now follows from (7.21) that for arbitrary $\epsilon > 0$, $\delta > 0$,

$$\begin{aligned}
&\limsup_{m,n\to\infty} \mu\left(\left\{\omega : \left|\int_a^b f_m(t,\omega)\,dw(t,\omega) - \int_a^b f_n(t,\omega)\,dw(t,\omega)\right| > \delta\right\}\right) \\
&\quad \leqslant \epsilon/\delta^2 + \lim_{m,n\to\infty} \mu\left(\left\{\omega : \int_a^b |f_m(t,\omega) - f_n(t,\omega)|^2\,dt > \epsilon\right\}\right) \\
&\quad = \epsilon/\delta^2.
\end{aligned}$$

Since $\epsilon > 0$ was arbitrary, for every $\delta > 0$

$$\limsup_{m,n\to\infty} \mu\left(\left\{\omega : \left|\int_a^b f_m(t,\omega)\,dw(t,\omega) - \int_a^b f_n(t,\omega)\,dw(t,\omega)\right| > \delta\right\}\right) = 0.$$

It follows from the above that the sequence of random variables $\int_a^b f_n(t,\omega)\,dw(t,\omega)$ converges in probability to a limit. The limit random variable is independent of the choice of the sequence of step functions in M_2 for which (7.22) holds. (We remark that if there are two sequences f_n and f'_n such that (7.22) obtains, then a single sequence can be formed by combining the two sequences, and, with probability one, the two sequences have the same limit). Finally, we define

$$\int_a^b f(t,\omega)\,dw(t,\omega) = \lim_{n\to\infty} \int_a^b f_n(t,\omega)\,dw(t,\omega) \qquad \text{(in probability).} \tag{7.23}$$

This limit is the *Itô stochastic integral* of the function $f(t,\omega) \in M_2$; and, as usual, this limit is uniquely defined except for its values on a negligible set $N \subset \Omega$.

The definite integral defined by (7.23) is a linear functional of $f(t,\omega)$ on M_2. Furthermore, for $c \in (a,b)$,

$$\int_a^b f(t,w)\,dw(t,\omega) = \int_a^c f(t,\omega)\,dw(t,\omega) + \int_c^b f(t,\omega)\,dw(t,\omega). \tag{7.24}$$

We remark that property (4), given by (7.21), is valid for all functions $f(t,\omega) \in M_2$. This can be verified by a limit process, using an approximating sequence of step functions in M_2. We now state some additional properties of the Itô integral.

5. If $f(t,\omega) \in M_2$, $f_n(t,\omega) \in M_2$, and $\int_a^b |f_n(t,\omega) - f(t,\omega)|^2\,dt \to 0$ in probability, then

$$\lim_{n\to\infty} \int_a^b f_n(t,\omega)\,dw(t,\omega) = \int_a^b f(t,\omega)\,dw(t,\omega), \tag{7.25}$$

in probability.

6. (Generalization of properties (2) and (3)). If the function $f(t,\omega)$ is such that

$$\int_a^b \mathscr{E}\{|f(t,\omega)|^2 \big| \mathfrak{F}_a\}\,dt < \infty$$

almost surely, then

$$\mathscr{E}\left\{\int_a^b f(t,\omega)\,dw(t,\omega) \middle| \mathfrak{F}_a\right\} = 0 \tag{7.26}$$

almost surely, and

$$\mathscr{E}\left\{\left(\int_a^b f(t,\omega)\,dw(t,\omega)\right)^2 \middle| \mathfrak{F}_a\right\} = \int_a^b \mathscr{E}\left\{|f(t,\omega)|^2 \middle| \mathfrak{F}_a\right\} dt \tag{7.27}$$

almost surely.

For applications of the Itô stochastic integral in the formulation of random integral equations we need to consider the integral as a function of the upper limit. Since for each t the integral

$$g(t,\omega) = \int_a^t f(\tau,\omega)\,dw(\tau,\omega) \tag{7.28}$$

is uniquely determined except for its values on a negligible set $N \subset \Omega$, we can assume that $g(t,\omega)$ is a separable random function. It can also be shown that $g(t,\omega)$ is continuous.

It is also possible to define the Itô stochastic integral for random functions with values in R_n. Let $w_1(t,\omega), w_2(t,\omega), \ldots, w_n(t,\omega)$ be n mutually independent real-valued Wiener processes; and assume that (i) each process $w_i(t,\omega)$, $i = 1,2,\ldots,n$ is $\mathfrak{F}_t$-measurable for every t, and $\mathfrak{F}_{t_1} \subset \mathfrak{F}_{t_2}$, for $t_1 < t_2$, and (ii) each $h > 0$ the processes $w_i(t+h,\omega) - w_i(t,\omega)$ $i = 1,2,\ldots,n$, are all independent of $\mathfrak{F}_t$. Then we can define the Itô integral

$$\int_a^b f(t,\omega)\,dw_i(t,\omega) \tag{7.29}$$

for every $i = 1,2,\ldots,n$ and every R_n-valued function $f(t,\omega)$ whose component functions $f_i(t,\omega) \in M_2$.

We now mention briefly some other studies on the Itô stochastic integral. Kallianpur and Striebel [41] have proved some theorems of Fubini type for Itô integrals; and have used these results in the study of certain random differential equations which arise in the study of estimation problems for continuous parameter stochastic processes.

The Itô stochastic integral was generalized by Doob (cf. Doob [19, pp. 436–451], Skorohod [67, Chap. 2], and Saaty [66, pp. 413–416]) by considering the integrator function to be a martingale process $\{z(t,\omega), t \in T\}$ with respect to the family of σ-algebras $\mathfrak{F}_t$. The resulting integral is referred to as the *Itô–Doob stochastic integral.* We refer to the above references for a precise definition of the Itô–Doob integral and discussions of its properties.

Within the framework of integration theory, the definitions of the integrals of Itô and Itô–Doob are completely self-consistent, and there is no reason for stochastic integrals to have any connection with ordinary (classical) integrals. For example, Doob (cf. Doob [19, p. 444]) has shown that if $x(t,\omega)$ is a Wiener process with unit variance, then

$$\int_a^b x(\tau,\omega)\,dx(\tau,\omega) = \tfrac{1}{2}[x^2(b,\omega) - x^2(a,\omega)] - \tfrac{1}{2}(b-a). \tag{7.30}$$

However, if we consider the evaluation of the analogous ordinary integral we obtain

$$\int_a^b x(\tau)\,dx(\tau) = \tfrac{1}{2}[x^2(b) - x^2(a)]. \tag{7.31}$$

This example shows that the stochastic integral calculus based on an Itô type integral cannot be entirely compatible with the classical integral calculus based on an ordinary integral.

In recent years an increasing number of applied mathematicians have been employing probabilistic methods; and many studies involve models formulated as Itô random differential and integral equations. Many workers have expressed concern over the fact that the theory of Itô random equations is self-consistent, and is not an extension of the classical theory of ordinary differential equations and integral equations.

Stratonovich [68; 69, Chap. 2] was one of the first mathematicians to consider the formulation of a stochastic calculus which would be compatible with the ordinary calculus. Let $\Phi(t,x)$ denote a function defined on $[a,b]$, which is continuous in t, with continuous derivative $\partial\Phi(t,x)/\partial x$, and satisfying the conditions

$$\int_a^b \mathscr{E}\{\Phi(\tau,x)\,a(\tau,x)\}\,d\tau < \infty$$

$$\int_a^b \mathscr{E}\{|\Phi(\tau,x)|^2\,b(\tau,x)\}\,d\tau < \infty.$$

In particular, Stratonovich considered functions of the form $\Phi(t,\omega) = F(t,x(t,\omega))$ and $\Phi(t,\omega) = F(t,y(t,\omega))$, where $y(t,\omega)$, $t \in [a,b]$ is a diffusion process related to $x(t,\omega)$ through a random differential equation.

Consider the partition of $[a,b]$, $a = t_1 < \cdots < t_n = b$; and put

$$\Delta = \max(t_{i+1} - t_i).$$

Using the above notation, the Itô stochastic integral can be defined as follows:

$$(\mathrm{I})\int_a^b \Phi(\tau,x(\tau,\omega))\,dx(\tau,\omega) = \operatorname*{l.i.m.}_{\Delta\to 0} \sum_{i=0}^{n} \Phi(t_i, x(t_i,\omega)) \times [x(t_{i+1},\omega) - x(t_i,\omega)]. \tag{7.32}$$

The *Stratonovich stochastic integral*, which is a symmetrized integral, can be defined as follows:

$$(\mathrm{S})\int_a^b \Phi(\tau,x(\tau,\omega))\,dx(\tau,\omega) = \operatorname*{l.i.m.}_{\Delta\to 0} \sum_{i=0}^{n} \Phi\left(t_i, \frac{x(t_{i+1},\omega) + x(t_i,\omega)}{2}\right)[x(t_{i+1},\omega) - x(t_i,\omega)]. \tag{7.33}$$

Stratonovich has shown that the relationship between the two stochastic integrals is given by the following formula:

$$(\mathrm{S})\int_a^b \Phi(\tau,x(\tau,\omega))\,dx(\tau,\omega) = (\mathrm{I})\int_a^b \Phi(\tau,x(\tau,\omega))\,dx(\tau,\omega) + \frac{1}{2}\int_a^b \frac{\partial\Phi(\tau,x(\tau,\omega))}{\partial x}\,b(\tau,x(\tau,\omega))\,d\tau \tag{7.34}$$

with probability one. For further discussion of the connection between the Itô and Stratonovich stochastic integrals, and their use in the study of random differential equations, we refer to Jazwinski [39, Chap. 4], Stratonovich [69, Chap. 2], and Wonham [78].

Wong and Zakai [74, 75, 77] (cf. also, Jazwinski [39, Chap. 4]) have considered the relationship between limits of sequences of Riemann–Stieltjes integrals and the corresponding Itô stochastic integrals. Their research was motivated by the fact that in many applied problems one is often interested in the limit of a sequence of Riemann–Stieltjes integrals which resembles an Itô integral, but with a sequence of smooth approximations $\{x_n(t,\omega)\}$ replacing the Wiener process $x(t,\omega)$. They showed, for example, that if the $x_n(t,\omega)$ have piecewise continuous derivatives, then

$$\int_a^b x_n(\tau,\omega)\,dx_n(\tau,\omega) = \tfrac{1}{2}[x_n^2(b,\omega) - x_n(a,\omega)] \to \tfrac{1}{2}[x^2(b,\omega) - x^2(a,\omega)],$$

which differs from (7.30) by a correction term equal to $\frac{1}{2}(b-a)$.

Consider the sum

$$\sum_{i=1}^{n-1} f(\tau_i,\omega)\,[z(t_{i+1},\omega) - z(t_i,\omega)]. \tag{7.35}$$

It is known that the limit of (7.35) in the usual Riemann sense which is obtained when $\tau_i = t_i$ differs from the limit obtained when $\tau_i = (t_i + t_{i+1})/2$, so that if $\tau_i \in [t_i, t_{i+1}]$ the limit fails to exist (cf. [19, p. 444]). This observation led McShane [57, 58, 58a] to introduce the notion of a belated partition and define a belated stochastic integral. A *belated partition* Δ of the interval $[a,b]$ is an ordered $(2n+1)$-tuple of real numbers (n an arbitrary positive integer), say $(t_1, t_2, \ldots, t_{n+1}, \tau_1, \tau_2, \ldots, \tau_n)$, such that $a = t_1 < t_2 < \cdots < t_{n+1} = b$, and such that for $i = 1, 2, \ldots, n$, $\tau_i \in D$ (where D is a set of real numbers with $[a,b] \subset D$) and $\tau_i \leqslant t_i$. The mesh of Δ, written $m(\Delta)$, is defined to be $\max(t_{i+1} - \tau_i)$, $i = 1, 2, \ldots, n$. Corresponding to Δ, form the sum

$$S(\Delta) = S(\Delta; f, z) = \sum_{i=1}^{n} f(\tau_i,\omega)\,[z(t_{i+1},\omega) - z(t_i,\omega)]. \tag{7.36}$$

We remark that Δ is belated in the sense that each τ_i is associated with an interval $[t_i, t_{i+1}]$ of times later than τ_i. We now define the belated or McShane stochastic integral as follows: If there exists a random variable $M(\omega)$ with $\mathscr{E}\{M^2\} < \infty$ such that

$$\lim_{m(\Delta)\to 0} \|S(\Delta; f, z) - M\|_2 = 0, \tag{7.37}$$

then $M(\omega)$ is the *belated* or *McShane stochastic integral* of $f(t,\omega)$, $t \in D$, $\omega \in \Omega$, with respect to $z(t,\omega)$ over $[a,b]$. We remark that $M(\omega)$ is not unique; that is, if $M(\omega)$ satisfies (7.37), then so does every $\tilde{M}(\omega)$ that is equivalent to $M(\omega)$. Any one of these versions of the McShane integral is denoted by

$$(\mathrm{M})\int_a^b f(\tau,\omega)\,dz(\tau,\omega). \tag{7.38}$$

The McShane integral exists under hypotheses which, compared with those for the Itô integral, are stronger with regard to continuity properties but weaker with regard to the probabilistic properties of the integrator process. When the hypotheses for the existence of both integrals are satisfied, the McShane and Itô integrals agree.

C. Existence and uniqueness of solutions of Itô random integral equations

In this subsection we consider the problem of the existence and uniqueness of solutions of the Itô random integral equation

$$x(t,\omega) = x_0(\omega) + \int_{t_0}^{t} a(\tau, x(\tau,\omega))\,d\tau + \int_{t_0}^{t} b(\tau, x(\tau,\omega))\,dw(\tau,\omega); \qquad (7.39)$$

which is, as we already know, the equivalent integral equation formulation of the Itô random differential equation

$$dx(t,\omega) = a(t, x(t,\omega))\,dt + b(t, x(t,\omega))\,dw(t,\omega). \qquad (7.40)$$

Equation (7.39) is to be solved under the initial condition $x(t_0,\omega) = x_0(\omega)$; and we assume that the random variable $x_0(\omega)$ is given, and is independent of the Wiener process $\{w(t,\omega),\ t \in [t_0, T]\}$.

In order for the integrals in Eq. (7.39), equivalently, the differentials in Eq. (7.40), to be well defined it is necessary to introduce an appropriate family of σ-algebras $\mathfrak{F}_t$. Let $\mathfrak{F}_t$ denote the σ-algebras of subsets of Ω generated by the initial condition $x_0(\omega)$ and the increments $w(\tau,\omega) - w(t_0,\omega)$ for $\tau \in [t_0, t]$; that is, $\mathfrak{F}_t$ is the minimal σ-algebra with respect to which $x_0(\omega)$ and $w(\tau,\omega) - w(t_0,\omega)$, for $\tau \in [t_0, t]$, are measurable.

Definition. A random function $\{x(t,\omega),\ t \in [t_0, T]\}$ is a *solution* of Eq. (7.39) if (i) $x(t,\omega)$ is $\mathfrak{F}_t$-measurable, (ii) the integrals in Eq. (7.39) exist, and (iii) Eq. (7.39) holds for each $t \in [t_0, T]$ almost surely.

In this subsection we will be concerned with only continuous solutions of the Itô integral equation. We remark that (7.21) of Sect. 7.2B implies that for equivalent random functions $f_1(t,\omega)$ and $f_2(t,\omega)$,

$$\int_{t_0}^{t} f_1(\tau,\omega)\,dw(\tau,\omega) = \int_{t_0}^{t} f_2(\tau,\omega)\,dw(\tau,\omega)$$

almost surely. This follows from that fact that if $f_1(\tau,\omega) = f_2(\tau,\omega)$ almost surely for every τ, then

$$\mu\left(\left\{\omega : \int_{t_0}^{T} |f_1(\tau,\omega) - f_2(\tau,\omega)|^2\,d\tau > 0\right\}\right) = 0.$$

It follows from the above that if $\{\tilde{x}(t,\omega)\}$ is a random function equivalent to a solution $\{x(t,\omega)\}$ of Eq. (7.39), then $\{\tilde{x}(t,\omega)\}$ is also a solution of Eq. (7.39). Now, the right-hand side of Eq. (7.39) is equivalent to the left-hand side, and is continuous almost surely. Hence for every solution of the Itô integral, equation there exists an equivalent continuous solution.

We now state and prove the following existence and uniqueness theorem for solutions of Itô random integral equations.

THEOREM 7.1. *Let the functions $a(t,x)$ and $b(t,x)$ be Borel measurable for $t \in [t_0, T]$, $x \in R$, satisfying the following conditions for some constant K:*

a. *(uniform Lipschitz condition) For all $x, y \in R$,*

$$|a(t,x) - a(t,y)| + |b(t,x) - b(t,y)| \leqslant K|x-y|,$$

b. *(growth condition) For all x,*

$$|a(t,x)|^2 + |b(t,x)|^2 \leqslant K^2(1+|x|^2).$$

Then, there exists a solution of Eq. (7.39), and it is unique in the following sense: If $x_1(t,\omega)$ and $x_2(t,\omega)$ are two continuous solutions (for a fixed initial condition $x_0(\omega)$) of Eq. (7.39), then

$$\mu\left(\left\{\omega : \sup_{t\in[t_0,T]} |x_1(t,\omega) - x_2(t,\omega)| = 0\right\}\right) = 1.$$

Proof. We first prove the uniqueness of a continuous solution. Let $x_1(t,\omega)$ and $x_2(t,\omega)$ be two continuous solutions of Eq. (7.39). We define a random variable $\chi_N(t,\omega)$ as follows:

$$\chi_N(t,\omega) = \begin{cases} 1 & \text{if } |x_1(\tau,\omega)| < N, \quad |x_2(\tau,\omega)| < N, \quad \text{for} \quad \tau \in [t_0, t] \\ 0 & \text{otherwise.} \end{cases}$$

Since $\chi_N(t,\omega)\chi_N(\tau,\omega) = \chi_N(t,\omega)$ for $\tau < t$, we have, using Eq. (7.39),

$$\begin{aligned} \chi_N(t,\omega)[x_1(t,\omega) - x_2(t,\omega)] = \chi_N(t,\omega)\Big\{\int_{t_0}^{T} &\chi_N(\tau)\,[a(\tau, x_1(\tau,\omega)) \\ &- a(\tau, x_2(\tau,\omega))]\,d\tau + \int_{t_0}^{T} \chi_N(\tau)\,[b(\tau, x_1(\tau,\omega)) \\ &- b(\tau, x_2(\tau,\omega))]\,dw(\tau,\omega)\Big\}. \end{aligned} \tag{7.41}$$

From condition (a) we have

$$\begin{aligned} &\chi_N(\tau,\omega)\,[|a(\tau,x_1(\tau,\omega)) - a(\tau,x_2(\tau,\omega))| + |b(\tau,x_1(\tau,\omega)) - b(\tau,x_2(\tau,\omega))|] \\ &\quad \leqslant K\chi_N(\tau,\omega)\,|x_1(\tau,\omega) - x_2(\tau,\omega)| \leqslant 2KN. \end{aligned}$$

Applying the above result to (7.41) we observe that the squares of the integrals on the right-hand side of (7.41) have expectations. Hence using the inequality $(x+y)^2 \leqslant 2(x^2+y^2)$, Cauchy's inequality, and (7.20) we obtain the inequality

$$\mathscr{E}\{\chi_N(t,\omega)\,[x_1(t,\omega)-x_2(t,\omega)]^2\}$$
$$\leqslant 2\mathscr{E}\Big\{\chi_N(t,\omega)\Big[\int_{t_0}^{t}\chi_N(\tau,\omega)\,(a(\tau,x_1(\tau,\omega))-a(\tau,x_2(\tau,\omega)))\,d\tau\Big]^2\Big\}$$
$$+\,2\mathscr{E}\Big\{\chi_N(t,\omega)\Big[\int_{t_0}^{t}\chi_N(\tau,\omega)\,(b(\tau,x_1(\tau,\omega))-b(\tau,x_2(\tau,\omega)))\,dw(\tau,\omega)\Big]^2\Big\}$$
$$\leqslant 2(T-t_0)\int_{t_0}^{t}\mathscr{E}\{\chi_N(\tau,\omega)\,(a(\tau,x_1(\tau,\omega))-a(\tau,x_2(\tau,\omega)))^2\}\,d\tau$$
$$+\,2\int_{t_0}^{t}\mathscr{E}\{\chi_N(\tau,\omega)\,(b(\tau,x_1(\tau,\omega))-b(\tau,x_2(\tau,\omega)))^2\}\,d\tau.$$

Using condition (a), we have

$$\mathscr{E}\{\chi_N(t,\omega)\,[x_1(t,\omega)-x_2(t,\omega)]^2\}\leqslant M\int_{t_0}^{t}\mathscr{E}\{\chi_N(\tau,\omega)\,|x_1(\tau,\omega)-x_2(\tau,\omega)|^2\}\,d\tau \tag{7.42}$$

for some constant M. In order to show that $x_1(t,\omega)$ and $x_2(t,\omega)$ are equivalent, we first use Gronwall's lemma in the following form: *If $\varphi(t)$, $t\in[t_0,T]$, is a nonnegative integrable function which satisfies the inequality*

$$\varphi(t)\leqslant C\int_{t_0}^{t}\varphi(\tau)\,d\tau+\xi(t),$$

where $\xi(t)$ is an integrable function and C is a nonnegative constant, then

$$\varphi(t)\leqslant C\int_{t_0}^{t}\exp\{C(t-\tau)\}\,\xi(\tau)\,d\tau+\xi(t).$$

Put $\varphi(t)=\mathscr{E}\{\chi_N(t,\omega)\,[x_1(t,\omega)-x_2(t,\omega)]^2\}$ and $\xi(t)=0$. Then

$$\mathscr{E}\{\chi_N(t,\omega)\,[x_1(t,\omega)-x_2(t,\omega)]^2\}=0;$$

that is,

$$\mu(\{\omega:x_1(t,\omega)\neq x_2(t,\omega)\})\leqslant\mu\Big(\Big\{\omega:\sup_t|x_1(t,\omega)|>N\Big\}\Big)$$
$$+\,\mu\Big(\Big\{\omega:\sup_t|x_2(t,\omega)|>N\Big\}\Big). \tag{7.43}$$

Now, the measures on the right-hand side of (7.43) approach zero, since the random functions $x_1(t,\omega)$ and $x_2(t,\omega)$ are continuous (and hence bounded) almost surely. Hence $x_1(t,\omega)$ and $x_2(t,\omega)$ are equivalent random functions. Finally, since both random functions are continuous almost surely, we have

$$\mu\Big(\Big\{\omega:\sup_t|x_1(t,\omega)-x_2(t,\omega)|>0\Big\}\Big)=0.$$

This completes the proof of uniqueness of the solution of the Itô integral equation (7.39).

We now consider the existence of a solution of Eq. (7.39). We first introduce the Banach space $\mathfrak{X}$ of measurable random functions $\xi(t, \omega)$ which are $\mathfrak{F}_t$-measurable for every t, and satisfy the condition

$$\sup_{t \in [t_0, T]} \mathscr{E}\{|\xi(t, \omega)|^2\} < \infty.$$

For $\xi \in \mathfrak{X}$, we introduce the norm

$$\|\xi(t, \omega)\| = \left(\sup_{t \in [t_0, T]} \mathscr{E}\{|\xi(t, \omega)|^2\}\right)^{1/2}.$$

We assume that the initial condition $\xi_0(\omega)$ satisfies the condition $\mathscr{E}\{|\xi_0(\omega)|^2\} < \infty$.

We define an operator T on $\mathfrak{X}$, which we will call the *Itô integral operator*, as follows:

$$T[\xi(t, \omega)] = \xi_0(\omega) + \int_{t_0}^{t} a(\tau, \xi(\tau, \omega))\, d\tau + \int_{t_0}^{t} b(\tau, \xi(\tau, \omega))\, dw(\tau, \omega). \quad (7.44)$$

Both integrals in (7.44) exist by condition (b); and the $\mathfrak{F}_t$-measurability of $T[\xi]$ follows from the $\mathfrak{F}_t$-measurability of $\xi(t, \omega)$.

Since $(x + y + z)^2 \leqslant 3(x^2 + y^2 + z^2)$, an application of condition (b) yields the following inequality:

$$\begin{aligned}
\mathscr{E}\{|T[\xi(t, \omega)]|^2\} &\leqslant 3\mathscr{E}\{|\xi_0(\omega)|^2\} + 3(T - t_0)\, \mathscr{E}\left\{\int_{t_0}^{t} K^2(1 + |\xi(\tau, \omega)|^2)\, d\tau\right\} \\
&\quad + 3\int_{t_0}^{t} \mathscr{E}\{K(1 + |\xi(\tau, \omega)|^2)\, d\tau\} \\
&\leqslant 3\mathscr{E}\{|\xi_0(\omega)|^2\} + 3K^2(T - t_0)^2 + (3K^2 + 6K^2\|\xi\|^2)(T - t_0) \\
&= 3\mathscr{E}\{|\xi_0(\omega)|^2\} + L_1 + L_2\|\xi\|^2,
\end{aligned}$$

where we have put $L_1 = 3K^2(T - t_0)(T_0 - t_0 + 1)$ and $L_2 = 6K^2(T - t_0)$. Hence $T[\xi] \in \mathfrak{X}$; that is, the Itô operator maps $\mathfrak{X}$ into itself. Furthermore, from (7.20) we obtain

$$\begin{aligned}
&\mathscr{E}\{|T[\xi_1(t, \omega) - \xi_2(t, \omega)]|^2\} \\
&\quad \leqslant 2(T - t_0)\int_{t_0}^{t} \mathscr{E}\{[a(\tau, \xi_1(\tau, \omega)) - a(\tau, \xi_2(\tau, \omega))]^2\, d\tau\} \\
&\quad\quad + 2\left(\mathscr{E}\left\{\int_{t_0}^{t} [b(\tau, \xi_1(\tau, \omega)) - b(\tau, \xi_2(\tau, \omega))]\, dw(\tau, \omega)\right\}\right)^2 \\
&\quad \leqslant L_0 \int_{t_0}^{t} \mathscr{E}\{|\xi_1(\tau, \omega) - \xi_2(\tau, \omega)|^2\, d\tau\} \\
&\quad \leqslant L_0(t - t_0)\, \|\xi_1 - \xi_2\|^2,
\end{aligned} \quad (7.45)$$

where $L_0 = 2K^2(T - t_0 + 1)$. The above shows that the Itô operator is continuous.

We now show that the Itô operator T is a contraction operator on $\mathfrak{X}$, hence there exists a fixed point of T which is a solution of the Itô integral equation. Iteration of the inequality (7.45) yields

$$\begin{aligned}
&\mathscr{E}\{|T^n[\xi_1(t,\omega)-\xi_2(t,\omega)]|^2\} \\
&\quad \leqslant L_0 \int_{t_0}^{t} \mathscr{E}\{|T^{n-1}[\xi_1(\tau,\omega)-\xi_2(\tau,\omega)]|^2\}\,d\tau \\
&\quad \leqslant L_0^n \int_{t_0}^{t}\int_{t_0}^{t_1}\cdots\int_{t_0}^{t_{n-1}} \mathscr{E}\{\|\xi_1(\tau,\omega)-\xi_2(\tau,\omega)\|^2\}\,d\tau\,dt_n\cdots dt_1 \\
&\quad \leqslant \frac{L_0^n(t-t_0)^n}{n!}\,\|\xi_1(t,\omega)-\xi_2(t,\omega)\|^2.
\end{aligned}$$

Therefore, for every $\xi \in \mathfrak{X}$,

$$\|T^n(T-I)\,\xi\| \leqslant \frac{L_0^n(T-t_0)^n}{n!}\,\|(T-I)\,[\xi]\|^2.$$

Now, from the convergence of the series $\sum_{n=1}^{\infty} \|T^n(T-I)\xi\|$ and the fact that $\mathfrak{X}$ is complete, it follows that $\lim_{n\to\infty} T^n[\xi]$ exists in $\mathfrak{X}$. Let

$$x(t,\omega) = \lim_{n\to\infty} T^n[\xi(t,\omega)].$$

Since T is a continuous operator, $T[T^n[\xi(t,\omega)]] \to T[x(t,\omega)]$. But

$$\lim_{n\to\infty} T[T^n[\xi(t,\omega)]] = \lim_{n\to\infty} T^{n+1}[\xi(t,\omega)] = x(t,\omega).$$

Therefore $\|T[x(t,\omega)] - x(t,\omega)\| = 0$; and $T[x(t,\omega)] = x(t,\omega)$ almost surely for every $t \in [t_0, T]$. Hence $x(t,\omega)$ is a fixed point of the Itô operator which is a solution of the Itô random integral equation (7.39).

We now establish the existence of a solution in the general case. Let

$$x_N(t_0,\omega) = \begin{cases} x_0(\omega) & \text{if } |x_0(\omega)| \leqslant N \\ 0 & \text{otherwise;} \end{cases}$$

and let the random function $x_N(t,\omega)$ denote the solution of Eq. (7.39) for the initial value $x_N(t_0,\omega)$. We will show that $\lim_{N\to\infty} x_N(t,\omega) = x(t,\omega)$ (in probability), where $x(t,\omega)$ is a solution of Eq. (7.39).

Let $N' > N$, and let

$$\lambda(\omega) = \begin{cases} 1 & \text{if } |x_0(\omega)| \leqslant N \\ 0 & \text{otherwise.} \end{cases}$$

Then $(x_N(t_0,\omega) - x_{N'}(t_0,\omega))\lambda(\omega) = 0$. Clearly $\lambda(\omega)$ is $\mathfrak{F}_{t_0}$-measurable. It now follows from the inequality

$$|x_N(t,\omega) - x_{N'}(t,\omega)|^2 \lambda(\omega)$$
$$\leqslant 2\Big(\int_{t_0}^{t} [a(\tau, x_N(\tau,\omega)) - a(\tau, x_{N'}(\tau,\omega))]\,\lambda(\omega)\,d\tau\Big)^2$$
$$+ 2\Big(\int_{t_0}^{t} [b(\tau, x_N(\tau,\omega)) - b(\tau, x_{N'}(\tau,\omega))]\,\lambda(\omega)\,dw(\tau,\omega)\Big)^2,$$

from condition (a), and previous estimates of the integrals, that for some constant H

$$\mathscr{E}\{|x_N(t,\omega) - x_{N'}(t,\omega)|^2 \lambda(\omega)\} \leqslant H \int_{t_0}^{t} \mathscr{E}\{|x_N(\tau,\omega) - x_{N'}(\tau,\omega)|^2 \lambda(\omega)\}\,d\tau.$$

Consequently

$$\mathscr{E}\{|x_N(t,\omega) - x_{N'}(t,\omega)|^2 \lambda(\omega)\} = 0,$$

so that

$$\mu(\{\omega : |x_N(t,\omega) - x_{N'}(t,\omega)| > 0\}) \leqslant \mu(\{\omega : |x_0(\omega)| > N\}).$$

It follows from the last inequality that $\lim_{N\to\infty} x_N(t,\omega) = x(t,\omega)$; and since the limit is uniform in t, we have

$$\lim_{N\to\infty} \int_{t_0}^{T} [x_N(\tau,\omega) - x(\tau,\omega)]^2\,d\tau = 0 \qquad \text{(in probability)}.$$

Utilizing condition (a) we obtain

$$\Big|\int_{t_0}^{t} a(\tau, x_N(\tau,\omega))\,d\tau - \int_{t_0}^{t} a(\tau, x(\tau,\omega))\,d\tau\Big|$$
$$\leqslant K\Big((t - t_0)\int_{t_0}^{t} [x_N(\tau,\omega) - x(\tau,\omega)]^2\,d\tau\Big)^{1/2};$$

and using property (7.21) we have

$$\mu\Big(\Big\{\omega : \Big|\int_{t_0}^{t} b(\tau, x_N(\tau,\omega))\,dw(\tau,\omega) - \int_{t_0}^{t} b(\tau, x(\tau,\omega))\,dw(\tau,\omega)\Big| > \delta\Big\}\Big)$$
$$\leqslant \epsilon^2/\delta^2 + \mu\Big(\Big\{\omega : K^2 \int_{t_0}^{t} [x_N(\tau,\omega) - x(\tau,\omega)]^2\,d\tau > \epsilon^2\Big\}\Big).$$

Hence, we can pass to the limit (in probability) on the right-hand side of the Itô integral equation for $x_N(t,\omega)$. Therefore, $x(t,\omega)$ is a solution of the Itô random integral equation (7.39). This completes the proof of the theorem.

For other results on the existence and uniqueness of solutions of Itô random integral equations we refer to Dynkin [21, Chap. 11], Gihman [30], Gihman and Skorohod [31, Chap. 8], Girsanov [32], Krylov [50], Maruyama [59], and Skorohod [67, Chap. 3]. In particular, in [67], existence and uniqueness was established for the case where $a(t,x)$ and $b(t,x)$ are continuous functions of their arguments, but satisfy less restrictive conditions than those of Theorem

7.1; and Krylov [50] considered existence and uniqueness of the solution in R_n ($n > 1$) of the equation

$$x(t, \omega) = x(t_0, \omega) + \int_{t_0}^{t} A(\tau, x)\, d\tau + \int_{t_0}^{t} B(\tau, x)\, dw(\tau, \omega) \tag{7.46}$$

under the following conditions:

1. The random n-vector $A(t,x)$ and the random $n \times n$ matrix $B(t,x)$ are Borel measurable; and
2. there exist constants $\gamma > 0$ and $M < \infty$ such that (a) $|A(t,x)| < M$ for all $x \in R_n$ and (b) $\gamma|\lambda|^2 \leqslant (B\lambda, \lambda) \leqslant \gamma^{-1}|\lambda|^2$ for all $x \in R_n$ and $\lambda = (\lambda_1, \lambda_2, \ldots, \lambda_n)$. In Eq. (7.46) $w(t, \omega)$ is an n-dimensional Wiener process.

Girsanov [33] (cf. also McKean [56, pp. 78–80]) has given an example of nonuniqueness of the solution of an Itô random integral equation. He considered the equation

$$x(t, \omega) = x_0(\omega) + \int_0^t |x(\tau, \omega)|^\alpha \, dw(\tau, \omega); \tag{7.47}$$

and showed that Eq. (7.47) has a unique solution for $\alpha \geqslant \frac{1}{2}$, but for $\alpha \in (0, \frac{1}{2})$ there exists many solutions.

Stroock and Varadhan [70] have recently studied the existence and uniqueness of solutions of Itô random integral equations; in particular, they considered a new general type of uniqueness theorem in which uniqueness is considered in the sense of the probability distributions.

Itô and Nisio [38] have studied the existence of stationary solutions of Itô random integral equations. A solution $x(t, \omega)$ is said to be a *stationary solution* if the probability distribution of the joint process

$$(x(t, \omega), w(t, \omega)) \equiv (x(t, \omega), t \in (-\infty, \infty), \quad w(t_2, \omega) - w(t_1, \omega), -\infty < t_1 < t_2 < \infty)$$

is invariant under the shift transformation. Clearly a stationary solution is a strictly stationary random function.

D. *Itô processes*

1. *Introduction.* In this section we consider the class of Markov processes, called *Itô processes*, determined by solutions of Itô random integral equations. This section contains two subsections, the first of which is devoted to a proof of the theorem that the solution process $\{x(t, \omega), t \in [t_0, T]\}$ determined by the solution of Eq. (7.39) is a Markov process. In the second subsection we present a brief discussion of the operator-theoretic formulation of Itô processes. For a detailed discussion of Itô processes and related topics we refer to Dynkin [21].

2. ***The solution process of the Itô random integral equation.*** We recall that a random function $\{x(t,\omega),\ t \in [t_0, T]\}$ with state space $(R, \mathfrak{B})$ is said to be a *Markov process* if

i. for all $t_1 < t_2 < \cdots < t_n < t$ (where $t,\ t_i \in [t_0, T],\ i = 1, 2, \ldots, n$) and Borel sets $B \subset R$,

$$\mathscr{P}\{x(t,\omega) \in B \,|\, x(t_1, \omega), \ldots, x(t_n, \omega)\} = \mathscr{P}\{x(t,\omega) \in B \,|\, x(t_n, \omega)\}$$

with probability one; and

ii. the transition function

$$P(s, x, t, B) = \mathscr{P}\{x(t,\omega) \in B \,|\, x(s,\omega) = x\}$$

defined for $t_0 \leqslant s \leqslant t \leqslant T$, $x \in R$, and $B \in \mathfrak{B}$, is a $\mathfrak{B}$-measurable function of x for every fixed s, t, B; and for fixed s, t, and x is a probability measure on $\mathfrak{B}$.

We now state and prove the following basic theorem.

THEOREM 7.2. *Let the functions $a(t,x)$ and $b(t,x)$ satisfy the hypotheses of Theorem* 7.1; *and let $\{x_{t,\xi}(s,\omega),\ s \in [t,T]\}$, where $t \geqslant t_0$, $\xi \in R$, be the solution process determined by the equation*

$$x_{t,\xi}(s,\omega) = \xi + \int_t^s a(\tau, x_{t,\xi}(\tau,\omega))\, d\tau + \int_t^s b(\tau, x_{t,\xi}(\tau,\omega))\, dw(\tau). \tag{7.48}$$

Then the solution process $\{x(t,\omega),\ t \in [t_0, T]\}$ determined by the Itô random integral equation (7.39) *is a Markov process with transition function*

$$\begin{aligned} P(t, \xi, s, B) &= \mathscr{P}\{x(s,\omega) \in B \,|\, x(t,\omega) = \xi\} \\ &= \mathscr{P}\{x_{t,\xi}(s,\omega) \in B\}, \quad B \in \mathfrak{B}. \end{aligned} \tag{7.49}$$

Proof. Throughout this proof $\mathfrak{F}_t$ denotes the σ-algebra introduced in Sect. 7.2A. Since (1) the solution $x(t,\omega)$ is $\mathfrak{F}_t$-measurable, and (2) $x_{t,\xi}(s,\omega)$ is completely determined by the differences $w(s,\omega) - w(t,\omega)$ for $s \in [t,T]$ (independently of $\mathfrak{F}_t$), it is clear that $x_{t,\xi}(s,\omega)$ is independent of $x(t,\omega)$ and of the events in $\mathfrak{F}_t$. It follows from the basic existence and uniqueness theorem (Theorem 7.1) that $x(s,\omega)$, $s \in [t,T]$ is the unique solution of the Itô random integral equation

$$x(s,\omega) = x(t,\omega) + \int_t^s a(\tau, x(\tau,\omega))\, d\tau + \int_t^s b(\tau, x(\tau,\omega))\, dw(\tau,\omega). \tag{7.50}$$

Now, the process $\{x_{t,x(t)}(s,\omega),\ s \in [t,T]\}$ is also a solution of Eq. (7.50); hence

$$x(s,\omega) = x_{t,x(t,\omega)}(s,\omega) \tag{7.51}$$

almost surely.

We now show that

$$\mathscr{P}\{x(s,\omega) \in B \,|\, x(t,\omega)\} = \mathscr{P}\{x(s,\omega) \in B \,|\, \mathfrak{F}_t\}. \tag{7.52}$$

In order to establish (7.52) it is sufficient to show that for any bounded, $\mathfrak{F}_t$-measurable function φ, and any bounded continuous function $g(x)$

$$\mathscr{E}\{\varphi g(x(s,\omega))\} = \mathscr{E}\{\varphi \mathscr{E}\{g(x(s,\omega))\,|\,x(t,\omega)\}. \tag{7.53}$$

Put $\Phi(\xi,\omega) = g(x_{t,\xi}(s,\omega))$. The function $\Phi(\xi,\omega)$ is clearly a measurable function of the pair (ξ,ω); and $\Phi(\xi,\omega)$ is also a bounded and continuous function of ξ for almost all ω. We can, therefore, write $\Phi(\xi,\omega)$ as the pointwise almost sure limit of bounded functions of the form

$$\sum_{i=1}^{n} \Phi_i(\xi)\, h_i(\omega), \tag{7.54}$$

where the $\Phi_i(\xi)$ are deterministic functions and the $h_i(\omega)$ are independent of $\mathfrak{F}_t$. Then

$$\mathscr{E}\left\{\varphi \sum_{i=1}^{n} \Phi_i(x(t,\omega))\, h_i(\omega)\right\} = \sum_{i=1}^{n} \mathscr{E}\{\varphi \Phi_i(x(t,\omega))\}\mathscr{E}\{h_i(\omega)\}$$

$$= \mathscr{E}\left\{\sum_{i=1}^{n} \varphi \Phi_i(x(t,\omega))\right\}\mathscr{E}\{h_i(\omega)\};$$

and

$$\mathscr{E}\left\{\sum_{i=1}^{n} \Phi_i(x(t,\omega))\, h_i(\omega)\,|\,x(t,\omega)\right\} = \sum_{i=1}^{n} \Phi_i(x(t,\omega))\,\mathscr{E}\{h_i(\omega)\}.$$

Since functions of the form (7.54) are dense, it follows that for any function $\Phi(\xi,\omega)$ depending on the differences $w(s,\omega) - w(t,\omega)$, $s \geqslant t$,

$$\mathscr{E}\{\Phi(x(t,\omega),\omega)|\mathfrak{F}_t\} = \mathscr{E}\{\Phi(x(t,\omega),\omega)|x(t,\omega)\}$$

$$= \mathscr{E}\{\Phi(\xi,\omega)\}|_{\xi = x(t,\omega)}\,.$$

Since $\Phi(\xi,\omega) = g(x_{t,\xi}(s,\omega))$, we have $\Phi(x(t,\omega),\omega) = g(x(s,\omega))$, and (using (7.51))

$$\mathscr{E}\{g(x(s,\omega))|\mathfrak{F}_t\} = u(x(t,\omega)),$$

where $u(\xi) = \mathscr{E}\{g(x_{t,\xi}(s,\omega))\}$. Hence

$$\mathscr{P}\{x(s,\omega) \in B\,|\,\mathfrak{F}_t\} = P_{t,x(t,\omega)}(s,B),$$

where

$$P_{t,\xi}(s,B) = \mathscr{P}\{x_{t,\xi}(s,\omega) \in B\}.$$

This completes the proof of the theorem.

3. *Operator-theoretic formulation of Itô processes.* Consider the measurable state space $(X,\mathfrak{B})$, where X is an arbitrary set and $\mathfrak{B}$ is a σ-algebra of subsets of X. A function $P(t,x,B)$, $t \in [0,\infty)$, $x \in X$, $B \in \mathfrak{B}$, is said to be a *transition function* if the following conditions are satisfied:

a. For fixed t and x, $P(t,x,B)$ is a probability measure on $\mathfrak{B}$.
b. For fixed t and B, $P(t,x,B)$ is a $\mathfrak{B}$-measurable function of x.
c. $P(t,x,X) \leqslant 1$.
d. $P(0,x,X-\{x\}) = 0$.
e. $P(t_1+t_2,x,B) = \int_X P(t_1,x,d\xi)P(t_2,\xi,B)$, $t_1,t_2 \geqslant 0$.

Let $\mathfrak{X}$ denote the Banach space of all bounded measurable functions $f(x)$, $x \in X$, with norm $\|f\| = \sup_{x\in X} |f(x)|$; and let $P(t,x,B)$ be a transition function on $(X,\mathfrak{B})$. We can define an operator on $\mathfrak{X}$ as follows:

$$T(t)[f(x)] = \int_X P(t,x,d\xi)f(\xi), \qquad f \in \mathfrak{X}; \tag{7.55}$$

hence (7.55) defines an integral operator on $\mathfrak{X}$ with a transition function as its kernel. It follows from properties (c) and (e) of a transition function that (7.55) defines an operator-valued function $T(t)$: $[0,\infty) \to \mathfrak{L}(X)$ such that $\{T(t), t \geqslant 0\}$ is a *contraction semigroup of operators* on $\mathfrak{X}$; that is,

$$T(t_1+t_2) = T(t_1)T(t_2),\ T(0) = I, \text{ and } \|T(t)\| \leqslant 1.$$

The *infinitesimal generator* of the semigroup $\{T(t), t \geqslant 0\}$ is defined by

$$Af = \lim_{h\to 0} A_h f, \qquad A_h = \frac{T(h)-I}{h}, \tag{7.56}$$

whenever the limit (in the strong sense) exists. The *domain* of A, written $\mathfrak{D}(A)$, is the set of all $f \in \mathfrak{X}$ for which this limit exists. It is clear that A is a linear operator.

A contraction semigroup $\{T(t), t \geqslant 0\}$ on $\mathfrak{X}$ is said to be a *Markov semigroup* (or *semigroup representation of a Markov process*) if (i) $f \geqslant 0$ implies $T(t)f \geqslant 0$ for all $t \in [0,\infty)$, and (ii) for all $x \in X$ and $t \in (0,\infty)$, $\sup_{0\leqslant f\leqslant 1} T(t)f \leqslant 1, f \in \mathfrak{X}$.

We now introduce an important class of Markov semigroups. Let $(X,\mathfrak{C},\mathfrak{B})$ be a topological measurable state space; that is, $(X,\mathfrak{C})$ is a topological space and $\mathfrak{B}$ is, as above, a σ-algebra of subsets of X, such that the sets $\{x: f(x) > 0\}$ (where f is a continuous, $\mathfrak{B}$-measurable function) generates $\mathfrak{B}$ and form a base of $(X,\mathfrak{C})$. The semigroup $\{T(t), t \geqslant 0\}$ defined by (7.55) is said to be a *Feller semigroup* if $T(t)$ maps the space $C(X,\mathfrak{C},\mathfrak{B})$ into itself.

The above definitions and concepts are all that are required for the present section. For expositions of the semigroup theory of Markov processes we refer to Dynkin [21] and Loève [55, Chap. XII].

We have shown (Theorem 7.2) that the solution process $\{x(t,\omega), t \in [0,T]\}$ of the Itô random integral equation

$$x(t,\omega) = \xi + \int_0^t a(\tau,x(\tau,\omega))\,d\tau + \int_0^t b(\tau,x(\tau,\omega))\,dw(\tau,\omega) \tag{7.57}$$

is a Markov process with transition function

$$P(t, \xi, B) = \mathscr{P}\{x(t, \omega) \in B \,|\, x(0, \omega) = \xi\},$$

which we will call an *Itô process*.

Following Doob [20], we now present a brief survey of the operator-theoretic formulation of Itô processes in the one-dimensional case. For the n-dimensional case we refer to Dynkin [21].

Let the state space $E = I$, where I is an interval. We will assume that (i) the coefficients $a(t, x)$ and $b(t, x)$ are functions of x alone, and are Baire functions on I, (ii) $b(x) \geqslant 0$, $x \in I$, and (iii) to every compact subset I_0 of I there is associated a constant K such that $|a(x)| < K$, $|b(x)| < K$, $|a(x_2) - a(x_1)| \leqslant K|x_2 - x_1|$, $|b(x_2) - b(x_1)| \leqslant K|x_2 - x_1|$, for $x_1, x_2 \in I_0$.

Let $\{x(t, \omega),\ t \in [0, \infty)\}$ be the Itô process generated by the solution of Eq. (7.57) on I; and let $C(-\infty, \infty)$ denote the Banach space of functions defined and continuous on $(-\infty, \infty)$, with norm

$$\|f\| = \sup_x |f(x)|\, e^{-|x|} < \infty. \tag{7.58}$$

We define an operator $T(t)$ on $C(-\infty, \infty)$ as follows:

$$\begin{aligned} T(t)\,[f(x)] &= \int_I P(t, x, d\xi) f(\xi), \qquad f \in C, \\ &= \mathscr{E}\{f(x_\xi(t, \omega))\} \\ &= \mathscr{E}\{f(x(t, \omega)) \,|\, x(0, \omega) = \xi\}. \end{aligned} \tag{7.59}$$

$T(t)$ as defined by (7.59) is a linear operator, bounded on every compact t-set; and $\{T(t),\ t \in [0, \infty)\}$ is a semigroup of operators on $C(-\infty, \infty)$. Doob has shown that

$$\|T(t)\,[f(x)]\| \leqslant M \|f\| e^{|x|},$$

where M is a constant which depends on t and K.

We now show that the Itô process is a Feller process; that is we will show that $T(t)[f]$ is continuous, from which it will follow that $T(t)$ maps C into itself. We first use the fact that $x_{\xi_1}(t, \omega) \to x_{\xi_2}(t, \omega)$ in probability as $\xi_1 \to \xi_2$ for all t; hence $f(x_{\xi_1}(t, \omega)) \to f(x_{\xi_2}(t, \omega))$ in probability also. To establish the continuity of $T(t)[f]$ it follows from (7.59) that we must show that

$$\mathscr{E}\{f(x_{\xi_1}(t, \omega))\} \to \mathscr{E}\{f(x_{\xi_2}(t, \omega))\}.$$

But

$$\mathscr{E}\{f^2(x_\xi(t, \omega))\} \leqslant \|f\|^2 \mathscr{E}\{e^{2|x_\xi(t, \omega)|}\} \leqslant \|f\|^2 \tilde{M} e^{2|\xi|},$$

where $\tilde{M}$ is a constant independent of ξ; hence we have shown that $T(t)$ maps C into itself. It can also be shown that if C_a is the subclass of C whose functions are absolutely continuous, with $\sup_x |f'(x)| e^{-|x|} < \infty$, then $T(t)$ maps C_a into C_a, and for $f \in C_a$, $T(t)[f]$ is a strongly continuous function of t on $[0, \infty)$.

We now compute the infinitesimal generator A of $\{T(t), t \in [0, \infty)\}$, and show that A coincides with the differential operator

$$L = \frac{b^2}{2}\frac{d^2}{d\xi^2} + a\frac{d}{d\xi} \tag{7.60}$$

on a subspace of C. Let $C_2(-\infty, \infty)$ denote the space of functions f with two continuous derivatives and such that outside of some finite interval f has a third continuous derivative with $\sup_x e^{-|x|}|f^{(3)}(x)| < \infty$. C_2 is a linear, but not closed, subspace of C. Let $f \in C_2$, and let ϵ be a positive number. Then

$$\begin{aligned}\frac{(T(h) - I)[f(x)]}{h} &= (1/h)\int_I [f(y) - f(x)]\,P(h, x, dy)\\ &= (1/h)\int_{|y-x|\leq\epsilon} [f(y) - f(x)]\,P(h, x, dy)\\ &\quad + (1/h)\int_{|y-x|>\epsilon} [f(y) - f(x)]\,P(h, x, dy).\end{aligned}$$

Now, since coefficients $a(x)$ and $b(x)$ are bounded, and the Itô process is continuous, the second term on the right (for a fixed x) tends to zero as $h \to 0$. The first term on the right can be written as

$$\begin{aligned}&\frac{1}{h}\int_{x-\epsilon}^{x+\epsilon}\left[(y - x)f'(x) + \frac{(y - x)^2}{2}f''(x) + (y - x)^2 g(x, y)\right]P(h, x, dy)\\ &\quad = L[f] + O(h^{1/2}) + \epsilon\tilde{M},\end{aligned}$$

where $\tilde{M}$ is a constant which depends on x. Since ϵ can be chosen arbitrarily small, it follows that

$$A[f] = \lim_{h\to 0}\frac{(T(h) - I)[f]}{h} = L[f]. \tag{7.61}$$

Hence, the infinitesimal generator A of the semigroup $\{T(t), t \in [0, \infty)\}$ is defined on C_2 and coincides with L there. For a more detailed discussion we refer to Doob [20].

7.3 Itô Random Integral Equations in Hilbert Spaces

A. Introduction

In this section we consider Itô random integral equations in Hilbert spaces. The study of the random integral equations equivalent to Itô random differential equations in Hilbert spaces requires (1) the basic concepts of probabilistic functional analysis in Hilbert spaces presented in Chaps. 1 and 2, and (2) the definition of the Itô stochastic integral for Hilbert space-valued integrands.

A number of mathematicians have considered stochastic integrals required for the study of random integral equations of Itô type in Hilbert spaces. Baklan [2, 3], Cabaña [12], Daletskiĭ [16], Falb (cf. Kalman *et al.* [42]), and Mortensen [61] have studied stochastic integrals with Hilbert space-valued integrands. In [16, 61] the integrator function is a real-valued Wiener process; and in [16, 42] the integrator function is a Hilbert space-valued Wiener process, but in [42] the integrand is an operator-valued function. Kunita [51] has defined stochastic integrals with Hilbert space-valued or operator-valued integrands, and the integrator function is a Hilbert space-valued martingale. Vahaniya and Kandelaki [71a] have considered stochastic integrals with operator-valued integrands and whose integrator processes are Hilbert space-valued. Using tensor product methods, Kannan and Bharucha-Reid [43] have defined a stochastic integral with a second order Hilbert space-valued random function as the integrand and a Hilbert space-valued Wiener process as the integrator function. The Kannan–Bharucha-Reid integral is operator valued.

This section is divided into two subsections. In Sect. 7.3B we consider, following Cabaña [12], Itô random integral equations in separable Hilbert spaces; and in Sect. 7.3C we consider, following Daletskiĭ [16], Itô random integral equations in scales of Hilbert spaces.

B. Itô random integral equations in separable Hilbert spaces

Let H and $\mathfrak{H}$ be separable Hilbert spaces with inner products $(\cdot,\cdot)$ and $\langle\cdot,\cdot\rangle$, respectively; and let $\tilde{H}=L_2(\Omega,H)$ and $\overline{\mathfrak{H}}=L_2(\Omega,\mathfrak{H})$ be the Hilbert spaces of second-order random variables with values in H and $\mathfrak{H}$, respectively. We denote by $[\cdot]_2$ the norm of an element in either $\tilde{H}$ or $\overline{\mathfrak{H}}$. Consider the measurable space $(\Theta,\mathfrak{T})$, where $\Theta=[0,T)$ (T might be $+\infty$) and $\mathfrak{T}$ is the σ-algebra of Borel sets of Θ; and let ν be a finite measure on $\mathfrak{T}$ which is absolutely continuous with respect to Lebesgue measure.

Let $x(t,\omega)$: $\Theta\to\overline{\mathfrak{H}}$; that is $x(t,\omega)$ is an $\overline{\mathfrak{H}}$-valued random function. We define

$$|||x||| = \operatorname*{ess\,sup}_{t\in\Theta}[x]_2 .$$

Now, let $L_\infty(\overline{\mathfrak{H}})$ be the Banach space of random functions $x(t,\omega)$ such that $|||x|||<\infty$; and let $C(\overline{\mathfrak{H}})\subset L_\infty(\overline{\mathfrak{H}})$ be the space of continuous random functions.

In this section we consider the existence and uniqueness of a solution of the Itô random integral equation in $\bar{\mathfrak{H}}$

$$x(t, \omega) = x_0(\omega) + \int_0^t a(\tau, x(\tau, \omega))\, d\nu(\tau) + \int_0^t b(\tau, x(\tau, \omega))\, dW(\tau, \omega), \quad (7.62)$$

which is the integral equation formulation of the Itô random differential equation

$$dx(t, \omega) = a(t, x(t, \omega))\, d\nu(t) + b(t, x(t, \omega))\, dW(t, \omega) \quad (7.63)$$

with initial condition $x(0, \omega) = x_0(\omega)$. In Eqs. (7.62) and (7.63), $W(t, \omega)$ is a Wiener operator, which was defined, and its properties discussed, in Sect. 2.5. Hence, $W(t, \omega)$ is a mapping of Θ into $L_2(\Omega, \mathfrak{L}(H, \mathfrak{H}))$—the space of second-order random operators on H to $\mathfrak{H}$. In order that Eq. (7.62) be meaningful, we must first define the stochastic integral of an $\bar{\mathfrak{H}}$-valued random function with respect to a Wiener operator.

A random function $x(t, \omega) \in \bar{\mathfrak{H}}$ will be said to have *property* (P) with respect to a Wiener operator W if when given τ, t_1 and t_2 with $\tau \leqslant t_1 \leqslant t_2$, $x(\tau, \omega)$ and $W(t_2, \omega) - W(t_1, \omega)$ are independent. Now, let I and I' denote subintervals of Θ. We will write $I' \leqslant I$ to indicate that no point in I' is greater than any point in I. Let $\bar{\mathfrak{H}}_W^0$ be the set of finite sums of terms of the form $W(I, \omega)x$ such that (a) $x \in \tilde{H}$ and (b) $\chi_{I'}x$ has property (P) with respect to W for each $I' \leqslant I$.

Now, let $\mathfrak{X}$ be the Hilbert space of equivalence classes of random functions $x(t, \omega): \Theta \to \tilde{H}$ such that

$$\int_\Theta [x]_2^2\, d\nu(t) < \infty$$

with norm

$$\|x\|_{\mathfrak{X}} = \left(\int_\Theta [x]_2^2\, d\nu(t)\right)^{1/2}$$

Given $x \in \mathfrak{X}$, let the mapping $\varphi_0(x): \bar{\mathfrak{H}}_W^0 \to C$ be defined by

$$(\varphi_0(x))(\xi) = \sum_{i=1}^n \int_{I_i} (f_i, x)_{\tilde{H}}\, d\nu(t),$$

where

$$\xi = \sum_{i=1}^n W(I_i, \omega) f_i,$$

and the terms in the above sum have properties (a) and (b) given above. Finally, let $\psi(x)$ be the unique element in $\bar{\mathfrak{H}}_W$ (the closure of $\bar{\mathfrak{H}}_W^0$) such that

$$\langle \xi, \psi(x) \rangle_{\bar{\mathfrak{H}}} = (\varphi_0(x))(\xi) \quad (7.64)$$

for all $\xi \in \bar{\mathfrak{H}}_W^0$. The mapping $\psi: x \to \psi(x)$ has domain $\mathfrak{X}$ and range a subset of $\bar{\mathfrak{H}}_W$; and it can be shown that ψ is linear and bounded.

The *Cabaña stochastic integral* can now be defined as follows†: Let $x(t, \omega)$ be a random function in $\mathfrak{X}$, and let $W(t,\omega)$ be a Wiener operator. Then the stochastic integral of x with respect to W is defined as

$$\psi(x) = \int_{\Theta} x(\tau, \omega)\, dW(\tau, \omega), \tag{7.65}$$

where ψ is defined by (7.64). For a detailed discussion of the above integral and its properties we refer to Cabaña [12].

We now state and prove the following existence and uniqueness theorem.

THEOREM 7.3. *Let the functions $a(t,x)\colon \Theta \times \tilde{\mathfrak{H}} \to \mathfrak{H}$ and $b(t,x)\colon \Theta \times \tilde{\mathfrak{H}} \to \tilde{H}$ satisfy the following conditions:*

a. *for any measurable function $x\colon \Theta \to \tilde{\mathfrak{H}}$, $a(t,x)$ and $b(t,x)$ are measurable;*
b. *there exists a constant M such that for almost all $t \in \Theta$ and every $\xi \in \tilde{\mathfrak{H}}$*

$$[a(t, \xi)]_2 \leqslant M^2(1 + ([\xi]_2)^2)$$
$$[b(t, x)]_2 \leqslant M^2(1 + ([\xi]_2)^2);$$

c. *there exists a constant L such that for almost all $t \in \Theta$ and $\xi_1, \xi_2 \in \tilde{\mathfrak{H}}$*

$$[a(t, \xi_2) - a(t, \xi_1)]_2 \leqslant L[\xi_2 - \xi_1]_2$$
$$[b(t, \xi_2) - b(t, \xi_1)]_2 \leqslant L[\xi_2 - \xi_1]_2.$$

Then, there exists a unique continuous random function $x(t,\omega)\colon \Theta \to \tilde{\mathfrak{H}}$ which satisfies Eq. (7.62) for all $t \in \Theta$, with $x_0(\omega) = x_0$.

Proof. We define an operator T on $C(\tilde{\mathfrak{H}})$ as follows:

$$T[x] = x_0 + \int_0^t a(\tau, x(\tau, \omega))\, d\nu(\tau) + \int_0^t b(\tau, x(\tau, \omega))\, dW(\tau, \omega). \tag{7.66}$$

We will show that under the conditions of the theorem that $T[C] \subset C$ and some power of T is a contraction. Hence the theorem will follow from the Banach contraction mapping theorem.

Condition (a) implies that

$$\begin{aligned} |||a(t, x)|||^2 &= \operatorname*{ess\,sup}_{t\in\Theta} ([a(t, x)]_2)^2 \\ &\leqslant M^2\Big(1 + \sup_{t\in\Theta} ([a(t, x)]_2)^2\Big) \\ &= M^2(1 + |||x|||^2), \end{aligned} \tag{7.67}$$

and, similarly,

$$|||b(t, x)|||^2 \leqslant M^2(1 + |||x|||^2), \tag{7.68}$$

† See also Cabaña [13].

where the norm on the left-hand side of (7.68) is taken in $L_\infty(\tilde{H})$.

Now, $x(t, \omega)$ is continuous on the compact set Θ; hence $|||x||| < \infty$. Therefore (7.67) and (7.68) imply that $a(t, x) \in L_\infty(\mathfrak{H})$ and $b(t, x) \in L_\infty(\tilde{H})$.

To establish the continuity of T, we note that

$$\left(\left[\int_{t_1}^{t_2} a(\tau, x(\tau, \omega))\, d\nu(\tau)\right]_2\right)^2 \leq \nu((t_1, t_2)) \int_{t_1}^{t_2} ([a(\tau, x)]_2)^2\, d\nu(\tau)$$
$$\leq \nu^2((t_1, t_2))|||a(t, x)|||^2$$
$$\leq \nu^2((t_1, t_2))\, M^2(1 + |||x|||^2);$$

and, using the fact that the norm of the Cabaña integral is less than or equal to one, we have

$$\left(\left[\int_{t_1}^{t_2} b(\tau, x(\tau, \omega))\, dW(\tau, \omega)\right]_2\right)^2 \leq \int_{t_1}^{t_2} ([b(\tau, x)]_2)^2\, d\nu(\tau)$$
$$\leq \nu((t_1, t_2))|||b(t, x)|||^2$$
$$\leq \nu((t_1, t_2))\, M^2(1 + |||x|||^2).$$

The continuity of T follows from the above calculations and the assumption that ν is absolutely continuous with respect to Lebesgue measure.

We now show that T^n is a contraction operator for some n. Let $x_1, x_2 \in C(\mathfrak{H})$; and put

$$\Delta_k(t) = T^k[x_2] - T^k[x_1]$$
$$\Delta a_k(t) = a(t, T^k[x_2]) - a(t, T^k[x_1])$$
$$\Delta b_k(t) = b(t, T^k[x_2]) - b(t, T^k[x_1]).$$

Condition (b) implies that

$$[\Delta a_k(t)]_2 \leq L[\Delta_k(t)]_2 \qquad \text{and} \qquad [\Delta b_k(t)]_2 \leq L[\Delta_k(t)]_2$$

for almost all t. In the above, the norms $[\Delta a_k(t)]_2$ and $[\Delta_k(t)]_2$ are taken in $\mathfrak{H}$ and the norm $[\Delta b_k(t)]_2$ is taken in $\tilde{H}$. Therefore,

$$([\Delta_k(t)]_2) = \left(\left[\int_0^t \Delta a_{k-1}(\tau)\, d\nu(\tau) + \int_0^t \Delta b_{k-1}(\tau)\, dW(\tau, \omega)\right]_2\right)^2$$
$$\leq 2\left(\left[\int_0^t \Delta a_{k-1}(\tau)\, d\nu(\tau)\right]_2\right)^2 + 2\left(\left[\int_0^t \Delta b_{k-1}(\tau)\, dW(\tau, \omega)\right]_2\right)^2$$
$$\leq 2\left\{\nu(\Theta) L^2 \int_0^t ([\Delta_{k-1}(\tau)]_2)^2\, d\nu(\tau) + L^2 \int_0^t ([\Delta_{k-1}(\tau)]_2)^2\, d\nu(\tau)\right\}$$
$$= 2(\nu(\Theta) + 1) L^2 \int_0^t ([\Delta_k(\tau)]_2)^2\, d\nu(\tau). \tag{7.69}$$

If we now put $k = 1, 2, \ldots$ in (7.69), and put $K = 2(\nu(\Theta) + 1)L^2$, we obtain

$$([\Delta_1(t)]_2)^2 \leqslant K \||\Delta_0(t)\||^2 \int_0^t d\nu(\tau_1)$$

$$([\Delta_2(t)]_2)^2 \leqslant K^2 \||\Delta_0(t)\||^2 \int_0^t d\nu(\tau_1) \int_0^{\tau_1} d\nu(\tau_2)$$

$$\vdots$$

$$([\Delta_k(t)]_2)^2 \leqslant K^k \||\Delta_0(t)\||^2 \int_0^t d\nu(\tau_1) \int_0^{\tau_1} d\nu(\tau_2) \cdots \int_0^{\tau_{k-1}} d\nu(\tau_k).$$

The iterated integral on the right-hand side of the above inequality equals

$$(k!)^{-1}\left[\int_0^t d\nu(\tau)\right] = (k!)^{-1}\, \nu^k((0,t)),$$

since ν is absolutely continuous with respect to Lebesgue measure. Therefore

$$\||\Delta_k(t)\|| \leqslant (k!)^{-1}\,[K\nu(\Theta)]^k\, \||\Delta_0(t)\||^2.$$

Now, given α, with $\alpha \in (0,1)$, we can pick n so large that $(n!)^{-1}[K\nu(\Theta)]^n \leqslant \alpha^2$. Hence

$$\||T^n[x_2] - T^n[x_1]\|| = \||\Delta_n\|| \leqslant \alpha \||\Delta_0\|| = \alpha \||x_2 - x_1\||.$$

Therefore, T^n is a contraction; and the Banach contraction mapping theorem implies the existence of a unique solution $x(t,\omega)$ of Eq. (7.62) in the space $C(\mathfrak{H})$.

C. Itô random integral equations in scales of Hilbert spaces

Let H be a Hilbert space with inner product $(\cdot,\cdot)$, and let $\mathfrak{S}_2$ denote the class of Hilbert–Schmidt operators on H. Let T be an unbounded, self-adjoint, positive-definite operator on H satisfying the condition $\|T^{-1}\| \leqslant 1$. The domain H_α of T^α $(\alpha > 0)$ is dense in H, and is a complete Hilbert space under the norm $\|x\|_\alpha = \|T^\alpha x\|$, $x \in H_\alpha$. It is also possible to introduce in H another norm, namely $\|x\|_{-\alpha} = \|T^{-\alpha} x\|$, $\alpha > 0$; and then consider the space $H_{-\alpha}$ obtained from H by completion with respect to the norm $\|\cdot\|_{-\alpha}$. The operator $T^{-\alpha}$ is bounded with respect to the norm $\|\cdot\|_{-\alpha}$; hence, after closure, it can be defined on the whole of $H_{-\alpha}$. The domain of $\tilde{T}^{-\alpha}$ (the closure of $T^{-\alpha}$) is H, and the inverse operator $\tilde{T}^\alpha$ (which is the closure in $H_{-\alpha}$ of T^α) maps H to $H_{-\alpha}$. The spaces H_α and $H_{-\alpha}$ are conjugate to one another in the sense of inner products.

The value of the functional $y \in H_{-\alpha}$ at $x \in H_\alpha$ is given by the formula $y(x) = (T^\alpha x, \tilde{T}^{-\alpha} y)$. Since $\tilde{T}^{-\alpha} y = T^{-\alpha} y \in H_\alpha$, $y \in H$, we have

$$y(x) = (T^\alpha x, \tilde{T}^{-\alpha} y) = (x, T^\alpha T^{-\alpha} y) = (x, y). \tag{7.70}$$

We are now able to define a scale of Hilbert spaces. A one-parameter family of Hilbert spaces $\{H_\alpha, \alpha \in (-\infty, \infty)\}$, with $H_0 = H$, is said to be a *scale of Hilbert spaces* if (i) $H_\alpha \subset H_\beta$, $-\infty < \beta < \alpha < \infty$, and $\|x_\beta\| \leqslant \|x_\alpha\|$, $x \in H_\alpha$; and (ii) $H_{-\alpha} = H_\alpha^*$. Scales of Hilbert spaces occupy a central position among scales of Banach spaces, and are of importance in the theory of partial differential equations (cf. Krein and Petunin [49]). The study of Itô random integral equations in scales of Hilbert spaces leads to more general results than those obtained by other workers who have restricted their attention to Itô equations in a single Hilbert space, rather than consider a system of Hilbert spaces.

In this subsection we consider the Itô random integral equation

$$x(t, \omega) = x_0(\omega) + \int_{t_0}^{t} a(\tau, x(\tau, \omega))\, d\tau + \int_{t_0}^{t} B(\tau, x(\tau, \omega))\, dw(\tau, \omega), \quad (7.71)$$

which is the integral equation formulation of the Itô random differential equation

$$dx(t, \omega) = a(t, x(t, \omega))\, dt + B(t, x(t, \omega))\, dw(t, \omega) \quad (7.72)$$

with initial condition $x(t_0, \omega) = x_0(\omega)$. In Eq. (7.71) $a(t,x) \in H_\alpha$, $B(t,x)$ is an operator-valued function, and $w(t, \omega)$ is a Wiener process with values in H_{-1}.

Let K_α denote the class of operator-valued functions satisfying the following conditions:

1. $B(t, \omega) \in \mathfrak{L}(H_{-1}, H_\alpha)$;
2. $B(t, \omega)$ is $\mathfrak{F}_t$-measurable;
3. $\int_{t_0}^{T} \mathscr{E}\{\|B(\tau, \omega)\|^2_{(H_{-1}, H_\alpha)}\}\, d\tau = \int_{t_0}^{T} \mathscr{E}\{\|T^\alpha B(\tau, \omega)\, T\|^2\}\, d\tau < \infty$.

For an operator-valued function $B(t, \omega) \in K_\alpha$, the integral of $B(t, \omega)$ with respect to an H_{-1}-valued Wiener process, that is,

$$I(B) = \int_{t_0}^{T} B(\tau, \omega)\, dw(\tau, \omega), \quad (7.73)$$

has been defined by Daletskiĭ as follows:

$$(I(B), y) = I(B^*(t, \omega)\, y), \quad (7.74)$$

for any $y \in H_{-\alpha}$. We remark that this definition is correct, since $B^* y \in H$ if $B \in K_\alpha$. The *Daletskiĭ stochastic integral* (7.73) has the following properties:

i. $\mathscr{E}\{I(B(t, \omega))\} = 0$;

ii. $\mathscr{E}\{\|I(B(t, \omega))\|_\alpha^2\} \leqslant \int_{t_0}^{T} \mathscr{E}\{\sigma^2(T^\alpha B(\tau, \omega))\}\, d\tau$.

Having defined the Daletskiĭ stochastic integral, Eq. (7.71) is well defined; and we can now consider the problem of the existence and uniqueness of the solution of Eq. (7.71). We first introduce some spaces of functions which will be referred to in the statement of the existence and uniqueness theorem.

Let $\mathfrak{X}_\alpha^p$ denote the space of H_α-valued functions which are $\mathfrak{F}_t$-measurable. Under the norm

$$|||x|||^p = \sup_{t \in [t_0, T]} \mathscr{E}\{\|x(t,\omega)\|_\alpha^p\},$$

$\mathfrak{X}_\alpha^p$ is a Banach space. Let $C_\alpha(H_\alpha)$ denote the space of continuous functions defined on H_α with range H_α; and let $C_{0,\alpha}(H_\alpha)$ denote the space of continuous operator-valued functions defined on H_α with range $\mathfrak{L}(H, H_\alpha)$. We also need the following definition: A Banach space-valued function $\xi(t,x)$, $t \in [t_0, T]$, $x \in H_\alpha$ is said to have *property* (L) if (i) $\|\xi(t,x)\| \leqslant C_1 + C_2\|x\|_\alpha$, and (ii) $\|\xi(t,x_1) - \xi(t,x_2)\| \leqslant C_2\|x_1 - x_2\|$ $(x_1, x_2 \in H_\alpha)$, where C_1 and C_2 are constants.

THEOREM 7.4. *Consider the operator*

$$S[x(t,\omega)] = \varphi(t) + \int_{t_0}^t a(\tau, x(\tau,\omega))\,d\tau + \int_{t_0}^t B(\tau, x(\tau,\omega))\,dw(\tau,\omega). \quad (7.75)$$

(a) *Let* $\varphi(t) \in \mathfrak{X}_\alpha^{2m}$, *and let the functions* $a(t,x)$ *and* $T^\alpha B(t,x)$ *have property (L) in the spaces* H_α *and* $\mathfrak{S}_2(H)$, *respectively. Then the operator S defined by* (7.75) *is a continuous operator on* $\mathfrak{X}_\alpha^{2m}$; *and for some n,* S^n *is a contraction operator on* $\mathfrak{X}_\alpha^2$.

(b) *Let* $\varphi(t) \in \mathfrak{X}_\alpha^2$, *and let the functions* $a(t,x) \in C_\alpha(H_\alpha)$ *and* $B(t,x) \in C_{0,\alpha}(H_\alpha)$ *have property (L). Then the Itô random integral equation*

$$x(t,\omega) = \varphi(t) + \int_{t_0}^t a(\tau, x(\tau,\omega))\,d\tau + \int_{t_0}^t B(\tau, x(\tau,\omega))\,dw(\tau,\omega), \quad (7.76)$$

where $\varphi(t) = x_0$ *is a continuous function independent of* $w(t,\omega)$, *has a unique continuous solution which is* $\mathfrak{F}_t$-*measurable*† *for every* $t \in [t_0, T]$.

Proof. (a) We first remark that the function $S[x(t,\omega)]$ is $\mathfrak{F}_t$-measurable. To show that $S[\mathfrak{X}_\alpha^p] \subset \mathfrak{X}_\alpha^p$, we consider the following estimate:

$$\begin{aligned} \mathscr{E}\{\|Sx(t,\omega) - \varphi(t)\|_\alpha^{2m}\} &\leqslant C_3 \int_{t_0}^t \mathscr{E}\{\|a(\tau, x(\tau,\omega))\|_\alpha^{2m}\,d\tau \\ &\quad + C_4 \int_{t_0}^t \mathscr{E}\{\sigma_2^2(T^\alpha B(\tau, x(\tau,\omega)))\}\,d\tau \\ &\leqslant 2^{m-1}(C_3 + C_4)\Big[C_1^{2m}(T - t_0) + C_2^{2m} \int_{t_0}^t \mathscr{E}\{\|x(\tau,\omega)\|_\alpha^{2m}\}\,d\tau\Big], \end{aligned}$$

where we have put $C_3 = (2(T - t_0))^{2m-1}$ and $C_4 = (2(T - t_0))^{m-1}(m(2m-1))^m$;

† The σ-algebra $\mathfrak{F}_t$ is taken to be generated by the random variables $w(\tau,\omega)$, $\tau \in [t_0, t]$ and $\varphi(\tau,\omega)$, $\tau \in [t_0, T]$.

and $\sigma_2(A)$ denotes the Hilbert–Schmidt norm of $A \in \mathfrak{S}_2$. Hence it follows that if $x \in \mathfrak{X}_\alpha^{2m}$, then $Sx \in \mathfrak{X}_\alpha^{2m}$.

To establish continuity, we consider the following estimate, which is analogous to the estimate obtained above. We have, using property (L),

$$
\begin{aligned}
|||Sx_2 - Sx_1|||^{2m} &= \sup_{t\in[t_0,T]} \mathscr{E}\{\|Sx_2(t,\omega) - Sx_1(t,\omega)\|_\alpha^{2m}\} \\
&\leqslant \sup_{t\in[t_0,T]} \Big[C_3 \int_{t_0}^{t} \mathscr{E}\{\|a(\tau, x_2(\tau,\omega)) - a(\tau, x_1(\tau,\omega))\|_\alpha^{2m}\}\, d\tau \\
&\quad + C_4 \int_{t_0}^{t} \mathscr{E}\{\sigma_2^{2m}(T^\alpha B(\tau, x_2(\tau,\omega)) - T^\alpha B(\tau, x_1(\tau,\omega)))\}\, d\tau\Big] \\
&\leqslant \sup_{t\in[t_0,T]} \Big[C_2^{2m}(C_3 + C_4) \int_{t_0}^{t} \mathscr{E}\{\|x_2(\tau,\omega) - x_1(\tau,\omega)\|_\alpha^{2m}\}\, d\tau\Big] \\
&\leqslant C_2^{2m}(C_3 + C_4) |||x_2(t,\omega) - x_1(t,\omega)|||^{2m}(T - t_0).
\end{aligned}
$$

The above estimate implies the continuity of S.

To prove that S^n is a contraction operator on $\mathfrak{X}_\alpha^2$ for some n, we must show that

$$|||S^n x_2 - S^n x_1||| \leqslant K|||x_2 - x_1|||, \qquad K < 1.$$

Using the definition of the norm in $\mathfrak{X}_\alpha^2$, taking the expectation of the last estimate obtained above, we obtain

$$
\begin{aligned}
&\mathscr{E}\{\|S^2 x_2(t,\omega) - S^2 x_1(t,\omega)\|_\alpha^2\} \\
&\quad\leqslant 2(T - t_0 + 1)\, C_2^2 \int_{t_0}^{t} \mathscr{E}\{\|Sx_2(\tau,\omega) - Sx_1(\tau,\omega)\|^2\}\, d\tau \\
&\quad\leqslant (2(T - t_0 + 1)\, C_2^2)^2 \int_{t_0}^{t} (\tau - t_0)\, d\tau |||x_2 - x_1|||^2 \\
&\quad= \tfrac{1}{2}(2(T - t_0 + 1)\, C_2^2)^2 (t - t_0)^2 |||x_2 - x_1|||^2.
\end{aligned}
$$

An induction argument yields

$$
\begin{aligned}
&\mathscr{E}\{\|S^n x_2(t,\omega) - S^n x_1(t,\omega)\|_\alpha^2\} \\
&\quad\leqslant 2(T - t_0 + 1)\, C_2^2 \int_{t_0}^{t} \mathscr{E}\{\|S^{n-1} x_2(\tau,\omega) - S^{n-1} x_1(\tau,\omega)\|\}^2\, d\tau \\
&\quad= \frac{(2(T - t_0 + 1)\, C_2^2)^n}{n!} (t - t_0)^n \|x_2(t,\omega) - x_1(t,\omega)\|.
\end{aligned}
$$

It is clear that we can pick an n large enough such that

$$(n!)^{-1}(2(T - t_0)(T - t_0 + 1)\, C_2^2)^n < 1;$$

hence S^n is a contraction for some $n \geqslant 1$.

(b) Since S^n is a contraction, the Banach contraction mapping theorem asserts the existence of a unique element $x \in \mathfrak{X}_\alpha^2$ such that $S^n x = x$. But $S^n S x = S^{n+1} x$; hence $Sx = x$. However, the last equation implies that $S^n x = x$; hence we can conclude that $x = x(t, \omega) \in \mathfrak{X}_\alpha^2$ is the unique solution of Eq. (7.76). Since, by hypothesis, $a(t,x) \in C_\alpha(H_\alpha)$, $B(t,x) \in C_{0,\alpha}(H_\alpha)$, and $\varphi(t)$ is a continuous, the continuity of $x(t,\omega)$ follows from the continuity of the right-hand side of Eq. (7.76).

7.4 Some Additional Studies on Itô Equations and Their Applications

A. Introduction

In this section we present a brief survey of some studies on Itô random equations and their applications. We will restrict our attention to (1) some concrete Itô random integral equations, and (2) recent studies which extend or generalize the basic theory of Itô random integral equations.

B. Some concrete Itô random integral equations

In this subsection we consider three Itô random integral equations which have been considered in connection with certain applied problems.

1. Åström [1] has considered the Itô random differential equation

$$dx(t,\omega) = x(t,\omega)\, dw_1(t,\omega) + dw_2(t,\omega), \tag{7.77}$$

where $w_1(t,\omega)$ and $w_2(t,\omega)$ are Wiener processes with

$$\begin{aligned} &\mathscr{E}\{\Delta w_1\} = -m_1 h, \qquad \mathrm{Var}\{\Delta w_1\} = 2a_1 h \\ &\mathscr{E}\{\Delta w_2\} = m_2 h, \qquad \mathrm{Var}\{\Delta w_2\} = 2a_2 h \\ &\mathrm{Cov}\{\Delta w_1, \Delta w_2\} = 2a_{12} h, \end{aligned}$$

where $\Delta w_i = w_i(t+h) - w_i(t)$, $i = 1,2$; and $m_1 > 0$, $m_2 \geqslant 0$, $a_1 \geqslant 0$, $a_2 \geqslant 0$, and $a_{12}^2 \leqslant a_1 a_2$. Equation (7.77) can be rewritten as

$$dx(t,\omega) = [-x(t,\omega) m_1 + m_2]\, dt + x(t,\omega)\, d\tilde{w}_1(t,\omega) + d\tilde{w}_2(t,\omega), \tag{7.78}$$

where $\tilde{w}_1(t,\omega)$ and $\tilde{w}_2(t,\omega)$ are Wiener processes whose increments have expectations zero and $\mathrm{Cov}(\Delta\tilde{w}_1, \Delta\tilde{w}_2) = \mathrm{Cov}(\Delta w_1, \Delta w_2)$. The Itô random integral equation equivalent to (7.78) is

$$\begin{aligned} x(t,\omega) = {}& \exp\{-m_1(t-t_0)\}\, x(t_0,\omega) + (m_2/m_1)(1 - \exp\{-m_1(t-t_0)\}) \\ &+ \int_{t_0}^t \exp\{-m_1(t-\tau)\}\, x(\tau,\omega)\, d\tilde{w}_1(\tau,\omega) \\ &+ \int_{t_0}^t \exp\{-m_1(t-\tau)\}\, d\tilde{w}_2(\tau,\omega), \qquad t \in [t_0, T]. \end{aligned} \tag{7.79}$$

Consider the space $L_2(\Omega)$, and let $x: [t_0, T] \to L_2(\Omega)$. The norm of f is given by

$$\|x(t, \omega)\| = \max_{t\in[t_0,T]} (\mathscr{E}\{x^2(t, \omega)\})^{1/2}.$$

As an operator equation in $L_2(\Omega)$, Eq. (7.79) is of the form

$$x(t, \omega) = Lx(t, \omega) + y(t, \omega), \tag{7.80}$$

where

$$y(t, \omega) = \exp\{-m_1(t - t_0)\} x(t_0, \omega) + (m_2/m_1)(1 - \exp\{-m_1(t - t_0)\}) + \int_{t_0}^{t} \exp\{-m_1(t - \tau)\} d\tilde{w}_2(\tau, \omega);$$

and the operator L on $L_2(\Omega)$ is defined as

$$L(\omega)[x] = \int_{t_0}^{t} \exp\{-m_1(t - \tau)\} x(\tau, \omega) d\tilde{w}_1(\tau, \omega).$$

A routine calculation yields

$$\|Lx\|^2 \leqslant \|x\|^2 (a_1/m_1)(1 - \exp\{-2m_1(t - t_0)\}).$$

For $x(t, \omega) = 1$, $\|L\| = [(a_1/m_1)(1 - \exp\{-2m_1(t - t_0)\})]^{1/2}$ almost surely; hence by choosing h sufficiently small we can satisfy the condition $\|L\| < 1$. If the above condition is satisfied, the solution of Eq. (7.80), equivalently Eq. (7.79), is given by the Neumann series

$$x(t, \omega) = y(t, \omega) + \sum_{n=1}^{\infty} L^n y(t, \omega). \tag{7.81}$$

2. Beutler [8] has studied the vector random integral equation

$$x(t, \omega) = \int_0^t A(\tau) x(\tau, \omega) d\tau + \int_0^t B(\tau) dy(\tau, \omega), \qquad t \in [0, \infty), \tag{7.82}$$

where $A(t)$ and $B(t)$ are $n \times n$ matrix-valued functions and $y(t, \omega)$ is a second-order homogeneous (temporally and spatially) vector process with orthogonal increments, and has shown that under certain conditions the solution process is a multivariate wide-sense Markov process.

Equation (7.82) is studied in a Hilbert space H, which can be obtained as follows: Let $x(t, \omega)$ be a vector random function with components $x_1(t, \omega, \ldots, x_n(t, \omega)$, each of which is a complex-valued second-order random function. $x(t, \omega)$ can be regarded as a matrix with elements $x_{1j}(t, \omega) = x_j(t, \omega)$, and $x_{ij}(t, \omega) = 0$, $i = 2, \ldots, n$, $j = 1, \ldots, n$. Let $x^*(t, \omega)$ denote the complex conjugate transpose of $x(t, \omega)$. Consider the pre-Hilbert space of all elements of the form $A(t)x(t, \omega)$ with inner product

$$(A(s) x(s, \omega), B(t) x(t, \omega)) = \mathrm{Tr}[A(s)(\mathscr{E}\{x(s, \omega) x^*(t, \omega)\} B^*(t)].$$

Completion of the above pre-Hilbert space yields the Hilbert space H.

We now assume the following:

i. $y(t,\omega)$ with $y(0,\omega)=0$, is a vector random function, each of whose components is a process with orthogonal increments;

ii. $\int_0^t \|B(\tau)\|^2 \, d\tau < \infty$ (hence the second integral on the right-hand side of Eq. (7.82) is a stochastic integral);

iii. the elements $a_{ij}(t)$ of $A(t)$ are measurable and $\|A(t)\|$ is locally integrable (hence the first integral on the right-hand side of Eq. (7.82) can be defined as a Bochner integral).

Under the above assumptions, Beutler has shown that the unique solution of Eq. (7.82) is of the form

$$x(t,\omega) = \int_0^t K(t,\tau)\, B(\tau)\, dy(\tau,\omega), \tag{7.83}$$

where $K(t,\tau) = \xi(t)\xi^{-1}(\tau)$, and $\xi(t)$ is the solution of the matrix differential equation $\xi'(t) = A(t)\xi$, $\xi(0) = I$. If H is the Hilbert space generated by the process $y(t,\omega)$, then $x(t,\omega) \in H$ for every $t \in [0,\infty)$.

3. Nagai [62] has studied the random integral equation

$$x(t,\omega) = x_0(\omega) + \alpha \int_0^t x(\tau,\omega)\, d\tau + \int_0^t \varphi(\tau,\omega)\, d\tau$$
$$+ \beta \int_0^t x(\tau,\omega)\, dw(\tau,\omega), \qquad t \in [0,T], \tag{7.84}$$

under the following assumptions:

i. $w(t,\omega)$ is a real-valued Wiener process;

ii. $x_0(\omega)$ has finite first and second moments;

iii. (a) $\varphi(t,\omega)$ is a measurable function of the pair (t,ω); (b) $\mathscr{E}\{|\varphi(t,\omega)|\} < \infty$ and $\mathscr{E}\{|\varphi(t,\omega)|^2\} < \infty$ for all $t \in [0,T]$, and (c) $\varphi(t,\omega)$ is continuous in mean-square.

iv. for any fixed t,τ, $(t \geqslant 0, \tau \leqslant T)$, the vector random variable $(x_0(\omega), \varphi(t,\omega))$ is independent of $w(\tau,\omega)$;

v. α and β are real constants.

Under the above assumptions, the unique solution of Eq. (7.84) admits the representation

$$x(t,\omega) = e^{-\alpha t} \sum_{k=0}^{\infty} \beta^k \Phi_k(t,\omega), \tag{7.85}$$

where

$$\Phi_0(t,\omega) = x_0(\omega) + \int_0^t e^{\alpha\tau} \varphi(\tau,\omega)\, d\tau$$

$$\Phi_{n+1}(t,\omega) = \int_0^t \Phi_n(\tau,\omega)\, dw(\tau,\omega), \qquad n = 0, 1, \ldots.$$

Nagai also studied the asymptotic behavior of the covariance function of the solution process $\{x(t,\omega),\ t \in [0,T]\}$ and the stability of the solution process.

C. The Bogoliubov-Mitropolskiĭ method of averaging for Itô random integral equations

Vrkoč [72] has applied the Bogoliubov–Mitropolskiĭ method of averaging† to Itô random integral equations. In particular, the integral equation

$$x(t,\omega) = x_0(\omega) + \epsilon \int_0^t a(\tau, x(\tau,\omega))d\tau + \epsilon^{1/2} \int_0^t b(\tau, x(\tau,\omega))\, d(x(\epsilon\tau,\omega)/\epsilon^{1/2}) \tag{7.86}$$

is studied, where $\epsilon > 0$, $|a(t,x_2) - a(t,x_1)| + |b(t,x_2) - b(t,x_1)| \leqslant K|x_2 - x_1|$ (independently of t), and $w(t,\omega)$ is a real Wiener process. Under the assumption that (i) there exists a function $\bar{a}(x)$ such that

$$\bar{a}(x) = \lim_{\xi\to\infty} (1/\xi) \int_0^\xi a(\tau, x)\, d\tau$$

uniformly in x; and (ii) there exists a function $\bar{b}(x)$ such that

$$\lim_{\xi\to\infty} \int_0^\xi |b(\tau,x) - \bar{b}(x)|^2\, d\tau = 0,$$

then, if $\bar{x}(t,\omega)$ is the solution of the "averaged" Itô random integral equation

$$\bar{x}(t,\omega) = x_0(\omega) + \int_0^t a(\bar{x}(\tau,\omega))\, d\tau + \int_0^t \bar{b}(\bar{x}(\tau,\omega))\, dw(\tau,\omega), \tag{7.87}$$

Vrkoč proved that for every $\lambda > 0$ there exists an ϵ_0 such that

$$\mathscr{E}\Big\{ \sup_{t\in[0,T/\epsilon]} |x(t,\omega) - \bar{x}(t,\omega)|^2 \Big\} < \lambda$$

for $\epsilon \in (0, \epsilon_0]$.

For applications of this method in the stability theory of Itô random integral equations we refer to Vrkoč [72, 73].

D. Second-order Itô processes and the associated random integral equations

The Itô random differential equation (7.2) is a first-order differential equation; and the solution process of either Eq. (7.2) or (7.1) can be referred to as a first-order Itô process. Although first-order Itô processes are of fundamental importance in many theoretical and applied areas, the systematic study of Itô processes of higher order is required in order to have a rigorous theory of differential equations of order n driven by Wiener processes. In

† We refer to Bogoliubov and Mitropolskiĭ [9].

particular, the simple harmonic oscillator driven by a Wiener process leads to the following random differential equation of second order:

$$dx'(t,\omega) + 2\alpha x'(t,\omega) + \beta x(t,\omega) = dw(t,\omega), \tag{7.88}$$

where $x'(t,\omega)$ is the derivative of the $x(t,\omega)$-process which describes the position of the particle. An equation of the above type leads to an Itô random differential equation of second order of the form

$$dx'(t,\omega) = a(t,x,x')\,dt + b(t,x,x')\,dw(t,\omega). \tag{7.89}$$

The study of second-order Itô equations was initiated by Borchers [10]; and Goldstein [34] has studied Eq. (7.89) and the equivalent Itô random integral equation

$$\begin{aligned} x(t,\omega) &= x(0,\omega) + \int_0^t x'(\tau,\omega)\,d\tau \\ x'(t,\omega) &= x'(0,\omega) + \int_0^t a(\tau, x(\tau,\omega), x'(\tau,\omega))\,d\tau \\ &\quad + \int_0^t b(\tau, x(\tau,\omega), x'(\tau,\omega))\,dw(\tau,\omega) \end{aligned} \tag{7.90}$$

Equation (7.90) can be rewritten as the vector random integral equation

$$\tilde{y}(t,\omega) = \tilde{y}(0,\omega) + \int_0^t A(\tau, \tilde{y}(\tau,\omega))\,d\tau + \int_0^t B(\tau, \tilde{y}(\tau,\omega))\,d\tilde{w}(\tau,\omega), \tag{7.91}$$

where

$$\tilde{y}(t,\omega) = \begin{pmatrix} y(t,\omega) \\ y'(t,\omega) \end{pmatrix}, \qquad A(t,\xi_1,\xi_2) = \begin{pmatrix} \xi_2 \\ a(t,\xi_1,\xi_2) \end{pmatrix},$$

$$B(t,\xi_1,\xi_2) = \begin{pmatrix} 0 & 0 \\ 0 & b(t,\xi_1,\xi_2) \end{pmatrix}, \qquad \tilde{w}(t,\omega) = \begin{pmatrix} w_0(t,\omega) \\ w(t,\omega) \end{pmatrix}.$$

In the above $w_0(t,\omega)$ is a (dummy) Wiener process, independent of $w(t,\omega)$ and the initial condition $\tilde{y}(0,\omega)$. Equation (7.90) has been studied under the following conditions:

i. $a(t,x)$, $b(t,x)$: $[0,\infty) \times R_2 \to R$ are Baire functions.

ii. For each $T > 0$ there is a constant $K(T)$ such that for $t \in [0,T]$ and $x, y \in R_2$

$$\begin{aligned} |a(t,x)| &\leq K(T)(1 + |x|^2)^{1/2}, \\ 0 \leq b(t,x) &\leq K(T)(1 + |x|^2)^{1/2}, \\ |a(t,x) - a(t,y)| &\leq K(T)\,|x - y|, \\ |b(t,x) - b(t,y)| &\leq K(T)\,|x - y|. \end{aligned}$$

iii. For every fixed t, $y(t,\omega)$ and $y'(t,\omega)$ are square-integrable random variables which are independent of all the increments Δw of the Wiener process $w(t,\omega)$.

Borchers has shown that if the above conditions are satisfied, then (1) there exists a unique solution $\tilde{y}(t,\omega)$, $t \in [0,\infty)$ of Eq. (7.91), (2) $\tilde{y}(t,\omega)$ is a vector Markov process, and (3) with probability one, $y(t,\omega)$ and $y'(t,\omega)$ are continuous on $[0,\infty)$. Furthermore, $y'(t,\omega)$ is the mean-square derivative (or the strong derivative in $L_2(\Omega)$) of $y(t,\omega)$.

Goldstein has given a detailed analysis of the behavior of the realizations of second-order Itô processes, and has developed the semigroup theory of second-order Itô processes which is the analogue of the operator-theoretic formulation of first-order Itô processes due to Doob [20].

E. Random operational integral equations

Bensoussan [6, 7] has introduced a class of random integral equations which generalize the Itô equations. These equations are referred to as random operational integral equations, where the term "operational" is used in the sense of Lions [54]. Let H and $\mathfrak{H}$ be two separable Hilbert spaces with $\mathfrak{H} \subset H$ (which means inclusion with continuous injection), and $\mathfrak{H}$ is dense in H. Let $\mathfrak{H}'$ denote the dual of $\mathfrak{H}$, and if H is identified with its dual, then $\mathfrak{H} \subset H \subset \mathfrak{H}'$. Let $\{A(t),\, t \in [0,T]\}$ denote a family of operators $A(t)\colon \mathfrak{H} \to \mathfrak{H}'$ satisfying the following conditions: (i) $\|A(t)\|_{\mathfrak{L}(\mathfrak{H},\mathfrak{H}')} \leqslant M$, (ii) the mapping $t \to (A(t)\xi_1, \xi_2)_{\mathfrak{H}}$ is measurable for all $\xi_1, \xi_2 \in \mathfrak{H}$, and (iii) $(A(t)\xi,\xi)_{\mathfrak{H}} + \lambda|\xi|^2 \geqslant \alpha\|\xi\|_{\mathfrak{H}}^2$ for all $\xi \in \mathfrak{H}$, where $\lambda, \alpha > 0$.

Let $y(t,\omega)$ be a H-valued random function, with $y(t,\omega) \in C([0,T], L_2(\Omega,H))$. It is assumed that $y(t,\omega)$ has orthogonal increments; that is,

$$\mathscr{E}\{(y(t_1,\omega) - y(t_2,\omega), B(y(t_j,\omega) - y(t_y,\omega)))_H\} = 0$$

for all $B \in \mathfrak{L}(H)$ and $t_4 \leqslant t_3 \leqslant t_2 \leqslant t_1 \in [0,T]$. Finally, consider the random variable $x_0(\omega) \in L_2(\Omega,H)$ which satisfies the condition

$$\mathscr{E}\{(Bx_0(\omega), y(t,\omega) - y(0,\omega))_H\} = 0$$

for all $B \in \mathfrak{L}(H)$ and all $t \in [0,T]$; and let $f(t)$ be a deterministic function with $f(t) \in L_2([0,T],H)$.

The random operational integral equation considered is of the form

$$x(t,\omega) + \int_0^t A(\tau)\,x(\tau,\omega)\,d\tau = x_0(\omega) + \int_0^t f(\tau)\,d\tau + y(t,\omega) - y(0,\omega), \quad (7.92)$$

which is the integral equation formulation of the random operational differential equation

$$\frac{dx}{dt} + A(t)\,x = x_0 + f + \frac{dy}{dt}, \quad (7.93)$$

where the derivative is taken in the sense of distributions with values in $L_2(\Omega, \mathfrak{H}')$.

Bensoussan has shown that there exists a unique solution $x(t,\omega)$ of Eq. (7.92), with $x(t,\omega) \in C_0([0,T],\ L_2(\Omega,\mathfrak{H}')) \cap L_\infty([0,T],\ L_2(\Omega,H)) \cap L_2([0,T],\ L_2(\Omega,\mathfrak{H}))$, and for all $t \in [0,T]$, Eq. (7.92) is satisfied with probability one. Furthermore, the solution process $x(t,\omega) \in C([0,T],\ L_2(\Omega,H))$ is continuous in probability in H.

F. Generalized Itô integrals and Itô equations

Dawson [17] has defined a stochastic integral with integrator function a generalized random function with independent values in the sense of Gel'fand and Vilenkin [29, Chap. III]. This extension of the Itô integral permits the formulation and study of generalized random equations of Itô type, the solutions of which are R_n-valued generalized random functions. Dawson's results are of great interest in the development of the theory of random equations, and should be of importance in applied fields which utilize Itô equations as models of physical processes.

References

1. Åström, K. J., On a first-order stochastic differential equation. *Internat. J. Control* **1** (1965), 301–326.

1a. Åström, K. J., "Introduction to Stochastic Control Theory." Academic Press, New York, 1970.

2. Baklan, V. V., A representation of the solution of the characteristic problem for the telegraph equation in the form of continuous integral (Ukrainian). *Dopovidi Akad. Nauk Ukrain. RSR* (1963), pp. 149–152.

3. Baklan, V. V., On the existence of solutions of stochastic equations in Hilbert space (Ukrainian). *Dopovidi Akad. Nauk Ukrain. RSR* (1963), pp. 1299–1303.

4. Baklan, V. V., Variational differential equations and Markov processes in Hilbert space (Russian). *Dokl. Akad. Nauk SSSR* **159** (1964), 707–710.

5. Barrett, J. F., Application of Kolmogorov's equations to randomly disturbed automatic control systems. *In* "Automatic and Remote Control," Vol. 2, pp. 724–733. Butterworth, London, 1961.

6. Bensoussan, A., Équations intégrales opérationnelles stochastiques. *C. R. Acad. Sci. Paris Ser. A* **269** (1969), 423–425.

7. Bensoussan, A., Sur les propriétés de la solution d'une équation intégrale opérationnelle stochastique. *C. R. Acad. Sci. Paris Ser. A* **269** (1969), 457–459.

8. Beutler, F. J., Multivariate wide-sense Markov processes and prediction theory. *Ann. Math. Statist.* **34** (1963), 424–438.

9. Bogoliubov, N. N., and Mitropolskiĭ, I., "Asymptotic Methods in the Theory of Non-linear Oscillations," translated from the Russian. Gordon & Breach, New York, 1961.

10. Borchers, D. R., Second order stochastic differential equations and related Itô processes. Ph.D. Dissertation, Carnegie Inst. of Technol., Pittsburgh, Pennsylvania, 1964.

11. Bucy, R. S., and Joseph, P. D., "Filtering for Stochastic Processes with Applications to Guidance." Wiley (Interscience), New York, 1968.

12. Cabaña, E. M., Stochastic integration in separable Hilbert spaces. *Publ. Inst. Mat. Estadist. Montevideo* **4** (1966), 49–80.
13. Cabaña, E. M., On stochastic differentials in Hilbert spaces. *Proc. Amer. Math. Soc.* **20** (1969), 259–265.
14. Chandrasekhar, S., Stochastic problems in physics and astronomy. *Rev. Modern Phys.* **15** (1943), 1–89.
15. Chuang, K., and Kazda, L. F., A study of nonlinear systems with random inputs. *Trans. Amer. Inst. Elec. Eng.* **78** (1959), 100–105.
16. Daletskiĭ, Yu. L., Infinite-dimensional elliptic operators and parabolic equations associated with them (Russian). *Uspehi Mat. Nauk* **22**(4) (1967), 3–54.
17. Dawson, D. A., Generalized stochastic integrals and equations. *Trans. Amer. Math. Soc.* **147** (1970), 473–506.
18. Doob, J. L., The Brownian movement and stochastic equations. *Ann. of Math.* **43** (1942), 351–369.
19. Doob, J. L., "Stochastic Processes." Wiley, New York, 1953.
20. Doob, J. L., Martingales and one dimensional diffusion. *Trans. Amer. Math. Soc.* **78** (1955), 168–208.
21. Dynkin, E. B., "Markov Processes," Vols. I and II, translated from the Russian. Academic Press, New York, 1965.
22. Fleming, W. H., The Cauchy problem for degenerate parabolic equations. *J. Math. Mech.* **13** (1964), 987–1008.
23. Fleming, W. H., Optimal continuous-parameter stochastic control. *SIAM Rev.* **11** (1969), 470–509.
24. Freĭdlin, M. I., On Itô's stochastic equations and degenerating elliptic equations (Russian). *Izv. Akad. Nauk SSSR Ser. Mat.* **26** (1962), 653–676.
25. Freĭdlin, M. I., The Dirichlet problem for elliptic second order differential equations with small parameters (Russian). *Dokl. Akad. Nauk SSSR* **144** (1962), 501–504.
26. Freĭdlin, M. I., Degenerate diffusion processes and the smoothness of solutions of boundary-value problems for degenerate elliptic equations (Russian). *Teor. Verojatnost. i Primenen.* **9** (1964), 757–758.
27. Freĭdlin, M. I., On small perturbations of coefficients of a diffusion process (Russian). *Teor. Verojatnost. i Primenen.* **12** (1967), 536–540.
28. Freĭdlin, M. I., On degenerating elliptic equations (Russian). *Teor. Verojatnost. i Primenen.* **14** (1969), 138–142.
29. Gel'fand, I. M., and Vilenkin, N. Ya., "Generalized Functions, Vol. 4: Applications of Harmonic Analysis," translated from the Russian. Academic Press, New York, 1964.
30. Gihman, I. I., On the theory of differential equations of random processes (Russian). *Ukrain. Mat. Ž.* **2**(4) (1950), 37–63.
31. Gihman, I. I., and Skorohod, A. V., "Stochastic Differential Equations" in Russian. Izdat. Naukova Dumka, Kiev, 1968.
32. Girsanov, I. V., On Itô's stochastic integral equations (Russian). *Dokl. Akad. Nauk SSSR* **138** (1961), 18–21.
33. Girsanov, I. V., An example of non-uniqueness of the solution of the stochastic equation of K. Itô (Russian). *Teor. Verojatnost. i Primenen.* **7** (1962), 336–342.
34. Goldstein, J. A., Second order Itô processes. *Nagoya Math. J.* **36** (1969), 27–63.
35. Itô, K., Stochastic differential equations in a differentiable manifold. *Nagoya Math. J.* **1** (1950), 35–47.
36. Itô, K., On stochastic differential equations. *Mem. Amer. Math. Soc.* **No. 4** (1951).
37. Itô, K., and McKean, H. P., "Diffusion Processes and Their Sample Paths." Springer-Verlag, Berlin and New York, 1965.

38. Itô, K., and Nisio, M., On stationary solutions of a stochastic differential equation. *J. Math. Kyoto Univ.* **4** (1964), 1–75.
39. Jazwinski, A. H., "Stochastic Processes and Filtering Theory." Academic Press, New York, 1970.
40. Kac, M., "Probability and Related Topics in Physical Sciences." Wiley (Interscience), New York, 1959.
41. Kallianpur, G., and Striebel, C., Stochastic differential equations occurring in the estimation of continuous parameter stochastic processes. *Teor. Verojatnost. i Primenen.* **14** (1969), 597–622.
42. Kalman, R. E., Falb, P. L., and Arbib, M. A., "Topics in Mathematical System Theory." McGraw-Hill, New York, 1969.
42a. Kannan, D., An operator-valued stochastic integral, II. *Ann. Inst. H. Poincaré Sect. B* **8** (1972), 9–32.
43. Kannan, D., and Bharucha-Reid, A. T., An operator-valued stochastic integral. *Proc. Japan Acad.* **47** (1971), 472–476.
43a. Kannan, D., and Bharucha-Reid, A. T., Random integral equation formulation of a generalized Langevin equation. *J. Statist. Phys.* **5** (1972), 209–233.
44. Khasminskiĭ, R. Z., Diffusion processes and elliptic differential equations degenerating at the boundary of the domain (Russian). *Teor. Verojatnost. i Primenen.* **3** (1958), 430–451.
45. Khasminskiĭ, R. Z., On positive solutions of the equation $Au + Vu = 0$ (Russian). *Teor. Verojatnost. i Primenen.* **4** (1959), 332–341.
46. Khasminskiĭ, R. Z., Ergodic properties of recurrent diffusion processes and stabilization of the solution to the Cauchy problem for parabolic equations (Russian). *Teor. Verojatnost. i Primenen.* **5** (1960), 196–214.
47. Khasminskiĭ, R. Z., On an averaging principle for parabolic and elliptic differential equations and Markov processes with small diffusion (Russian). *Teor. Verojatnost. i Primenen.* **8** (1963), 3–25.
48. Khazan, E. M., Evaluation of the one-dimensional probability densities and moments of a random process in the output of an essentially nonlinear system (Russian). *Teor. Verojatnost. i Primenen.* **6** (1961), 130–138.
49. Krein, S. G., and Petunin, Yu. I., Scales of Banach spaces (Russian). *Uspehi Mat. Nauk* **21**(2) (1966), 89–128.
50. Krylov, N. V., On Itô stochastic integral equations (Russian). *Teor. Verojatnost. i Primenen.* **14** (1969), 340–348.
51. Kunita, H., Stochastic integrals based on martingales taking values in Hilbert space. *Nagoya Math. J.* **39** (1970), 41–52.
52. Kushner, H. J., "Stochastic Stability and Control." Academic Press, New York, 1967.
52a. Kushner, H. J., "Introduction to Stochastic Control." Holt, New York, 1971.
53. Langevin, P., Sur la théorie du mouvement Brownien. *C. R. Acad. Sci. Paris* **146** (1908), 530–533.
54. Lions, J. L., "Équations différentielles opérationnelles et problèmes aux limites." Springer-Verlag, Berlin and New York, 1961.
55. Loève, M., "Probability Theory," 3rd ed. Van Nostrand-Reinhold, Princeton, New Jersey, 1963.
56. McKean, H. P., Jr., "Stochastic Integrals." Academic Press, New York, 1969.
57. McShane, E. J., Stochastic integrals and stochastic functional equations. *SIAM J. Appl. Math.* **17** (1969), 287–306.
58. McShane, E. J., Toward a stochastic calculus, I. *Proc. Nat. Acad. Sci. U.S.A.* **63** (1969), 275–280.

58a. McShane, E. J., Stochastic differential equations and models of random processes. *Proc. 6th Berkeley Symp. Math. Statist. Probability*, (*1970*), to be published.
59. Maruyama, G., Continuous Markov processes and stochastic equations. *Rend. Circ. Mat. Palermo* **4**(2) (1955), 48–90.
60. Middleton, D., "An Introduction to Statistical Communication Theory." McGraw-Hill, New York, 1960.
61. Mortensen, R. E., Existence and uniqueness of measure-valued solutions of a stochastic integral equation. *In* "Mathematical Theory of Control" (A. V. Balakrishnan and L. W. Neustadt, eds.), pp. 441–449. Academic Press, New York, 1967.
62. Nagai, T., Stability problem of random linear system of the first order. *Mem. Fac. Sci. Kyushu Univ. Ser. A* **16** (1962), 47–59.
63. Nelson, E., "Dynamical Theories of Brownian Motion." Princeton Univ. Press, Princeton, New Jersey, 1967.
64. Papoulis, A., "Probability, Random Variables, and Stochastic Processes." McGraw-Hill, New York, 1965.
65. Pontryagin, L., Andronov, A., and Vitt, A., On the statistical investigation of dynamic systems (Russian). *Ž. Eksper. Teoret. Fiz.* **3** (1933), 165–180.
66. Saaty, T. L., "Modern Nonlinear Equations." McGraw-Hill, New York, 1967.
67. Skorohod, A. V., "Studies in the Theory of Random Processes," translated from the Russian. Addison-Wesley, Reading, Massachusetts, 1965.
68. Stratonovich, R. L., A new representation for stochastic integrals and equations (Russian). *Vestnik Moscov. Univ. Ser. I Mat. Meh.* (1964), pp. 3–12; Engl. transl. *SIAM J. Control* **4** (1966), 362–371.
69. Stratonovich, R. L., "Conditional Markov Processes and Their Application to the Theory of Optimal Control," translated from the Russian. Amer. Elsevier, New York, 1968.
70. Stroock, D. W., and Varadhan, S. R. S., Diffusion processes with continuous coefficients, I. *Comm. Pure Appl. Math.* **22** (1969), 345–400.
71. Tanaka, H., Local solutions of stochastic differential equations associated with certain quasilinear parabolic equations, *J. Fac. Sci. Univ. Tokyo Sect. I* **14** (1967), 313–326.
71a. Vahaniya, N. N., and Kandelaki, N. P., A stochastic integral for operator-valued functions (Russian). *Teor. Verojatnost. i Primenen.* **12** (1967), 582–585.
72. Vrkoč, I., Extension of the averaging method to stochastic equations. *Czechoslovak Math. J.* **16** (1966), 518–544.
73. Vrkoč, I., The exponential stability and periodic solutions of Itô stochastic equations with small stochastic terms. *Czechoslovak Math. J.* **18** (1968), 301–314.
73a. Wong, E., "Stochastic Processes in Information and Dynamical Systems." McGraw-Hill, New York, 1971.
74. Wong, E., and Zakai, M., On the convergence of ordinary integrals to stochastic integrals. *Ann. Math. Statist.* **36** (1965), 1560–1564.
75. Wong, E., and Zakai, M., On the relation between ordinary and stochastic differential equations. *Internat. J. Engrg. Sci.* **3** (1965), 213–229.
76. Wong, E., and Zakai, M., On the relation between ordinary and stochastic differential equations and applications to stochastic problems in control theory. *Proc. Internat. Congr., IFAC, 3rd, London, 1966.*
77. Wong, E., and Zakai, M., Riemann-Stieltjes approximations of stochastic integrals. *Z. Wahrscheinlichkeitstheorie und Verw. Gebiete* **12** (1969), 87–97.
78. Wonham, W. M., Random differential equations in control theory. *In* "Probabilistic Methods in Applied Mathematics" (A. T. Bharucha-Reid, ed.), Vol. 2, pp. 131–212. Academic Press, New York, 1970.

Author Index

Numbers in parentheses are reference numbers and indicate that an author's work is referred to, although his name is not cited in the text. Numbers in italics show the page on which the complete reference is listed.

Subject Index